8th International Meeting of the Union for Compact Accelerator-Driven Neutron Sources (UCANS-8)

EPJ Web of Conferences Volume 231 (2020)

Paris, France
8 - 11 July 2019

Editors:

Frederic Ott
Alain Menelle
Christiane Alba-Simionesco

ISBN: 978-1-7138-0829-9

Printed from e-media with permission by:

Curran Associates, Inc.
57 Morehouse Lane
Red Hook, NY 12571

Some format issues inherent in the e-media version may also appear in this print version.

For additional information, please contact EDP Sciences – Web of Conferences at the address below.

EDP Sciences – Web of Conferences
17, Avenue du Hoggar
Parc d'Activité de Courtabœuf
BP 112
F-91944 Les Ulis Cedex A
France

Phone: +33 (0) 1 69 18 75 75
Fax: +33 (0) 1 69 28 84 91

contact-edps@webofconferences.org

Additional copies of this publication are available from:

Curran Associates, Inc.
57 Morehouse Lane
Red Hook, NY 12571 USA
Phone: 845-758-0400
Fax: 845-758-2633
Email: curran@proceedings.com
Web: www.proceedings.com

TABLE OF CONTENTS

EPJ Web of Conferences **231**, 01001 (2020)
UCANS-8
https://doi.org/10.1051/epjconf/202023101001

Development of compact neutron sources

John M. Carpenter[1,*]

[1]Nuclear Technologies Division of the Neutron Science Directorate, Oak Ridge National Laboratory, Oak Ridge, TN.

Abstract. Many small research reactors, neutron sources built in the 1950s and 1960s are being shut down. The activities that these supported have therefore ended, which motivates their replacements with new facilities. Compact accelerator-based neutron sources under development are able to take up many of those activities. This paper paints the background for these developments, summarizes the replacement technologies, relates recent work that supports them, and alludes to some of the new activities that they enable.

1 Small research reactors

During the 1950s and 1960s, numerous laboratories and universities built and operated several dozens of research and test reactors (distinguished from power reactors). That was the optimistic era of "Atoms for Peace" programs around the world. The reactors were of many different types, operating at total thermal powers between 1 and ~10 MW and providing steady thermal neutron fluxes between 10^{12} and 10^{13} n/cm^2-sec. The table below presents a representative list.
Most have offered materials irradiation (neutron damage) and activation analysis facilities, beams for neutron cross section measurements, and capabilities for medical and industrial isotope production. They also provided for reactor physics tests and included arrays of instruments for slow-neutron scattering research. In addition, they supported nuclear data and nuclear astrophysics measurements and still more applications such as the following: materials science neutron scattering studies of atomic structures and motions, neutron-induced radiation damage in power reactor materials, cross sections for design and production of medical isotope production systems, and cross sections for neutron-modulated nuclear reactions relevant to astrophysical isotope production. Around most of these reactors grew programs of nuclear science education, small-source technology (especially moderator development relevant to both small and large sources), reactor physics tests, neutron physics and materials science instrumentation development, neutron detector improvements, radiation applications, and training. Many of these became full-fledged university Nuclear Engineering departments, which furnish undergraduate and graduate education in broad programs in radiation physics and power reactor development. About 15 neutron summer schools established in those times, as well as new ones coming into being (see Table of Neutron Summer Schools, below), train students in the use of the local facilities and for entering research facilities at the major (~ 10^{15} – 10^{17} n/cm^2-sec) neutron sources based on fission (notably, for many years starting in the 1970s and still so, ILL) and spallation reactions (notably, SNS at ORNL).

Table 1. Atom for Peace-era reactors

FACILITY NAME	LOCATION	KW	OPERATION DATES
Ford Nuclear Reactor [5]	Ann Arbor, MI	2000	1957-2003
Breazeale Nuclear Reactor, PSBR	University Park, PA	1000	1955- …
Berliner Experimentier Reaktor II, BER II	Helmholz-Zentrum, Berlin	10000	1990-2019
CP-5	Argonne, IL	5000	1954-1979
Kyoto University Research Reactor, πKURR	Kyoto, Japan	5000	1964- …
PULSTAR	Raleigh, NC	1000	1964- …
ORPHEE	Gif-sur-Yvette, France	14000	1980-2019
McMaster University Reactor, MNR	Hamilton, Ont., Canada	3000	1959- …

2 Neutron summer schools

The first of the neutron scattering schools seems to have been the one at the Atomic Energy Research Establishment (AERE) at Harwell, UK, established in the 1960s and related to the early research reactors and the Harwell electron-linac-based bremsstrahlung photoneutron source there. The school supported and was

* Corresponding author: carpenterjohnm@yahoo.com

peopled largely by a community of users mostly in UK universities and the AERE research staff. That school and the UK Neutron Beam Research Unit of the UK Ministry of Science provided a model system that served the large user community, which is the same as is widely in use today.

Table 1. A list of a few past neutron schools

CETS, Central European Neutron School. Budapest.
HERCULES, HIGHER EUROPEAN RESEARCH COURSE FOR USERS OF LARGE EXPERIMENTAL SCIENCE. GRENOBLE, PARIS, VILLIGEN.
Oxford School of Neutron Scattering. Oxford.
Berlin Neutron School. HZB Berlin.
SONS School of Neutron Scattering. Scuola Neutroni, Erice, Italy.
LANSCE School on Neutron Scattering, Los Alamos, NM.
JCNS Laboratory Course on Neutron Scattering. Jülich.
AONSA Muon and Neutron School. Facilities of AONSA (Asia-Oceanica Neutron Scattering Association) member laboratories.
The All-Russian School for Young Scientists and Students and International Student Practice Summer School, Всероссийская Школя Молодых Учёнz, Dubna.
SwedNESS Swedish Neutron School (NNSP). Swedish Foundation for Strategic Research (SSF);
Nordic Neutron School, NordForsk (a Nordic collaboration). First meeting, Tartu, Estonia.
Argonne-Oak Ridge Neutron and X-ray School. ANL and ORNL.
PSI Summer School on Condensed Matter Research. Zuoz, Switzerland.
NIST Summer School on Fundamentals of Neutron Scattering. Gaithersburg, MD.
Bilbao Neutron School. Bilbao, Spain.

3 What's being lost

A large fraction of the original complement of small neutron sources has now been or will soon be shut down. Some have been dismantled on the basis of fear of reactor-related incidents, none of which ever involved any laboratory- or university-based reactors. And some mid-scale reactors have been shut down to be replaced. Capabilities are being lost as the small reactors are shutting down, especially for device prototyping, techniques development, and education and practice in the field, as well as opportunities for hands-on student projects and laboratory experiments.

Also being lost as some of the larger facilities are closing are medical isotope production facilities (especially for Mo^{99} production), a significant fraction of materials-science neutron-scattering capabilities, and facilities for neutron-induced transmutation doping of high-purity silicon to produce precisely controlled volumetric n-type P^{31}-doped material for use, for example, in photovoltaic cells (the reaction is $Si^{30} + n \longrightarrow Si^{31} \longrightarrow P^{31} + e^{+}$). Also diminished are the concentrations of expertise of engineers and scientists residing there and the consequent loss of bases for education and training of young scientists.

4 New sources

All these losses motivate development of new neutron facilities, probably compact, small, accelerator-driven neutron sources, CANS, and based on low-energy (typically less than 100 MeV particle energies) neutron-producing charged-particle reactions. Simply put, it's not that CANS are much better than reactors, it is that they are different, with a broad range of applications that are now being elaborated. Most CANS operate on d,d reactions (~ 2 MeV neutrons), d,t reactions (~ 14 MeV neutrons), p,Be reactions (~ 2 MeV neutrons), or p,Li reactions (~ 2.4 MeV neutrons). Operation can be either steady or pulsed. Slowing down in room-temperature moderators, these produce Maxwellian neutron spectra with temperatures ~ 300 K, mean energies 0.025 eV. As sources of neutron beams, accelerator-driven sources compensate for losing some of the reactor-based installations. Furthermore, the accelerator-based facilities are more than neutron sources; they are also sources of charged-particle beams useful for other purposes.

By comparison, reactors are steady sources, operating continuously, wherein fission neutrons have average energies of about 2 MeV and moderated neutron spectra have average energy 0.025 eV (corresponding to the moderator temperature ~ 300 K). Both CANS and reactor types adapt to cold moderators, producing neutron spectra with temperatures as low as 20 K, depending on the moderator temperature. For further information consult Carpenter and Loong 2015 [1].

5 Compact Accelerator-driven Neutron sources

5.1 UCANS – Union for Compact Accelerator-driven Neutron sources

UCANS is a loosely organized collaboration of international laboratory and university personnel that meets about once a year or year-and-a-half. Laboratories volunteer their sponsorship, and leadership passes to a representative from the sponsor of the next meeting. Scientists from about 22 laboratories now participate in UCANS, some from institutions with currently operating CANS, such as LENS at Indiana University, Bloomington, and CPHS at Tsinghua University, Beijing. Many come from where CANS are under development, such as LENOS at the Legnaro Laboratory, Italy; the ESS-BILBAO source in Bilbao, Spain; SONATE in France; and Era-Nova and HBS in Germany. Details will be found in the proceedings of this (Paris) meeting, UCANS VIII, where about 100 scientists took part. About 50 scientists took part in the 2018 meeting, UCANS VII, in Bariloche, Argentina. Several active, small ex-UCANS groups, for example in Japan and Korea, meet frequently in local venues. The next UCANS meeting (UCANS IX) will be held in Japan, in late 2020. Consult the website ucans.org.

The topics covered in the 2018 UCANS VII meeting were Accelerator and beam optics, Neutron detection, Nuclear astrophysics and nuclear data, Target radiation damage and heat removal, CANS projects and innovative instrumentation, Medical applications, Material characterization, Computer simulations and data analysis software, Optical devices, Other applications of CANS, and Education. See Carpenter, et al. 2018 [2]. The 2019 meeting in Paris, UCANS VIII, is the first UCANS meeting recorded in a published proceedings. Consult www.ucans.org for current information.

5.2. A conceptual CANS installation

The figure is a schematic representation of a compact accelerator-based neutron source. Such systems are either purchased from existing suppliers (smaller systems) or designed and built in-house (larger ones) to local specifications.

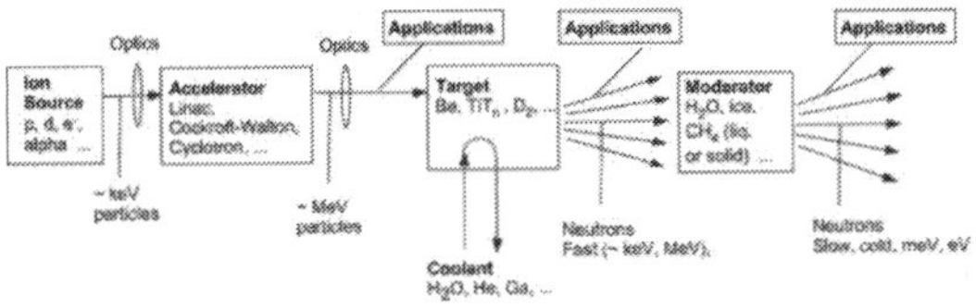

Fig. 1. Caption of the Figure 1. Below the figure.

5.3 Applications

Existing CANS support a wide variety of applications (Anderson, et al. 2017 [3]), for example, in the developing field of characterizing cultural artifacts (Kardjilov and Festa 2017 [4]). Several groups are developing and deploying portable neutron radiography technology for nondestructive infrastructure (roads and bridges) inspection and contraband nuclear materials (cargo) detection. Fissile materials give themselves away by emitting fission-generated neutrons that return with a time and energy signature detectably distinct from those scattered from common nonfissionable materials. CANS cargo inspection facilities are probably inexpensive enough to deploy permanently at ports where they are needed and to justify mobile inspection CANS stationed to move around when called for.

CANS are well-suited for simulations of single-event upsets (SEUs) caused by fast-neutron scattering in semiconductor devices in the high atmosphere. For these purposes, it would be easier in CANS to tailor neutron spectra than in the large facilities. Also, some of materials-science investigations that researchers carry out with higher resolution and at higher rates of throughput at the major sources rely on preliminary tests at a small source. Furthermore, a long-foreseen and hoped-for application of CANS in boron capture neutron therapy, BNCT, awaits development of boron-containing cancer-tumor-seeking pharmaceuticals. And there is pressing need for alternatives to reactor-produced medical isotopes, notably Mo^{99}, but other isotopes also, where hopes lie with accelerator-based production. Table 4.3 of Anderson, et al. 2016 summarizes characteristics of CANS useful for many categories of applications and includes detailed descriptions of those applications and facilities.

5.3 An example

The 2-MW Ford Nuclear Reactor (criticality 1957) at the University of Michigan's Phoenix Memorial Laboratory is an example of a response to the shutting down of an Atoms for Peace reactor. There, the Nuclear Engineering and Radiological Sciences Department has installed a d,d (2.2 MeV neutrons) and a d,t (14 MeV neutrons) neutron generator in its new Neutron Sciences Laboratory and an array of several low-energy charged-particle accelerators (p, d, α, …) in its Ion Beam Laboratory. These serve both teaching and research purposes involving students and faculty (see [5] for more details).

Throughout, these applications follow from the intrinsic advantages of CANS over reactors and the large, multiuser neutron facilities: low acquisition and maintenance costs, flexible scheduling, lighter safety concerns, easier access, local management, and more.

Acknowledgement

The work of Oak Ridge National Laboratory is supported by the U.S. Department of Energy, Office of Science, Office of Basic Energy Sciences, under contract 4000149663, and of Argonne National Laboratory under Contract DE-AC02-06CH11357.

About the author

John M. Carpenter is a Senior Physicist, retired, Emeritus Scientist, and Distinguished Fellow of Argonne National Laboratory, now attached to the APS Engineering Services Division. Until 1975, he served for ten years as Professor of Nuclear Engineering at the University of Michigan. He is the spokesperson for UCANS.

Carpenter serves as a consultant to the Nuclear Technologies Division of the Neutron Science Directorate of Oak Ridge National Laboratory, Oak Ridge, Tennessee.

References

[1] J. M. Carpenter, and C.-K. Loong, 2017. *Elements of Slow-Neutron Scattering*, Cambridge University Press.

[2] J. M. Carpenter, D. Baxter, and R. Granada (2018). UCANS VII Meeting Summary, Neutron News, **29**, Number 2.

[3] I.S. Anderson, C. Andreani, J. M. Carpenter, G. Festa, G. Gorini, C.- K. Loong, and R. Senesi (2016). Research opportunities with compact accelerator-based neutron sources, **Physics Reports** doi:10.1016/j/physrep.2016.07.007.

[4] N. Kardjilov and G. Festa, (2017). *Neutron methods for archeology and cultural heritage*, Springer International Publishing Switzerland, 10.1007/978-3-319-33163-8.

[5] J. J. Duderstadt (2018). *The University of Michigan Nuclear Engineering and Radiological Sciences: A History*, The Millennium Project, Ann Arbor.

See also an earlier version of this paper, (2019) J. M. Carpenter, "Development of Compact Neutron Sources."***Nature Reviews Physics* 1**, published online 28 January 2019, pp.177-179. Issue date March 2019.

 https://doi.org/10.1051/epjconf/202023101002

Newly constructed compact accelerator-based neutron facility at AIST

Koichi Kino[1,2,*], *Takeshi* Fujiwara[1,2], *Michihiro* Furusaka[1,2], *Noriyosu* Hayashizaki[1,2,3], *Hidetoshi* Kato[1,2], *Ryunosuke* Kuroda[1,2], *Koji* Michishio[1,2], *Takemi* Muroga[2], *Hiroshi* Ogawa[1,2], *Brian E.* O'Rourke[1,2], *Nagayasu* Oshima[1,2], *Daisuke* Satoh[1,2], *Norihiro* Sei[1,2], *Tamao* Shishido[2], *Ryoichi* Suzuki[1,2], *Masahito* Tanaka[1,2], *Yo* Tomota[1,2], *Hiroyuki* Toyokawa[1,2], *Akira* Watazu[1,2], *Kazuro* Furukawa[1,4], *Kazuyuki* Nigorikawa[1,4], and *Takashi* Obina[1,4]

[1]National Institute of Advanced Industrial Science and Technology (AIST), Tsukuba, Ibaraki, Japan
[2]Innovative Structural Materials Association (ISMA), Chiyoda-ku, Tokyo, Japan
[3]Tokyo Institute of Technology (Tokyo Tech), Meguro-ku, Tokyo, Japan
[4]High Energy Accelerator Research Organization (KEK), Tsukuba, Ibaraki, Japan

Abstract. We have constructed a compact accelerator-based neutron facility at the National Institute of Advanced Industrial Science and Technology (AIST) in Japan aiming at development of structural materials of transportation vehicles for weight reduction. The facility consists of electron linear accelerator, neutron source, and a neutron beamline with a measurement hutch. Currently we are commissioning the accelerator and expecting a first neutron beam in 2019.

1 Introduction

An energy-saving society is required recently and regulations of CO_2 exhaust are getting tighter. One of the solutions is reducing weight of transportation vehicles like automobiles by "multi-material design". Multi-material design involves choosing and applying lightweight materials, such as high strength steel, aluminum or carbon-fiber reinforced plastic (CFRP). The key is the joining technology of these dissimilar materials. For practical realization of such designs, it is important to analyze not only test pieces but also assembled structural materials non-destructively.

A neutron beam is effective for nondestructive analysis due to its high material penetrability. A particularly useful technique for materials analysis is wavelength-resolved neutron transmission imaging. This imaging technique requires a white (i.e. a beam with a broad energy distribution) pulsed neutron beam. The neutron beam is irradiated on a sample and transmitted neutrons are detected by a position sensitive neutron detector with the time of flight (TOF) between the neutron source and the detector of each neutron measured. The wavelength of a neutron is proportional to the TOF as the wavelength is inversely proportional to the neutron's velocity. If the sample has a crystalline nature, neutron diffractions are seen in the transmission spectrum as a function of a neutron wavelength. By analyzing the transmission spectra for all neutron detection positions, macroscopic images on the crystalline information can be obtained. This method is called Bragg edge imaging. Many pioneering measurements have been reported for crystalline phase, crystalline strain, crystalline size, and crystalline orientation especially with steel samples [1-7]. Such imaging data is easily understood by people in industrial fields, who are not neutron analysis experts.

Furthermore, a compact accelerator-based neutron source (CANS) is useful for industrial usages by on-demand measurements. Therefore, we think a combination of a CANS and the wavelength-resolved neutron transmission imaging matches to our purpose. We have designed and constructed a CANS at the National Institute of Advanced Industrial Science and Technology (AIST) in Tsukuba, Japan [8]. In this paper, the main components, the expected properties of the neutron beam, and the current status of the new CANS are described.

2 Main components

Figure 1 shows the overview of the CANS at AIST. It mainly consists of an electron linear accelerator, neutron source, and neutron beamline.

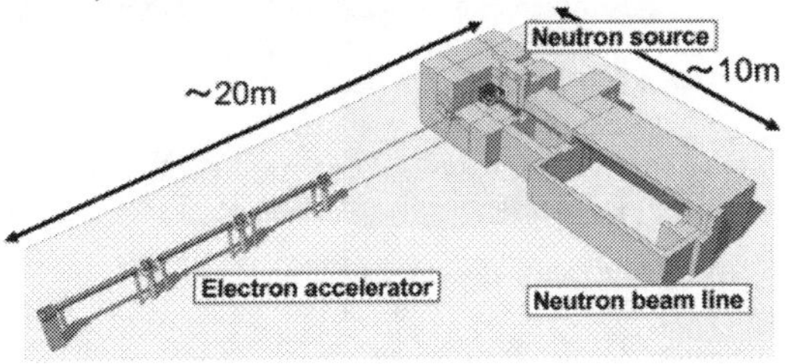

Fig. 1. Overview and features of the CANS at AIST [8].

* Corresponding author: koichi.kino@aist.go.jp

 https://doi.org/10.1051/epjconf/202023101002

An electron beam of about 35 MeV, with a maximum pulse width of 10 μs, and a maximum repetition rate of 100 Hz is produced by the electron linear accelerator [9]. The pulse width is optimized for a short pulse neutron beam by a decoupled solid methane moderator described later. The maximum repetition rate is optimized for the neutron wavelength region needed for the Bragg edge imaging. The designed maximum electron beam power is about 10 kW and this value is about 10 times larger than that of the HUNS (Hokkaido University Neutron Source)-I.

In the neutron source, the electron beam is irradiated on a neutron production target, which consists of several layers of thin tantalum plates with cooling water flowing between them. Evaporation neutrons generated in the tantalum plates are slowed down in a moderator positioned directly above the neutron production target. The moderator is a block of solid methane with a thickness of 3 cm cooled to 20 K. The dimensions and the decoupled designs (i.e. the moderator is surrounded by cadmium except for the neutron extraction hole) were chosen for a short pulse length of the neutron beam, which is needed for a high neutron wavelength resolution due to the limited length of the flight path.

The neutron beamline is surrounded by a radiation hutch made of concrete. The flight path length is 8 m due to the size of the hutch which is limited by the size of the experimental room. Super mirror guide tubes of $3Q_c$ are set along the neutron beamline to increase the neutron flux at the sample position. The inner dimensions and total length of the guide tubes are 130 mm×130 mm and 5 m. Furthermore, we are preparing a novel position sensitive neutron detector, which has a high detection efficiency and counting rate capability. The detector is based on gaseous detector (Glass GEM) with Boron foil neutron converter [10]. In order to improve the detection efficiency, micro-structured aluminium substrate were coated with ^{10}B [11]. The combination of high neutron production rate, with efficient transport and detection will allow measurements with good statistics within practical measurement time.

3 Properties of the neutron beam

We have calculated the expected properties of the neutron beam using a Monte-Carlo simulation [8]. Figure 2 shows the neutron wavelength resolution at L (flight path length) = 8 m (i.e. at the sample position).

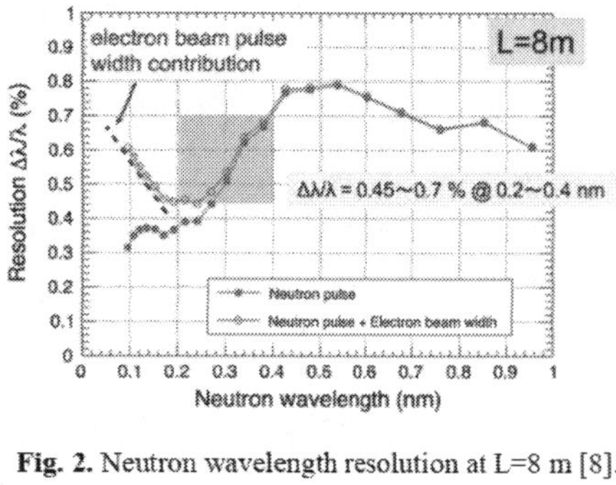

Fig. 2. Neutron wavelength resolution at L=8 m [8].

The maximum Bragg edge wavelength for structural materials such as steel and aluminium is about 0.4 nm. In the wavelength region from 0.2 to 0.4 nm the resolution is 0.45 ~ 0.7 %. This value is about 3 times larger than that of the J-PARC neutron imaging beamline (BL22) [12] mainly due to a difference of the flight path lengths. The wavelength resolution deteriorates below 0.2 nm as the wavelength becomes shorter. This is due to the pulse width of the electron beam (10 μs), which also contributes to the sharpness of the neutron beam in addition to the time width by the slowing down process in the moderator.

The neutron flux at L=8 m is shown in Fig. 3. The simulation includes a super mirror guide tube along the neutron beam extraction hole in the beam shutter. Super mirror guide tubes can also be placed downstream of the beam shutter, which consist of two parts (labelled upstream and downstream). We calculated the neutron flux at L=8 m for three configurations of the super mirror guide tubes; 1. shutter only, 2. shutter and upstream and 3. shutter, upstream and downstream mirrors. The results show a thermal neutron peak around 0.4 nm and the intensity of this peak increases as the number of super mirror guide tubes is increased. In the thermal and cold neutron energy region, the calculated instantaneous maximum neutron flux ranges from 2.5×10^4 ~ 1.0×10^5 /cm²/s depending on the mirror configuration. The maximum counting rate of currently available position sensitive neutron detectors, which have an effective area of 10 cm x 10 cm, is 10^5 ~ 10^6 /s. If a typical neutron detection efficiency of 10 % at 0.4 nm is assumed, these detectors will reach maximum counting rate for a neutron beam with a flux of 10^4 ~ 10^5 /cm²/s. This indicates that the expected neutron flux of the present CANS corresponds to the maximum detection rate of existing position sensitive detectors.

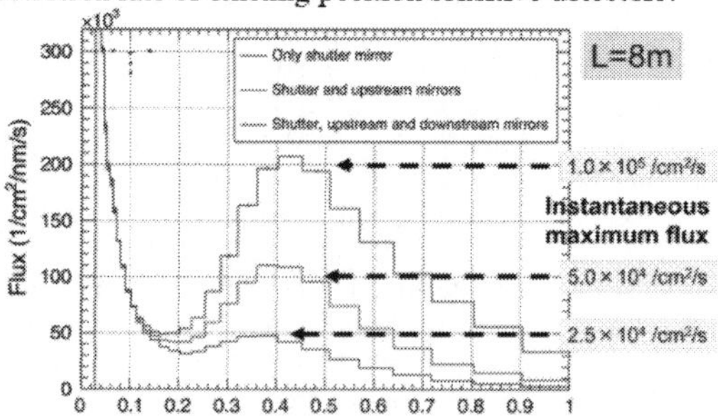

Fig. 3. Neutron flux at L=8 m [8].

4 Current status of the CANS

Figure 4 is a picture of the electron linear accelerator. An electron beam is accelerated from electron gun through three 2.7 m long accelerator cavities. The accelerated electron beam is then directed onto the neutron production target, which is located in an experimental room separated from the accelerator room by a 5 m thick concrete wall.

https://doi.org/10.1051/epjconf/202023101002

Currently we are commissioning the accelerator. A beam loading of the accelerator cavity 1 was observed in an RF dummy load signal when the electron gun was tuned on. We also observed a beam spot on a screen monitor set just downstream the accelerator cavity 1. The beam size was about 30 mm in diameter but focusing magnets were not used at this time.

Fig. 4. Photo of the electron linear accelerator.

The electron beam comes from the electron linear accelerator in the neighboring room (to the left hand side in this picture) and is irradiated on the neutron production target shown in Fig. 5 at the center of the neutron source. A cooling water test of the target has been done. Tests of the solid methane moderator test are ongoing. Solid methane is produced by condensation of methane gas by cooling down the aluminum moderator vessel. Condensation is confirmed not only by the methane gas pressure and temperature of the vessel but also by X-ray imaging of the liquid and solid methane. Liquid and solid methane were clearly observed in X-ray images.

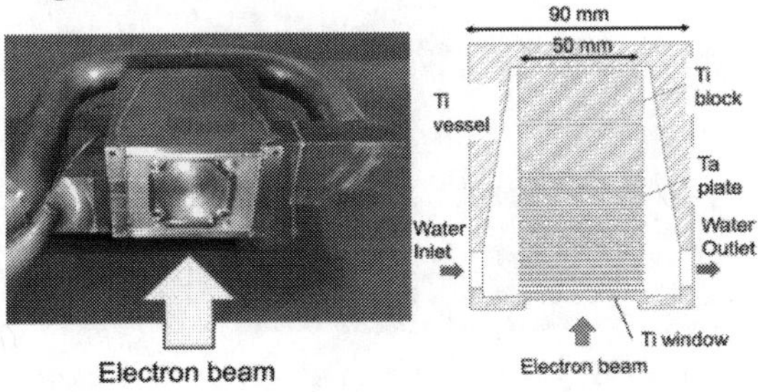

Fig. 5. Neutron production target.

Access into the beamline hutch is controlled by a door at the hutch entrance. The neutron beam can be blocked off using a shutter, preventing neutrons from entering the hutch when the shutter is closed. The shutter controller and door controller panel is located just outside the entrance. If the shutter is closed, the door can be unlocked and opened and workers can enter the hutch even when the accelerator is in operation. Before the beam shutter can be opened, all workers must evacuate from the hutch and the door must be locked within 20 seconds after the evacuation button is pushed. The status of the shutter, door and evacuation condition are controlled by a personal protection system. Status of the beam shutter is shown by green and red lights just above the entrance. Furthermore, this procedure is very similar to that of the J-PARC neutron beamlines. Therefore, neutron users will not feel difficulties between the present CANS and J-PARC.

Currently there are only two stages around the sample position in the hutch. We are preparing a neutron detector, sample goniometer, background shielding and so on for a first neutron beam.

5 Summary

We have constructed a CANS for wavelength resolved neutron transmission imaging of structural materials. The electron linear accelerator is now under commissioning. The neutron production target and solid methane moderator of the neutron source are undergoing testing. The sample position in the neutron beamline hutch is in preparation. The first neutron beam is expected later in 2019.

The new neutron facility is being developed under the auspices of the Innovative Structural Materials Association (ISMA), which is promoting weight reduction in automobiles and other transportation vehicles. This paper is based on results obtained from a project commissioned by the New Energy and Industrial Technology Development Organization (NEDO).

References

1. J.R. Santisteban, L. Edwards, M.E. Fitzpatrick, *et al.*, Nucl. Instrum. Methods Phys. Res., Sect. A **481**, 765 (2002)
2. H. Sato, T. Kamiyama, and Y. Kiyanagi, Mater. Trans., **52**, 1294 (2011)
3. K. Iwase, H. Sato, S. Harjo, *et al.*, J. Appl. Crystallogr., **45**, 113 (2012)
4. Y. Su, K. Oikawa, S. Harjo, *et al.*, Mater. Sci. Eng., A **675**, 19 (2016)
5. A.S. Tremsin, S. Ganguly, S.M. Meco, *et al.*, J. Appl. Crystallogr., **49**, 1130 (2016)
6. A.S. Tremsin, T.Y. Yau, and W. Kockelmann, Strain, **52**, 548 (2016)
7. K. Oikawa, Y.H. Su, R. Kiyanagi, *et al.*, Physica B, **551**, 436 (2018)
8. K. Kino, T. Fujiwara, M. Furusaka, *et al.*, Nucl. Instrum. Methods Phys. Res., Sect. A **927**, 407 (2019)
9. B.E. O'Rourke, T. Fujiwara, K. Furukawa, *et al.*, Nucl. Instrum. Methods Phys. Res., Sect. B **464**, 41 (2020)
10. T. Fujiwara, Y. Mitsuya, H. Takahashi, *et al.*, J. Instrum. **9** P11007 (2014)
11. T. Fujiwara, U. Bautista, Y. Mitsuya, *et al.*, Nucl. Instrum. Methods Phys. Res., Sect. A **838**, 124 (2016)
12. Y. Kiyanagi, T. Kamiyama, H. Sato, *et al.*, Nuclear Instrum. Methods Phys. Res., A **651**, 16 (2011)

Neutron beam performance of iBNCT as linac-based neutron source for boron neutron capture therapy

Hiroaki Kumada[1,*], *Susumu* Tanaka[1], *Fujio* Naito[2], *Toshikazu* Kurihara[2], *Takashi* Sugimura[2], *Hideyuki* Sakurai[1], *Akira* Matsumura[1], and *Takeji* Sakae[1]

[1]Faculty of Medicine, University of Tsukuba, 1-1-1, Tennodai, Tsukuba, Ibaraki, 305-8575, Japan
[2]High Energy Accelerator Research Organization, 1-1, Oho, Tsukuba, Ibaraki, 305-0801, Japan

Abstract. The linac-based neutron source "iBNCT" developed by the Tsukuba team has been shown to generate a large intensity of neutrons. To confirm the applicability of the device to BNCT, several characteristic measurements have been implemented. In a water phantom experiment, when the accelerator was operated with an average current of 1.4 mA, the maximum thermal neutron flux was approximately 7.8×10^8 (cm^{-2}s^{-1}). From estimation of the stability of the linac, the results demonstrate its applicability to actual treatment.

1 Introduction

In recent years, the Boron Neutron Capture Therapy (BNCT) has attracted attention as a treatment method for intractable cancers and recurrent cancers. BNCT is a binary system, based on the nuclear reaction between low-energy neutrons and the boron isotope boron-10 accumulated in the tumor cells [1]. In principle, BNCT can destroy cancer cells selectively while preventing serious damage to normal tissues, even if the cancer cells invade them. BNCT requires the availability of high-intensity neutrons. Thus far, clinical studies of BNCT have been conducted by using research reactors. And recent treatments have used epithermal neutrons (0.5 eV < Energy < 10 keV) to deliver therapeutic doses to deeper regions within the human body. The clinical results using reactors have demonstrated the effectiveness of BNCT for treating intractable cancers [2]. However, the therapy has not been established yet since its implementation requires the use of reactors.

The recent progress in the development of technologies for accelerator-based neutron sources have rendered it possible to produce enough neutrons for BNCT using a compact accelerator, which can be installed in hospitals. Therefore, it is expected that BNCT could be established as a general cancer treatment in the near future. Based on this background, many development projects for the accelerator-based neutron source applicable to BNCT are being undertaken in the world [3]. The iBNCT project, a development project for the linac-based neutron source for BNCT, is currently underway [4]. The accelerator of the iBNCT has adopted a Radio-Frequency Quadrupole Linac (RFQ) and a Drift Tube Linac (DTL) type linac. Fig. 1 shows the linac of the iBNCT. The linac accelerates protons to 8 MeV with pulsed mode operation and the average current that can be accelerated has been designed to be 5 mA or more. When the repetition cycle is 50Hz, the duty is 5.0% and the average current is approximately 1.4 mA. For the target material for neutron generation, iBNCT has adopted beryllium. Table 1 gives the main specifications for the iBNCT. Each value in parentheses in the table is a target value in the design stage.

Fig. 1. An RFQ and a DLT-type linac-based neutron source for BNCT at the University of Tsukuba, Japan.

Table 1. Main specifications for the demonstrator of iBNCT.

Accelerator type	RFQ + DTL proton linac
Proton energy	8 MeV
Peak proton current	> 28 mA (50mA)
Ave. proton current	> 1.4 mA (> 5 mA)
Pulse width	Max. 1 millisecond
Repetition cycle	50Hz (> 125 Hz)
Duty	> 5.0% (> 12.5%)
Target material	Beryllium
Accelerator size	< 8 m

The construction of the demonstrator of the iBNCT was completed in 2016. It has been successful for generating

* Corresponding author: kumada@pmrc.tsukuba.ac.jp

https://doi.org/10.1051/epjconf/202023101003

neutrons at that time, albeit having low intensity. Subsequently, various improvements and modifications have been carried out to enhance neutron intensity. At present, in order to conduct clinical trials using the device in the near future, several characteristic measurements of the neutron beam of the iBNCT device are being carried out.

2 Materials and Methods

2.1. Measurement of the distributions thermal neutron flux and gamma-ray dose rate in a water phantom

It is required to grasp the dose distribution administered around the tumor region when the neutron beam emitted from the iBNCT is irradiated to a human body. For this, several neutron beam irradiation experiments using a water phantom have been performed. In the experiments, a 200 × 200 × 200 mm3 rectangular phantom made of polymethyl methacrylate has been used. The thickness of each wall of the phantom except for the beam side wall is 10 mm, and the thickness of the beam-side wall is 3 mm. In the experiments, several detectors for radiation measurements have been set up inside the phantom that is filled with water. The phantom is positioned to irradiation position in the irradiation room of the iBNCT facility and is irradiated by the neutron beam. In Fig. 2-(a), the phantom is shown and in Fig. 2-(b), the experimental setup is shown with the phantom located at the irradiation position in the irradiation room.

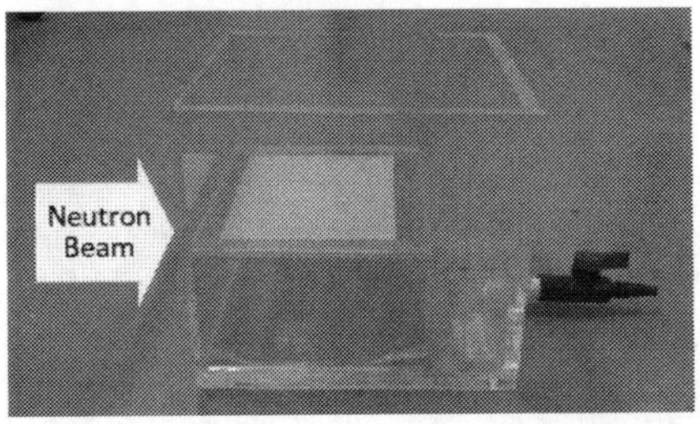

(a) 20 × 20 × 20 cm^3 rectangular water phantom

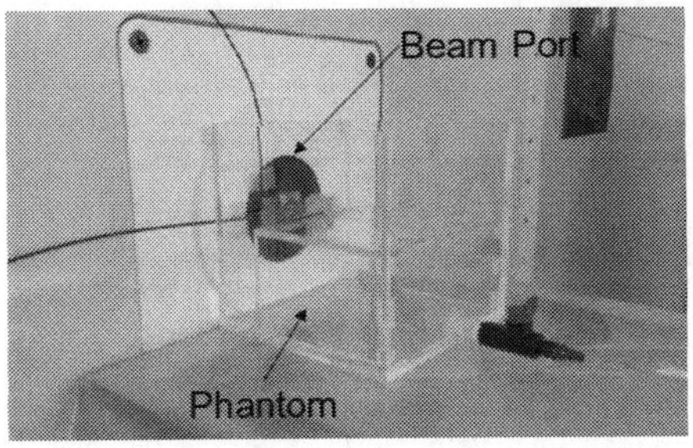

(b) Experimental situation with the phantom

Fig. 2 (a) Cubic water phantom (b) experimental setup with the phantom at the irradiation position

To estimate the dose distribution for BNCT, distributions for thermal neutron flux and gamma-ray dose rate, which are the fundamental dose elements of BNCT, have been measured.

For the measurement of thermal neutron flux, the activation foil method [5], has been adapted in which pure gold (Au-197) is used. A few Au wires of 10-mm length each have been placed on the beam axis inside the phantom. Fig. 3-(a) shows the schematic image of the arrangement of the Au wires in the phantom.

In the experiments, the linac is operated with an average current of 1.4 mA. After irradiation, the radioactivity of each gold wire has been measured using a germanium detector and finally, the distributions of the thermal neutron flux have been determined.

For the measurement of the distribution for the gamma-ray dose rate, a thermoluminescent dosimeter (TLD) has been used as the gamma-ray dose detector [6]. Several TLDs have been placed on the beam axis in the phantom as shown in Fig. 3-(b). After imparting neutron beam irradiation to the phantom, measurements have been taken from each TLD and finally, the distribution for the gamma-ray dose rate has been determined.

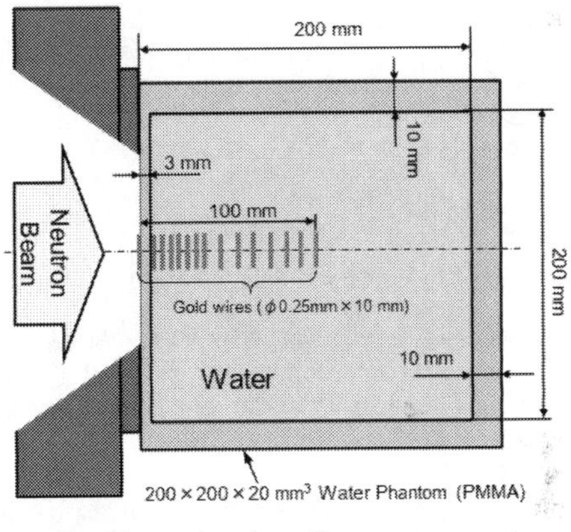

(a) Thermal neutron flux measurement

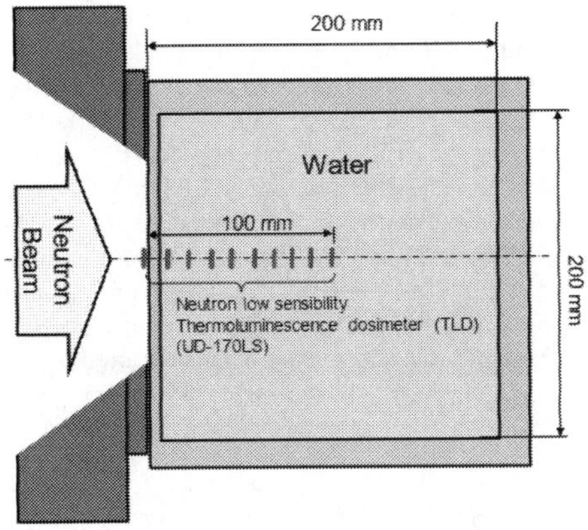

(b) Gamma-ray dose rate measurement

Fig. 3 Experimental conditions for the water phantom irradiation to measure the distributions for (a) thermal neutron flux and (b) gamma-ray dose rate

2.2 Dose estimation using Monte Carlo transport code

To estimate distributions of the equivalent doses for a tumor and for normal tissues in the phantom by neutron irradiation, Monte Carlo solusion has been carried out. For estimation, the Particle and Heavy Ion Transport Code System (PHITS) has been used as a Monte Carlo code. And for the nuclear data library required for the Monte Carlo calculation, JENDL-4.0 has been used. In the calculation, the specifications of the phantom experiment such as the geometry of the iBNCT facility and the phantom and the neutron beam characteristics have been reconstructed properly using PHITS. In the PHITS calculation, two-dimensional (2D) distributions for multiple absorbed doses consisting of boron-10 dose, nitrogen dose, hydrogen dose, and gamma-ray dose have been determined. The four absorbed doses are the basis of the equivalent dose [7]. In addition to the absorbed doses, the distributions for thermal neutron flux and for gamma-ray dose rate in the phantom have also been calculated in order to determine a normalization factor for the calculation values.

The 2D distributions for equivalent doses (Gy-Eq) in the phantom have been determined by combining with the calculated absorbed doses. Equation (1) is generally used for BNCT dosimetry [7].

$$ED\ (\text{Gy-Eq}) = (C_B \cdot D_{B,ppm} \times CBE_B) + (D_N \times RBE_N) + (D_H \times RBE_H) + (D_\gamma) \quad (1)$$

where ED is an equivalent dose rate. $D_{B,ppm}$ is the absorbed dose for boron-10 per ppm. Similarly, D_N, D_H, and D_γ indicate the absorbed doses for nitrogen, hydrogen, and gamma-ray, respectively. C_B which is required to calculate the boron dose in each organ is the boron concentration (ppm). CBE_B is the compound relative biological effectiveness for boron-10. In the dose evaluation, hypothetical values have been set according to the general protocol in BNCT. Each parameter used in the evaluation is summarized in Table 2.

The University of Tsukuba plans to implement a clinical study using the iBNCT for malignant melanoma. Therefore, the equivalent doses for normal tissue, skin, and tumors have been estimated by applying the protocol for malignant melanoma. In the simulation for the BNCT irradiation, the maximum equivalent dose of the skin is set as 15 Gy-Eq and the concentration of boron-10 in blood is 25 ppm. In the estimation condition, irradiation time and maximum tumor dose have been estimated.

Table 2. Parameters required to the equivalent dose.

	Tumor	Normal Tissue	Skin
B-10 Ratio with Blood	3.5	1.0	1.2
B-10 concentration (ppm)	88	25	30
CBE for BPA	3.8	1.35	2.5
RBE_N (Thermal Neutron)	3.0		
RBE_H (Fast Neutron)	3.0		
Gamma-ray	1.0		

3 Results and discussions

3.1. Measurement results of the thermal neutron flux and gamma-ray dose rate

Thermal neutron flux in the phantom has been measured by detecting the activation of the irradiated gold wires. Fig. 4 shows the distribution of the thermal neutron flux on the beam axis in the phantom. The maximum value is approximately 7.8×10^8 ($cm^{-2}s^{-1}$) at 20 mm depth of the phantom. The error margin of the measurement for the radio-activity of gold wire is approximately ± 7%.

Fig. 5 shows the distribution of the gamma-ray dose rate on the beam axis. The irradiation condition for the average current of the proton beam is the same as the thermal neutron flux measurement. The maximum gamma-ray dose rate is approximately 2.8 (Gy/h) at depth of 25 mm from the surface of the phantom. The error margin of the measurement with TLD is approximately ± 12%.

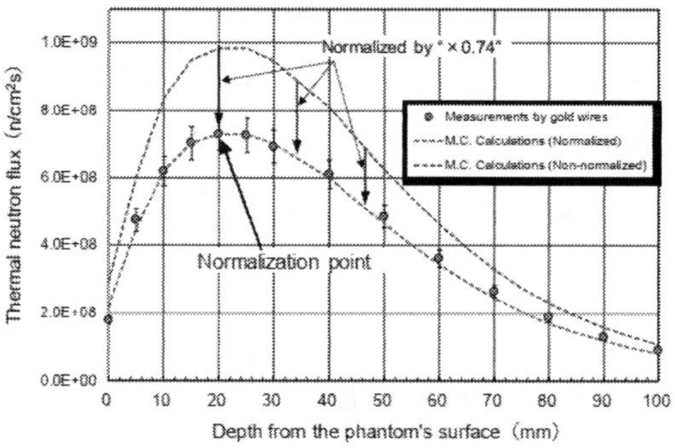

Fig. 4 Thermal neutron flux on the beam axis

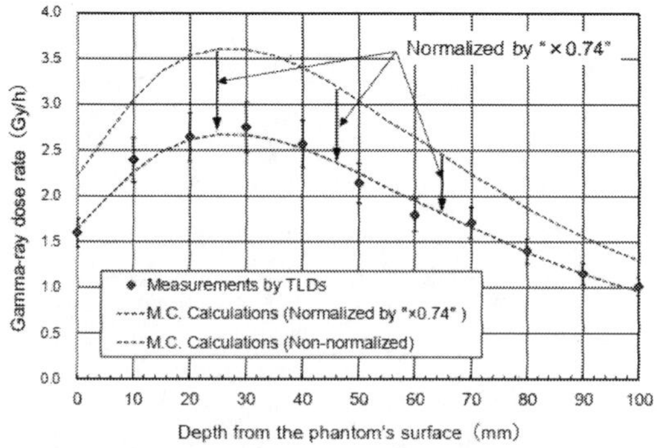

Fig. 5 Gamma-ray dose rate on the beam axis

3.2. Dose estimations and irradiation time

As described previously, the distributions for thermal neutron flux and gamma-ray dose rate in the phantom have been calculated by using PHITS. The value of the normalization factor was determined as 0.74 by dividing the experimental value of the maximum thermal flux (7.80×10^8) by the calculated value (10.51×10^8) at the same point (20-mm depth from the surface on the axis). The calculated values for the thermal neutron flux and gamma-ray dose rate have been normalized by the factor value. The calculated distributions for the thermal neutron flux and the gamma-ray dose rate are in good agreement with the measured distributions, as shown in Fig. 4 and Fig. 5. Based on these results, the distributions for four absorbed doses in the phantom have been determined by multiplying each calculated value for the absorbed doses with the normalization factor. Finally, the distributions for the equivalent doses for normal tissue, skin, and tumor regions have been estimated by applying the values of absorbed doses and each parameter listed in Table 2 to Equation 1.

From the estimation results, the irradiation time is approximately 44.10 min, when 15.00 Gy-Eq is administered to normal skin. According to this estimation condition, the maximum dose for the tumor and for normal tissue are approximately 54.64 Gy-Eq and 8.65 Gy-Eq, respectively. The estimation results prove that the iBNCT facility provides a neutron flux high enough for the treatment of malignant melanoma. This is because the irradiation time is less than 1 h, and the prescribed dose administered to the tumor is more than 25.00 Gy-Eq.

3 Conclusions

The development of the iBNCT, a linac-based neutron source for BNCT, is being carried out. The demonstration device of the iBNCT has been proven to succeed in generating high-intensity neutrons. Several experiments and measurements have been performed to evaluate the performance of the neutron source. The thermal neutron flux and the gamma-ray dose rate on the beam axis in a water phantom have been measured by phantom irradiation experiments. Furthermore, Monte Carlo simulations have been carried out to determine dose distribution administered to the irradiation field. The results of the experiments have demonstrated that the iBNCT can generate neutrons with sufficient intensity required for BNCT. To improve the stability and safety of the device further and to adopt the iBNCT device for actual BNCT clinical study, the device is being continuously advanced. Further experiments and analyses have been planned to understand the physical characteristics and biological effects of the neutron beam.

This work was supported by MEXT/JSPS KAKENHI (Grant number JP16H05236).

References

1. W. H. Sweet, J. Neuro-Oncol. **33**, 19-26 (1997)
2. T. Aihara, N. Morita, N. Kamitani, H. Kumada, K. Ono, J. Hiratsuka and T. Harada, Int. J. Oncol. **19**, 1437-1444 (2014)
3. A. J. Kreiner, M. Baldo, J.R. Bergueiro, D. Cartelli, W. Castell, V. Thatar Vento, et al., Appl. Radiat. Isot. **88**, 185-189 (2014)
4. H. Kumada, F. Naito, K. Hasegawa, H. Kobayashi, T. Kurihara, K. Takada, T. Ohnishi, H. Sakurai, A. Matsumura and T. Sakae, Plasma and Fusion Research, **13** (2406006), 1-6 (2018)
5. H. Tanaka, Y. Sakurai, M. Suzuki, S. Masunaga, Characteristics comparison between a cyclotron-based neutron source and KUR-HWNIF for boron neutron capture therapy. Nucl. Instrum. Methods. Phys. Res. B **267**, 1970–1977 (2009)
6. K. Yamamoto, H. Kumada, Y. Torii, T. Kishi, T. Yamamoto and A, Matsumura, Simple estimation method of gamma-ray dose using low neutron-sensitive TLD (UD-170LS) for intra-operative boron neutron capture therapy (IOBNCT). Research and Development in Neutron Capture Therapy, (Monduzzi), 499-503 (2002)
7. H. Kumada and K. Takada, Therapeutic Radiology and Oncology, **2**(50), 1-11 (2018)

EPJ Web of Conferences **231**, 01004 (2020) https://doi.org/10.1051/epjconf/202023101004
UCANS-8

The SONATE project, a French CANS for Materials Sciences Research

Frédéric Ott[1,*], *Alain* Menelle[1], *Christiane* Alba-Simionesco[1]

[1]Laboratoire Léon Brillouin CEA/CNRS, Université Paris Saclay, 91191 Gif sur Yvette Cedex, France

Abstract. We describe the Compact Accelerator-based Neutron Source SONATE which we are aiming for to replace the closed Orphée reactor at Saclay, France. The SONATE source would serve an instrumental suite of about 10 instruments. The instruments would be split into low resolution instruments and higher resolution instruments. Our reference design is based on a proton accelerator operating at an energy in the range 20-30 MeV. The accelerator would serve 2 target stations. The first one operating at 20Hz with 2ms long pulses serving low resolution instruments (SANS, reflectivity, imaging, spin-echo) and the second one operating at 100Hz, 200µs long pulses serving higher resolution instruments (powder diffraction, Direct Time-of-flight spectroscopy, Indirect geometry spectroscopy). The 2 operation modes would be interlaced. The peak current on the target is aimed at 100 mA with an average power on the target on the order of 50-80 kW. Numerical Monte-Carlo simulations show that we may expect instrument performances equivalent to the current instruments around Orphée or ISIS.

1 Introduction

The reactor Orphée at the CEA Saclay has been serving the French neutron scattering community for 38 years. The reactor will however stop operation in 2019. The Laboratoire Léon Brillouin has been operating up to 24 neutron scattering instruments. The reactor was also serving other purposes such as industrial radiography, silicon doping and irradiation. We are considering replacing the reactor neutron source with a compact accelerator-based source which would serve an instrumental suite of about 10 instruments. In the following we describe the design parameters of a Compact Accelerator-based Neutron Source which would be suitable to replace the reactor Orphée for neutron scattering purposes. This design is referred to as "SONATE".

2 The SONATE design parameters

We will consider the different degrees of freedom in the design of a Compact Accelerator-based Neutron Source and derive machine parameters we think are suitable for a CANS dedicated to neutron scattering which would provide sufficient performances to be operated as a user facility. The discussion is limited to neutron produced by the stripping reaction corresponding to proton energies below 50 MeV impinging a Lithium or Beryllium target.

2.1. Technological boundaries

2.1.1 Accelerator technology

Accelerators can operate in continuous mode or in pulsed mode. For neutron scattering the most efficient operation is in pulsed mode to benefit from the time-of-flight techniques. Most spallation sources are nowadays operating in pulsed mode which allows making use of most of the produced neutrons. In such pulsed machines, the key figure of merit is the peak flux which is given by the proton peak current.

In the ESS design [1], the peak current was set to a conservative value of 62 mA. New facilities such as IFMIF/EVEDA have been operating at currents of up to 125 mA [2]. We will thus assume that it is possible to operate reliably a LINAC accelerator at a peak current of 100 mA even though such high intensity accelerators are not common.

The choice of the ion particle is still under debate. The nuclear data are rather scarce in the 3-60 MeV range and the cross sections are poorly known. Experimental work is under way to fill the database gap in the energy range of interest [3]. First results suggest that the neutron yield gains obtained with deuterons are no higher than 50% at best which is hardly enough to justify using deuterons over protons. Protons are significantly easier to handle than deutons and also they do not induce activation issues in the accelerator parts as deuterons do. Hence in the following we will only consider protons.

2.1.2 Target material

The choice of the target material is non-trivial and would require an extensive discussion. The situation can be roughly summarized as follows:

* Corresponding author: frederic.ott@cea.fr

- For proton energies below 3 MeV, lithium should be preferred due to the low stripping reaction threshold. This is typically the choice made for low energy Boron Neutron Capture Therapy (BNCT) accelerators for example.
- For proton energies in the range 3 MeV- 60 MeV, beryllium and lithium are roughly equivalent in terms of neutron yield. However using significant power levels (kW) on the target will require to handle molten lithium which is challenging but is justified for very high power since it makes the cooling more efficient (see IFMIF@5MW). Beryllium on the other hand has a very high melting point (1287°C) and allows operating a solid target up to power densities in the range 0.5-1 kW/cm².
- From 60-100 MeV, there are several candidate materials (among which carbon for example). Again the lack of nuclear data makes choices non trivial.
- Above 100MeV, the neutron production via spallation channels starts being efficient and a heavy material target becomes the best choice (e.g. tantalum).

A key choice in a facility design is thus the choice of the proton energy. This choice must be weighted by different factors: (i) the neutron yield, (ii) the accelerator cost, (iii) the power deposited on the target.

2.2 Choice of the proton energy

In the case of beryllium, there is a threshold of 2 MeV for neutron production and above 20 MeV the yield is roughly proportional to the proton energy: Yield ~ E_p - 12 MeV.

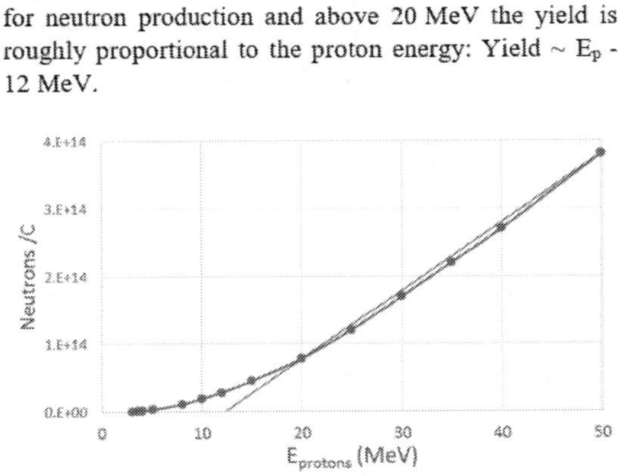

Fig. 1. Neutron yield as a function of the proton energy on a beryllium target (ENDF database).

The design of an efficient TMR requires that the target is as small as possible. A typical (maximum) size is on the order of 100 cm². Thermo-hydraulics requires that the maximum power density is in the range 0.5-1 kW/cm² which sets a limit of the ion beam energy at 50-100 kW. The Fig. 2 shows the power deposited on the target as a function of the proton energy. The orange line at P= 50 kW corresponds to a "safe" limit where thermo-hydraulics design are not too challenging. The limit at P=100 kW is probably a hard limit above which a fixed target cannot be used and a rotating target is necessary. The different load lines correspond to increasing duty cycle of the source.

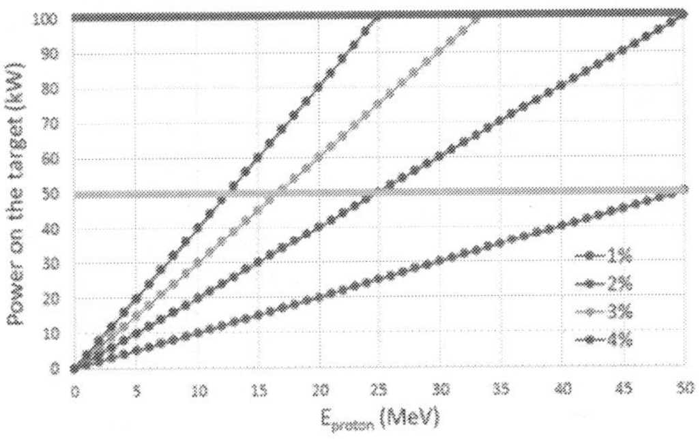

Fig. 2. Power on the target for increasing proton energies and various duty cycles from 1% to 4%.

As a reference, the long pulse source ESS is operating with 2.86ms pulses at 14 Hz corresponding to a duty cycle of 4%. In this operation scenario, the maximum proton energy which can be used is about 20-25 MeV. If the proton energy is increased to 50 MeV, the duty cycle has to be reduced below 2% otherwise the power density on the target would exceed reasonable levels. These figures are of course only indicative since the 100 kW limit may be either difficult to achieve or may be overcome depending on the technology used. Nevertheless it illustrates that compromises have to be made between the proton energy and the duty cycle for a given power on the target. This is illustrated on Fig. 3a. For example, for a proton energy of E_p= 20MeV, a peak current of 100 mA and a 4% duty cycle, the power on the target would be 80 kW and the neutron yield would be 3.1×10^{14} n/s. If the proton energy is increased by a factor 2 to 40 MeV while the power on the target remain limited at 80kW, the duty cycle has to be reduced to 2% and the neutron yield is 5.4×10^{14} n/s. For a given power deposited on the target, the neutron yield is roughly proportional the proton energy.

It is difficult to define a simple figure of merit since a number of "soft" parameters also play a role:

- What is the most suitable duty cycle? This depends on the aimed applications.
- For lower energy protons, the fast neutron spectrum is less energetic and thus easier to moderate.
- The gamma background is also less energetic for lower energy protons.
- A lower energy accelerator is cheaper to build.
- Above E_p = 30MeV, new activation channels open and lead to activation of accelerators parts which can make the maintenance more complicated.

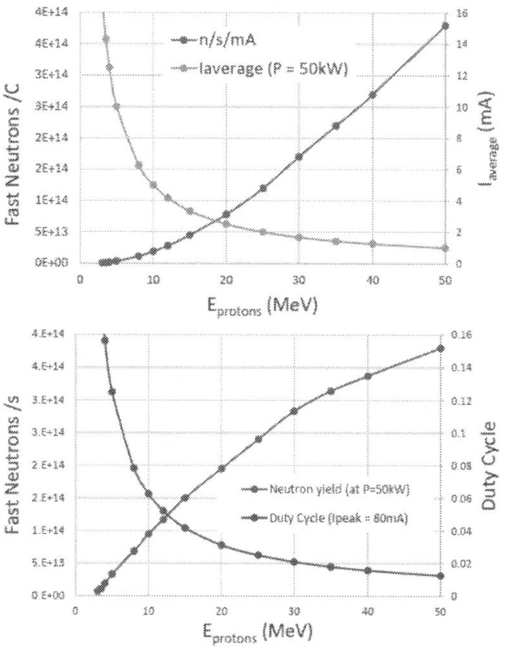

Fig. 3. (a) Neutron yield per Coulomb versus proton energy and maximum average current to limit the power on the target at 50 kW. (b) Actual yield with a power limit on the target set at 50 kW and corresponding duty cycle (for I_{peak} = 80mA). While the neutron yield per proton doubles from E_P = 20 to 30 MeV, power limits on the target reduce the gain to 45%. In parallel, the duty cycle on the target is reduced (from 3% to 2% if I_{peak} = 80 mA) which makes the neutron pulses easier to exploit for scattering experiments. From 20 MeV to 30 MeV the gain in flux is roughly proportional to the accelerator energy. From 30 MeV to 50 MeV, the gain in flux is 36% while the proton energy is increased by 67%.

Assuming that the accelerator cost and operation is proportional to its energy (which is rather crude), a simple figure of merit could be defined as FOM = [Neutrons Yield/ $E_{protons}$] (see Fig. 4) which reflects the cost per produced neutron. It is clearly efficient to work above 10 MeV. However, even though the neutron yield per proton increases quickly with the proton energy, the figure of merit nevertheless decreases slowly above 20 MeV.

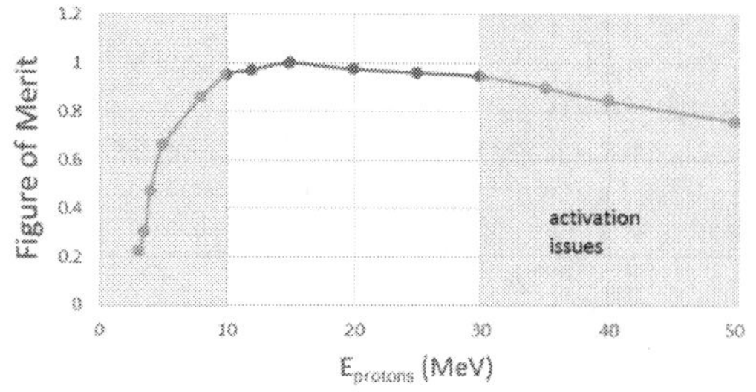

Fig. 4. Figure of merit defined as [Neutron Flux / Cost] (a.u.)

Below E_p = 15 MeV, the maximum duty cycle is above 4% which becomes difficult to exploit efficiently for ToF neutron scattering. Above 30 MeV, proton activation channels open up and make the accelerator maintenance more complicated. Hence an optimal operation region for CANS seems to be in the range 10 – 30 MeV.

2.2 The SONATE design parameters

From the above considerations, we decided to set the SONATE reference parameters as:

- E_p = 20MeV,
- I_{peak} = 100mA,
- duty cycle = 4%, P = 80kW,
- fixed Be target.

These parameters were partly chosen because they correspond to the first 20 m of the ESS Linac (out of 600m). Hence the components (Source, RFQ and DTL) are in principle available with no R&D developments.
From a French nuclear regulation perspective, installations producing ionizing particles may be classified as "Installations Classées pour la Protection de l'Environnement" (ICPE) or "Installations Nucléaires de Base" (INB) [4]. In the latter case, fall of course research reactors (such as Orphée@Saclay) and particles accelerators (such as SPIRAL2@Caen) producing radionuclides above some legal thresholds. The INB categories are subject to very stringent rules which make their exploitation difficult.
Calculation of the activation products produced within the SONATE indicate that the facility would be considered as a simple ICPE.
In other countries, other rules and thresholds apply.

3 The expected performances for neutron scattering

In order to have an estimate of the performances a source such as SONATE could provide, a moderator design using polyethylene as moderating medium and beryllium as a reflector was considered. Moderation calculations were performed both with MCNP and GEANT4 [5]. In the case of the SONATE design parameters, a brilliance of $1.2x10^{11}$ $n/cm^2/s/sr$ was calculated and was used as an input in the instruments Monte-Carlo simulations (using McSTAS).
The neutron flux at the sample position for various neutron scattering techniques was calculated using simple ToF instrument designs [6] or even considering existing instruments around Orphée which were simply "moved" around SONATE. The results are summarized in the table below and compared with "reference" instruments at sources such as Orphée@Saclay or ISIS. The figures for the inelastic instruments (Direct TOF and Backscattering) have been taken from [7]. The orange figures correspond to various types of instrumental upgrades which are either in progress at the LLB or being implemented on some of the ESS instruments. In the case of reflectivity, the SELENE@ESS design could

be implement to increase reflectivity measurement efficiency by an order of magnitude on small samples. In the case of SANS, focussing SANS is a rather simple technique which can be implemented to increase the flux at the sample position.

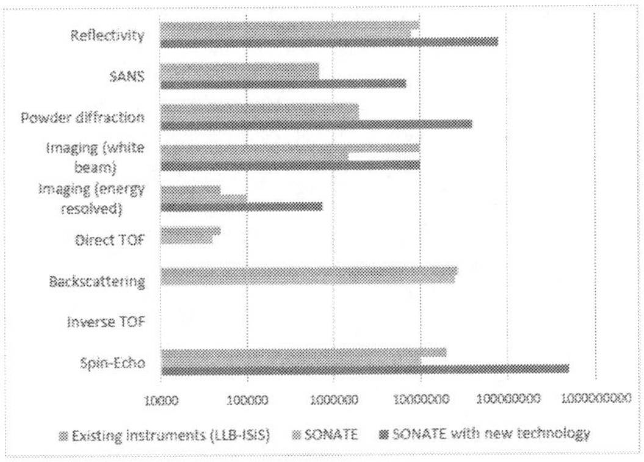

Fig. 5. Comparison of the performances of different scattering techniques in terms of flux at the sample position (n/cm²/s). (green) reference instruments at various facilities (LLB or ISIS) ; (yellow) move of the existing instruments from Orphée to SONATE ; (orange) performances after technical upgrades. The figures for the inelastic instruments (Direct TOF and Backscattering) have been taken from [7].

While it may be surprising that a CANS may achieve performances on par with existing medium scale research reactors or spallation sources, the origin for these performances can be qualitatively explained by the following reasons:

- The proton energy remaining low in a CANS, using a very high peak current is not detrimental to the overall (electrical) energy consumption of the source. While the ISIS TS2 is operated with an average current of 60μA, SONATE would be operated at an average current of 4mA. This large current compensates a large fraction of the neutron production efficiency difference between stripping and spallation.
- The moderator design is such that it is almost fully coupled to the fast neutron source. This is possible due to the small size of the primary fast neutron source, to the rather small heat load of the source but also from the rather low radiative gamma heating which is low is CANS especially compared to spallation. In CANS, the gamma spectrum is limited to the proton energy (a few 10 MeV) while in a spallation source high energy gammas are present in the source. The tight coupling between the source and the moderator leads to a gain of approximately a factor 5 compared to the coupling of the ESS moderator.
- The moderator design can make use of modern moderator design such as tube moderators as proposed recently [8]. A gain of a factor 5 relative to moderators at current sources is also expected.
- For each type of instrument, the source time structure has been optimized to fully fill the phase space, that is, the pulse length and the repetition rates have been considered as free parameters and hence for each instruments almost all the neutrons can be used for scattering. Since the shielding constrains are very low, the chopper systems can be in principle set very close to the source (as close as 1 m in theory) leading to additional gains compared to larger facilities.

Around existing sources, the time structure is usually fixed and better suited for specific instruments. For examples, short pulse spallation sources are well suited to high resolution experiments while short pulses are inefficient for low resolution experiments such as SANS, reflectometry or imaging. In the case of ESS, the operation parameters (2.6 ms, 14 Hz) were chosen as a weighted compromise to serve a very wide range of instruments. It may be argued that a CANS would face similar choices. The key difference is that the design and construction of a Target – Moderator – Reflector (TMR) assembly should have a rather limited cost (<1M€). It is even considered that the best option would be provide each instrument with its dedicated tube moderator [9]. A rather easy way to optimize the source time structure is to build several target stations with optimized instrumentation. The proton beam structure could have the time structure illustrated on Fig. 6 which interleaves long pulses at a low repetition rate with short pulses with a high repetition rate.

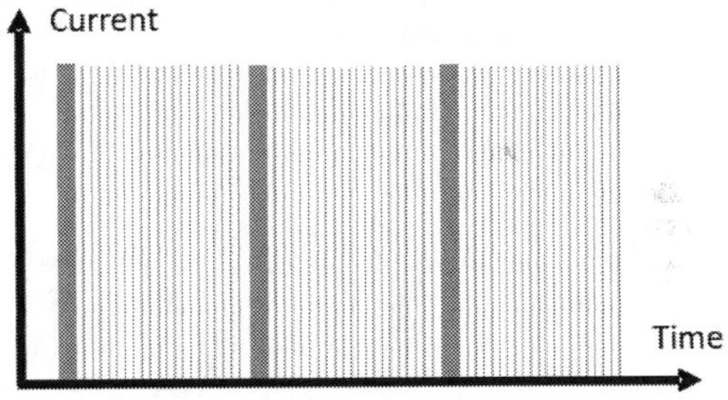

Fig. 6. Example of the proton pulse structure on SONATE. Long pulses (2ms, 20Hz) are interleaved with short pulse high repetition rate pulses (200μs, 100Hz). The long pulses are directed to a TMR station for low resolution instruments while the short pulses are directed to a second TMR station for higher resolution instruments. The first target would use a fraction of 4% of the duty cycle while the second target would use 2% of the duty cycle. Hence the accelerator should be designed to handle a 6% duty cycle.

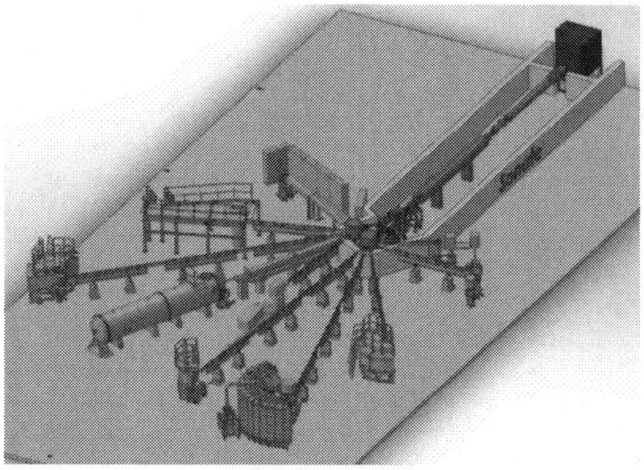

Fig. 7. A possible sketch of SONATE. A first long pulse / low repetition rate target station (green) would serve low resolution instruments such as an imaging station, a SANS, a reflectometer, a spin-echo and a low resolution powder diffractometer. A second target station with short pulses and high repetition rate would serve higher resolution instruments such as a direct TOF instrument, a high resolution powder diffractometer, an inverse TOF diffractometer.

3 Conclusion

With the foreseen loss of neutron capacity in Europe due to the closure of aging neutron research reactors, alternative solutions must be found to continue providing neutrons to neutron scattering users. We think that current accelerator technology is mature enough to build CANS which can provide neutrons for scattering instruments and have performances on par with current state of the art medium scale research reactors or medium scale spallation sources. Beyond providing an alternative to produce neutrons, these CANS represent an investment which is only a fraction of the cost of new nuclear reactors or spallation facilities together with reduced operation costs. Hence there is the possibility that the CANS technology may even allow an easier access to neutron scattering compared to the current situation. In which case, a network of CANS across Europe could support an extended user community. This would be beneficial for the efficient use of the future European Spallation Source.

References

1. ESS Technical Design Report (2013). https://docdb01.esss.lu.se/DocDB/0002/000274/006/TDR_final_130423_online_ch1.pdf
2. Successful acceleration of a 125 mA deuteron beam @5 MeV in LIPAc (July 2019). https://www.ifmif.org/
3. T. Gutberlet, HBS meeting (Unkel, 2018)
4. INB are defined in France by the Décret n° 2007-830
5. H.N. Tran, A. Marchix, A. Letourneau, N. Chauvin, A. Menelle, F. Ott, J. Schwindling, IOP Conf. Series: Journal of Physics: Conf. Series **1021** (2018) 01200. doi :10.1088/1742-6596/1021/1/012008
6. F. Ott, H. N. Tran , X. Fabrèges, A. Menelle, N. Chauvin, J. Schwindling, A. Letourneau, A. Marchix, C. Alba-Simionesco, IOP Conf. Series: Journal of Physics: Conf. Series **1021** (2018) 012007. doi :10.1088/1742-6596/1021/1/012007
7. J. Voigt, S. Böhm, J.P. Dabruck, U. Rücker, T. Gutberlet, T. Brückel, NIM A **884** (2018) 59-63. doi.org/10.1016/j.nima.2017.11.085.
8. F. Mezei et al , Journal of Neutron Research **17** (2014) 101–105. DOI 10.3233/JNR-140013.
9. S. Böhm, T. Cronert, J.-P. Dabruck, X. Fabrèges, T. Gutberlet, F. Mezei, A. Letourneau, A. Menelle, F. Ott, F. Porcher, U. Rücker, Hoang Tran, J. Voigt, P. Zakalek, *Neutron Scattering Instrumentation at Compact Neutron Sources (Gif sur Yvette, 2018).* https://hal-cea.archives-ouvertes.fr/cea-01870227

From micro- to macro- neutron sources: The Lund Broad-band Neutron Facility

Hanno Perrey[1,*], *Mikael* Elfman[1], *Kevin* Fissum[1], *Robert J. W.* Frost[1], *Nathaly* De La Rosa[1], *Markus* Kristensson[1], *Per* Kristiansson[1], *Nicholai* Mauritzson[1], *Francesco* Messi[1], *Charlotta* Nilsson[1], *Jan* Pallon[1], *Emil* Rofors[1]

[1] Lund University, Division of Nuclear Physics, Professorsgatan 1, 22363 Lund, Sweden

Abstract. The Lund Broad-band Neutron Facility provides access to a variety of neutron sources together with a well-established user infrastructure. Neutrons from radioactive sources have been successfully employed for the research and development of both detectors and materials for the European Spallation Source. A recently procured d-t neutron generator delivers higher neutron fluxes than those provided by the radioactive sources, and further allows for pulsed operation. With the currently on-going construction of a dedicated neutron beam-line at our 3 MeV Pelletron accelerator, the facility is anticipated to produce approximately 10^{10} n/s. Cost-effective access to neutrons as well as a platform for educational purposes are the ultimate goals of the project.

1 Introduction

Neutrons of all energies are important probes of matter and are crucial to an increasingly number of applications in both scientific and industrial fields. For many novel developments and for training purposes, dedicated local facilities offering straightforward and affordable access to neutrons are a vital need of the community. The Lund Broad-band Neutron Facility (LBNF) has been designed with this and a low barrier-of-entry in mind. Hosted by the Division of Nuclear Physics at Lund University and operated by the Neutronics Group, the LBNF provides well-established and user-focused infrastructure, expertise in nuclear physics and detection techniques and access to neutrons from both accelerator-based neutron sources and neutron-emitting radioactive sources. Since 2014, the facility has functioned as infrastructure for ESS-related detector and material research and development as well as an educational platform [1]. Currently, the 3 MeV Pelletron accelerator at the LBNF is being upgraded with a dedicated neutron production beam-line. The anticipated increase in neutron flux from approximately 10^6 n/s (radioactive source) or 10^8 n/s (neutron generator) to 10^{10} n/s will open up new possibilities for applications in other fields.

2 Source-based neutrons

Source-based continuous neutron beams from an ^{241}Am/Be source provide stable fluxes of typically 10^6 n/s. At the LBNF, these beams have been successfully used for response characterizations of neutron detectors [2,3,4] and their data-acquisition systems [5]. Using a technique called "tagging" and the infrastructure present at the LBNF, it is even possible to study attenuation and detector responses as function of the neutron energy even with standard radioactive sources [6,7,8]. Based on coincidence measurements between prompt gamma-ray emissions and the neutron in the reaction $\alpha + ^9Be \rightarrow ^{12}C^* + n \rightarrow ^{12}C + \gamma + n$ which are present in around 55% of all decays, one can determine the time of flight of the neutron on an event-by-event basis.

As radioactive sources are a stable source of neutrons, they are also a cost-effective tool for educational purposes. At the LBNF, students have used the neutron tagging setup in laboratory exercises in nuclear physics courses. As the setup consists of a water-filled Plexiglas tank, all the key components of the experiment are visible and accessible while keeping the dose-rate on the external surface of the tank well below 0.5 μSv/h. Coupled with modern teaching techniques such as flipped classroom and blended learning, the lab makes key concepts such neutron interactions in matter, neutron detection and neutron time of flight tangible and thus facilitates learning. Figure 1 shows a photograph taken during one of the lab exercises where students test the effect of shielding materials on the measured neutron flux.

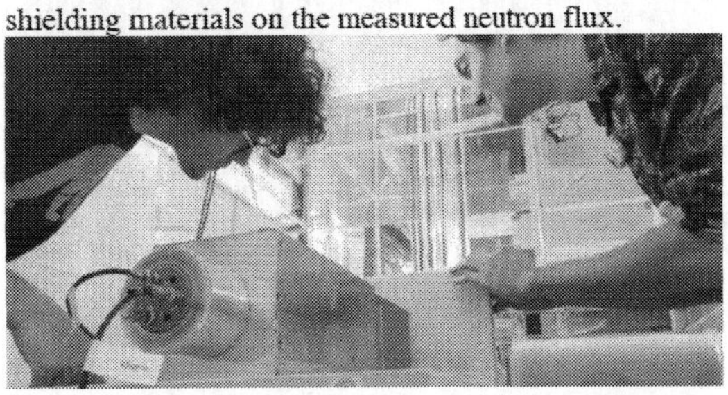

Figure 1: Photograph taken during neutron time-of-flight lab exercise showing students interacting with the setup. Visible are the water-filled tank hosting the radioactive source (background), the neutron detector (foreground) and shielding material being introduced into the neutron beam.

3 Accelerator-based neutrons

For applications requiring higher or variable neutron fluxes, the LBNF features a SONDERN Genie 16 neutron generator. Based on the d+t reaction, calibrated fluxes of up to 10^8 n/s can be achieved in DC operation mode. Even

* Corresponding author: hanno.perrey@nuclear.lu.se

higher fluxes are anticipated from the currently ongoing upgrade of the 3 MeV Pelletron accelerator with a dedicated beam-line for neutron production. A 10 μA, 2.5 MeV deuterium beam will be directed onto a beryllium target which, when run in continuous mode, is anticipated to produce approximately 10^{10} n/s in 4π sr.

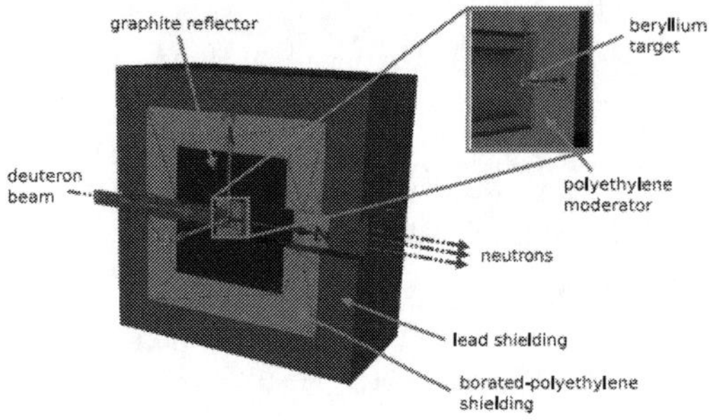

Figure 2: Geometry of the moderator and shielding assembly as simulated in GEANT4

A polyethylene moderator, graphite reflector and lead/borated-polyethylene shielding system will then thermalize the neutrons and direct them to the experimental area. The design for the moderator/shielding assembly is illustrated in Figure 2. The functionality of each of its material layers are as follows: The innermost polyethylene block, which surrounds the end of the beam-line, moderates fast neutrons exiting the beryllium target. From this moderator block there is an exit path cut through the surrounding layers in the forward direction to allow neutrons to escape. Surrounding the moderator is a graphite layer which acts as a neutron reflector, redirecting neutrons that escape the moderator in all but the forward direction. The graphite layer is then surrounded by borated-polyethylene to capture escaping neutrons that are not fully reflected. Finally, an outer casing of lead is placed to absorb gamma-rays produced in the target and by thermal neutron capture on boron.

At present the beam-line required to transport the deuterons, including all magnets, the electrostatic deflector system, a beam-dump and water-cooled target-holder, has been constructed and is ready for operation. Extensive simulations have been performed in GEANT4 [9] to model the performance of the moderator and shielding assembly.

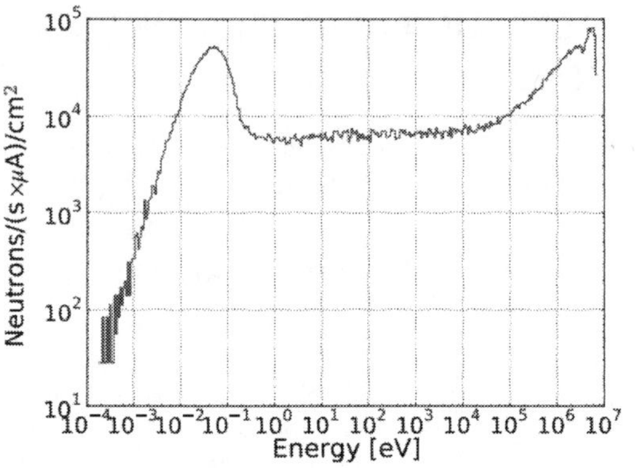

Figure 3: Simulated neutron energy spectrum at the exit port of the current moderator/shielding concept.

With the deuteron beam run in continuous mode the neutron flux at the exit port in the moderator/shielding assembly is anticipated to be on the order of 5×10^5 n/cm²/s, of which the thermalized-neutron flux will be around 5×10^4 n/cm²/s. Figure 3 shows the simulated neutron flux per μA of deuteron beam as function of neutron energy directly at the exit port of the currently favoured moderator concept. The spectrum shows a peak at neutron energies between 20 – 100 meV and contributions of similar order of magnitude from fast (>1MeV) as well as neutrons of intermediate energies. Further control of the neutron energy will be implemented by pulsing the beam using an electrostatic deflector and using time of flight to discriminate against unwanted neutron energies. Such an increase in neutron flux compared to the other sources available at the LBNF opens up many possibilities for neutron-detector characterizations or fast-neutron activation analysis targeting fast decay activation products using the available high-sensitivity gamma-ray spectrometer. Further experimental instrumentation at the neutron beam-line is foreseen to be developed and depends on future users' and collaborators' needs.

4 Summary

The LBNF provides its users with a variety of options for neutron sources of different intensities and beam characteristics. An on-going upgrade of the Pelletron accelerator with a dedicated neutron beam-line will extend this suite of sources and is expected to produce approximately 10^{10} n/s.

This cost-effective access to a wide range of neutron energies and beam types reduces the barrier of entry to novel neutron-related developments and applications in both science and industry significantly. Furthermore, such a facility located at a University provides an ideal site for the education of the next generation of neutron scientists.

References

1. Messi, Perrey et al, accepted by IAEA (2018)

2. Issa, Khaplanov, Hall-Wilton, Llamas, Riktor, Brattheim & Perrey, *Phys. Rev. Accel. Beams*, **20**, 092801 (2017). doi.
3. Mauri et al, *Journal of Instrumentation*, **13(03)**, P03004-P03004 (2018). doi.
4. Messi, Mauri & Piscitelli, *NIM A*, **936**, 499-500 (2019). doi.
5. Christensen, et al, *Journal of Instrumentation*, **13(11)**, (2018). doi.
6. Scherzinger, et al, *NIM A*, **840**, 121 - 127 (2016). doi.
7. Scherzinger et al, *Appl. Rad. and Isotopes*, **128(C)**, 270 - 274 (2017). doi.
8. DiJulio, Cooper-Jensen, Perrey, Fissum, Rofors, Scherzinger & Bentley, *NIM A*, **859**, 41 - 46 (2017). doi.
9. J. Allison et al, *NIM A*, **835** 186-225 (2016). doi

Development of a Setup for Material Identification Based on Laser-Driven Neutron Resonance Spectroscopy

Marc Zimmer[1*], *Stefan* Scheuren[1], *Annika* Kleinschmidt[2,3], *Alexandra* Tebartz[1], *Tina* Ebert[1], *Johannes* Ding[1], *Daniel* Hartnagel[1], and *Markus* Roth[1]

[1]Technische Universität Darmstadt, Institut für Kernphysik, 64289 Darmstadt, Germany
[2]GSI Helmholtzzentrum für Schwerionenforschung, 64291 Darmstadt, Germany
[3]Helmholtzinstitut Jena, 07743 Jena, Germany

Abstract. With the phasing out of many research reactors over the upcoming years, a shortcoming of small and medium sized neutron sources is to be expected. Laser-driven neutron sources have the potential to fill this void, with enormous progress being made in laser technology over the past years. Upcoming petawatt lasers with high repetition rates up to 10 Hz promise a tremendous increase in neutron flux. In this paper, a setup is developed and optimized to conduct neutron resonance spectroscopy at a laser-driven neutron source. This setup is then evaluated at an experimental campaign at the PHELIX laser system. Laser intensities up to 10^{21} W/cm² with a ns pre-pulse contrast of 10^{-7} were used for ion acceleration, resulting in $(1.8\pm0.7)\times10^{8}$ N/sr per pulse corresponding to $(2.3\pm1.0)\times10^{9}$ N in a 4 π equivalent. These pulses were moderated, collimated and investigated via the time of flight method in order to characterize the thermal neutron spectrum as well as the signal to noise ratio.

1 Introduction

Neutrons provide a unique tool for material analysis, from neutron diffraction [1] over prompt gamma neutron activation analysis [2] to *N*eutron *R*esonance *S*pectroscopy (NRS) [3]. These techniques can be used to identify the elemental and isotopic compositions of unknown materials without compromising the structural integrity of the sample. A key feature of neutrons is the large propagation range into materials. In combination with the unique distribution of scatter and capture resonances for every isotope, neutrons provide a particularly valuable analysis tool, complementing currently existing x-ray techniques.

In previous years, neutrons were in many cases provided by research reactors. With recent developments, a large fraction of these sources have been or will be shut down in the near future and a need for small and medium-sized neutron sources is growing [4]. *L*aser-*D*riven *N*eutron *S*ources (LDNS) promise to fulfill these needs and provide a compact short pulsed neutron source with high peak flux [5]. Recent improvements in laser technology enable high-intensity lasers to be operated up to 10 Hz [6], therefore, tremendously increasing the average neutron flux of these sources.

LDNS are especially well suited for material analysis utilizing NRS since it relies on the *T*ime *o*f *F*light (ToF) method which requires short initial pulse durations to decrease the energy uncertainty. LDNS have sub-ns production pulse lengths and consequently the dominant energy uncertainty is the moderation time which is below 2 µs for neutrons above 1 eV. Other Compact Accelerator Driven Neutron Sources (CANS) or reactors have longer pulse lengths above 100 µs [7] and therefore require longer flight paths to maintain the same energy uncertainty [3].

Spallation sources can operate similarly to LDNS in a regime only limited by the moderation time but they need longer flight paths due to the larger shielding required for higher neutron energies from the spallation process [8]. A longer flight path L results in a reduction in neutron flux at the detector since the flux decreases with $1/L^2$. This enables a more efficient utilization of the neutron beam in comparison to most other sources, except those based on electron linacs which can reach similar short pulse lengths and neutron energies.

NRS exploits the unique resonance distribution for most isotopes in their neutron cross section $\sigma(E)$. Therefore, when a neutron pulse with a broad energy spectrum, as it is produced by LDNS, travels through a sample, resonances with a high cross section will absorb parts of the spectrum, imprinting this information on the neutron distribution. A detector behind the sample can, therefore, identify the presence and the amount of each isotope in relation to the depth and position of the absorption peaks. The following part of this publication will describe the optimization and the characterization of an LDNS for NRS.

* Corresponding author: m.zimmer@gsi.de

2 Setup

For this experiment, the PHELIX laser at GSI was used. The (600±100) fs laser pulse [9] was focused to a FWHM diameter of (4±0.5) μm with (180±12) J per pulse. This setup allows peak intensities of up to 10^{21} W/cm² enabling ion acceleration via the Target Normal Sheath Acceleration (TNSA) [10] and via relativistic induced transparency that can commonly deliver ion energies between 60 and 85 MeV [11]. During this campaign, the ns laser contrast was reduced to 10^{-7} and consequently, maximum ion energies were limited to around (33±10) MeV. As target material, deuterated polystyrene foils with 500 nm thickness as well as 10, 25 and 50 μm silicon targets were used. Some of them with a micro-needle structure on the laser facing surface and a flat rear side for better conversion efficiency.

The neutron production is carried out using a pitcher-catcher design [5] where ions are directed from the target (pitcher) onto a beryllium cylinder (catcher). There, the ions are converted into neutrons via ^{9}Be(d,n)^{10}B, ^{9}Be(p,n)^{9}B as well as stripping reactions. Emitted neutrons have energies in the MeV regime, but for most NRS applications, neutrons in the epithermal (~1 to 100's of eV) regime are required. Therefore, the neutrons are directed into a moderator made out of *H*igh *D*ensity *P*oly*E*thylene (HDPE). The high hydrogen content slows neutrons down until they are emitted via diffusion at the moderator surface and propagate towards the detector.

To ensure only neutrons with a direct flight path from the moderator to the detector are counted, a series of borated HDPE plates are installed. These plates have an opening in the center with a converging diameter towards a borated *M*ulti-*C*hannel *P*late (MCP) detector to act as a collimator as seen in Fig. 1.

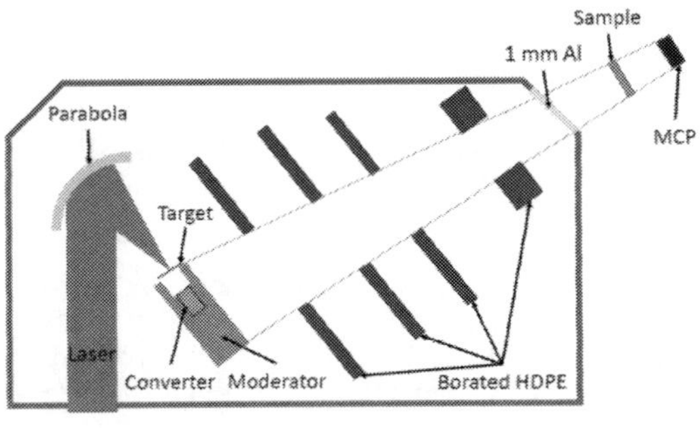

Fig. 1. Setup of the experiment. Neutrons are produced inside the converter, slowed down in the moderator and are emitted to the MCP via the flight path indicate by the light-blue lines.

The sample is placed at 10 cm distance in front of a MCP that is capable of single neutron detection. In this configuration, absorption gamma photons produced inside the sample have a smaller chance of reaching the detector in comparison to placing the sample directly in front of the detector.

The MCP itself is placed at a distance of (180±1) cm and is additionally shielded with 10 cm borated HDPE in all other directions which is not shown in Fig. 1 to improve clarity. The detector used in this experiment is a custom made design, consisting of a borated MCP from Nova Scientific encapsulated in a vacuum housing by the ProxiVision GmbH with a spatially integrated anode for a faster time response. This detector is capable of single neutron detection with a maximal time resolution of 10 ns. The neutron energy was calculated via ToF. The neutron flight time t was determined by:

$$t = t_d - t_0 - t_c - t_s \quad (1)$$

With the detection time t_d, the arrival time of the gamma flash t_0, the time delay of light to propagate from the target to the detector t_c = 6 ns and the average delay neutrons of a certain energy need to cross the surface of the moderator t_s.

3 Design and Simulations

3.1 Catcher

During the operation of an LDNS, the beryllium catcher has to be shielded against ablation by the expanding hot laser plasma, since beryllium dust is highly toxic and the integrity of the catcher has to be maintained. Therefore, a blast shield in front of the beryllium is required. In previous experiments [12], a 1 mm Al plate was used. Inside this blast shield deuterons with an energy below 18 MeV are stopped and do not reach the converter. Al itself has a low neutron yield and consequently, a large amount of ions does not contribute to the neutron production. Hence a material with low stopping power, high thermal and ablation resistivity, as well as relatively high neutron yield from deuteron bombardment should be chosen as a blast shield. Cirlex consists of a polyimide material with a high thermal resistivity and has a high carbon content which has a relatively high (d,n) cross section of 200 mb between 2 and 10 MeV [13]. It is comprised of only low Z materials, resulting in a low stopping power thus more deuterons (> 14 MeV) arrive at the catcher. Therefore, it fulfills the requirements as laser-plasma blast shield. In addition, the reaction products of d/p irradiation produce mostly short-lived radioactive isotopes like ^{13}N with a half-life of 10 minutes, ^{11}C with 20 minutes and stable ^{14}N. This reduces the radiation inside the chamber and enables fast target handling.

3.2 Moderator

Designing a moderator for an epithermal neutron source is fundamentally different from thermal or cold neutron sources. While these sources use reflectors to keep neutrons inside the moderator as long as possible until they diffuse out at the surface [8], an epithermal source has to emit neutrons before they reach thermal equilibrium. On average, neutrons undergo 18 collisions inside HDPE until they are thermalized. For 1 eV they have to be emitted after around 16 collisions. This can be achieved when an optimal thickness is found, where the average scattering number at the surface is corresponding to the desired energy for NRS. To find an optimal moderator configuration, a PHITS [14] simulation was used to scan the length height and width of the moderator while measuring the epithermal neutron flux at a distance of 50 cm to the moderator center. The neutron spectrum was produced by directing a TNSA ion spectrum with up to 32 MeV protons on to a beryllium catcher. Fig. 2 shows the results for the length and the height. It can be seen, that for both cases a saturation effect starts to appear for values larger than 10 cm. The preferred dimensions of this cuboid moderator were chosen to be at a length of 17 cm and a height of 10 cm towards the center.

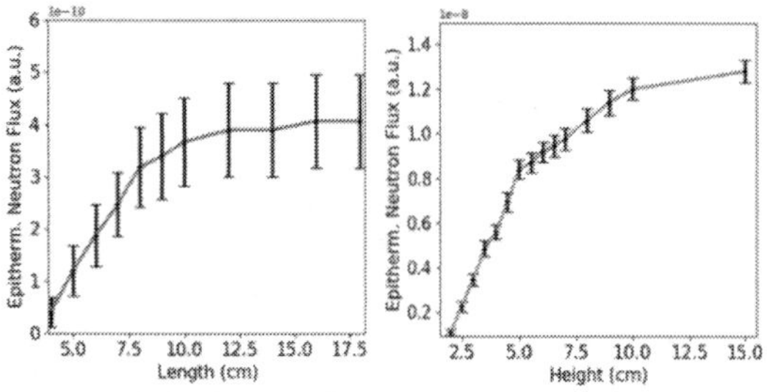

Fig. 2. Epithermal neutron flux as a function of moderator length and height. For both parameters a saturation effect is visible. Simulation performed with the PHITS code.

Fig. 3 shows the increase in epi-thermal flux as a function of moderator width. The flux increases with the thickness until a maximum of 4 cm towards the moderator center is reached. Afterwards the flux decreases until a width of 5 cm where the flux increases again until it reaches a second maximum at 5.5 cm. Further increasing the width causes the flux to decrease. The second peak is associated with neutrons that have an initial propagation direction towards the moderator surface as these neutrons pass through less material as those emitted in other directions.

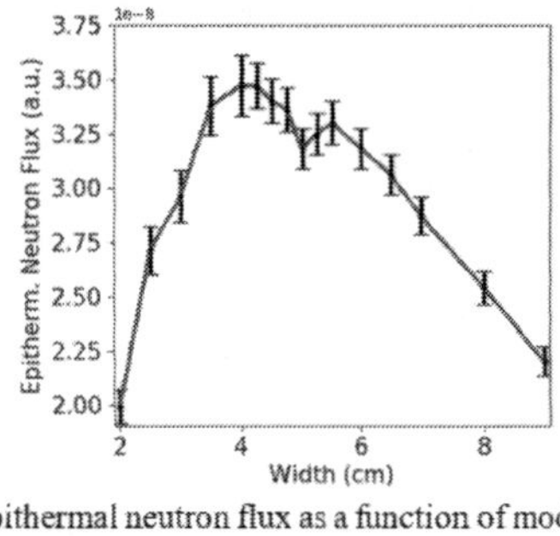

Fig. 3. Epithermal neutron flux as a function of moderator width. First, the flux is increasing with the moderator width until it reaches a maximum at 4 cm. Afterwards, the flux decreases until 5 cm where it starts increasing again. At 5.5 cm a second maximum is visible followed by a constant decrease in neutron flux. Simulation performed with the PHITS code.

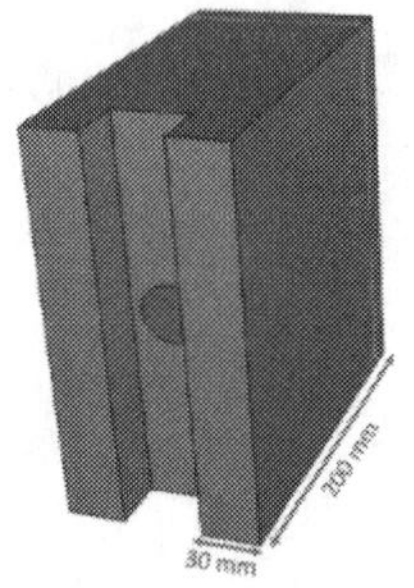

Fig. 4. Optimized moderator design for NRS. The dark circle in the middle represents the Be catcher. In the front the wings are visible.

In addition, two 3x3x20 cm wings were added in front of the moderator as seen in Fig. 4. These wings have a chance to capture neutrons that are emitted into laser backwards direction up to an angle of 56° measured from the converter front surface center. PHITS simulations have shown, that this can lead to an increase in epithermal neutron flux at the detector facing moderator side of 72% and an increase of thermal flux of 46% as seen in Fig. 5. The presence of the wings leads to a new optimal moderator thickness of 5 cm. Fig. 4 shows a CAD drawing of the optimized moderator design.

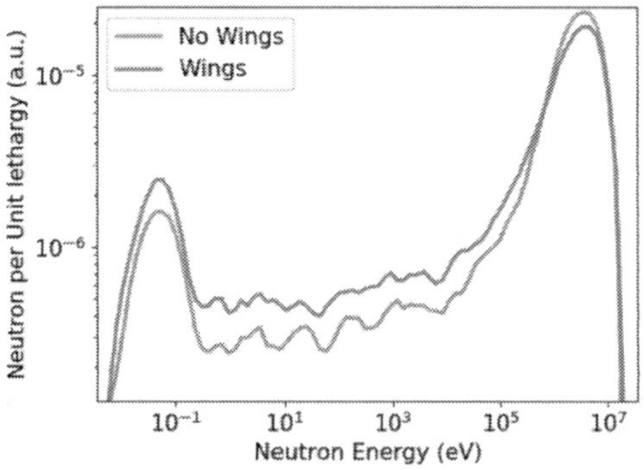

Fig. 5. PHITS simulation for the moderator with and without moderation wings. The epithermal yield is increased by 72% as well as the thermal yield by 46%

With a PHITS simulation, the flux of neutrons between 1 and 10 eV is investigated at the detector facing moderator surface. The result can be seen in Fig.6. Most neutrons in this energy regime are emitted in close proximity to the converter, indicated by the black square.

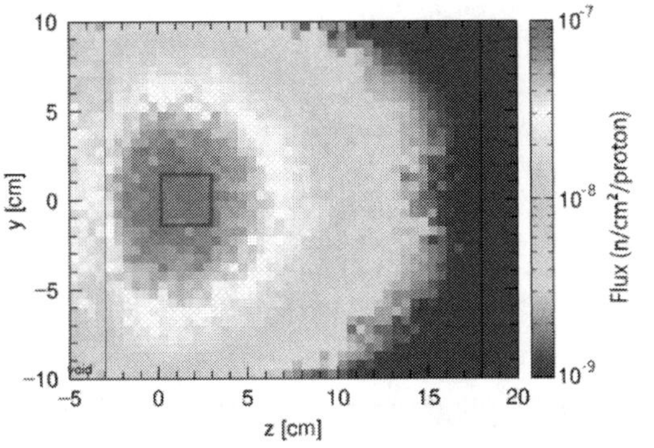

Fig. 6. Flux of neutrons with energies between 1 and 10 eV at the moderator surface. Most neutrons are emitted close to the catcher, indicated by the black square.

The energy resolution of an NRS measurement and, therefore, the minimal detector distance is limited by the uncertainty of the arrival time Δt. There are two main contributing factors to this uncertainty for an LDNS. The first one is the moderation time, e.g. the time a neutron propagates inside the moderator until it reaches its final energy. The second one is the time after the last collision until the neutron reaches the moderator surface. The sum of these two factors is estimated in a PHITS simulation for this moderator design for various energies from 1 eV to 1 keV. The results can be seen in Fig. 7. It is visible that for higher energies the average arrival time shifts to earlier times. Also, the width of each pulse is decreasing with higher energies. The average arrival time can be treated as an offset in the energy determination, while the width of each peak represents the uncertainty for each energy $\Delta t_s(E)$. For 1 eV this width is (2±0.5) μs, for 100 eV (200±50) ns and for 1 keV this width has decreased to (60±10) ns. If this is compared to the minimal ion pulse width at the LANSCE spallation source of 125 ns [8] convoluted with their moderation time, it becomes evident that an LDNS has the potential to achieve a higher energy resolution for neutron energies above 100 eV.

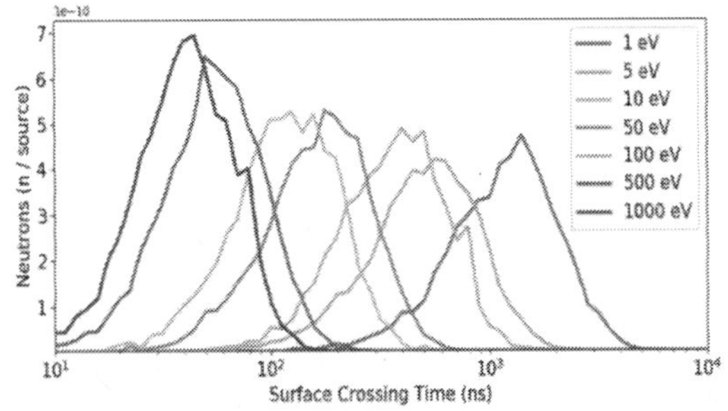

Fig. 7. Time needed for neutrons with different energies to cross the moderator surface. Faster neutrons leave the moderator quicker and have a lower moderation time uncertainty. Energy intervals are ±1%.

3.3 Collimation System

For NRS, it is crucial that neutrons detected at the MCP have traveled in a straight line from the moderator through the sample. Therefore, neutrons have to be prevented from scattering at other parts of the experimental setup, the 3 cm thick steel target chamber wall, the floor or the ceiling. Therefore, three 5 cm thick and 50 cm wide borated HDPE sheets were placed in the beam path with an inner recess fitting the converging projection of the moderator onto the MCP as seen in Fig. 1 and 8. In addition to the three large collimators, four smaller collimators with converging round openings to the detector were used. Two of them before the 1 mm aluminium flange inside the target chamber and two additional collimators directly in front of the detector. This geometry enables to view the entire moderator from the detector to increase the amount of usable neutrons. The increase in energy uncertainty for this setup from the larger visible surface area is below 1% (at 4 eV). This is smaller than the corresponding uncertainty from Δt_s with 3%.

The detector itself was surrounded by a 1 mm copper Faraday cage against the EMP and additional 10 cm borated HDPE in all other directions. The setup was reconstructed in a PHITS simulation to evaluate the collimation and the signal to noise ratio. The result can be seen in Fig. 8 for neutrons with energies between 1 and 10 eV. The collimator achieves, in this simulation, a signal to background ratio of (15±4):1 inside the detector shielding. An analysis of the arrival time and the energy of the neutrons revealed the origin of the background neutrons to be from fast neutrons that have been moderated in the last borated HDPE layer. Additional contributions to the background have been identified to come from the prompt gamma emission from neutron absorption in boron, hydrogen and carbon. This background can be significantly reduced in future experiments by adding additional steel collimation layers directly in front of the last HDPE layers and a cadmium layer around the detector can be used to filter out thermal neutrons. In addition, a lead shielding around the detector can reduce the gamma contribution.

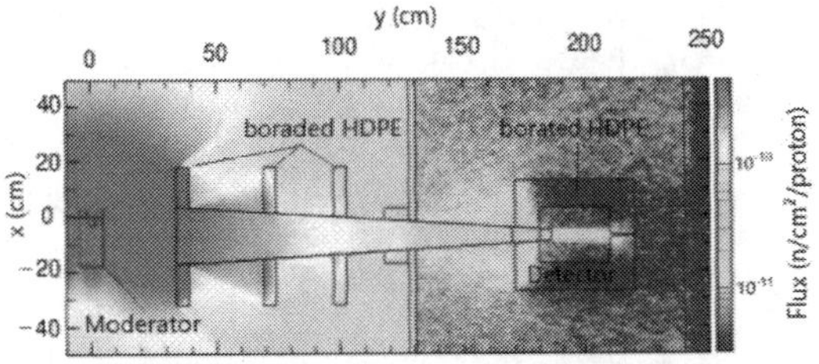

Fig. 8. Simulated flux of neutrons with energies between 1 and 10 eV. Neutrons are collimated by four layers of borated HDPE and enter the detector shielding through a tunnel. The signal to background ratio inside the detector shielding is (15±4):1.

4 Experimental Results

4.1 Determination of the Signal to Background Ratio

In the experimental campaign at PHELIX, two methods were used to determine the signal to noise ratio. For the first method, the thermal neutron spectrum is compared with and without a moderator. The result can be seen in Fig. 9. The ratio between these shots is (16±1):1 for thermal neutrons and (6±1):1 for epi-thermal neutrons at 1 eV. The lower ratio for faster neutrons is linked to the contribution of thermalized background neutrons which are generated in the shielding and their arrival time coincides with epi-thermal neutrons from the moderator.

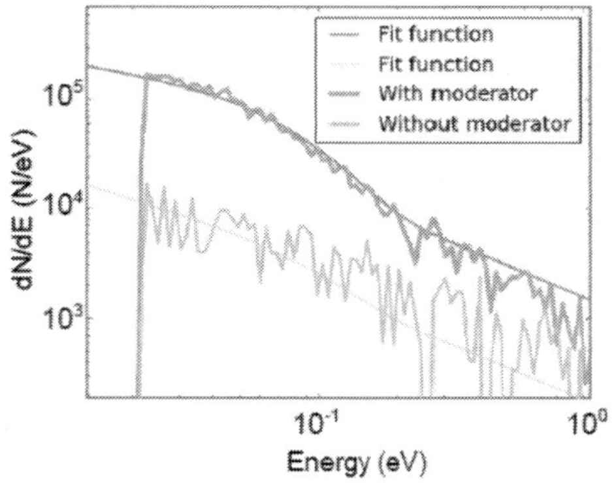

Fig. 9. Measurement of a thermal neutron distribution of a laser-based NRS campaign at PHELIX. Orange lines represent data taken without a moderator and green lines are single shot spectra with the moderator in place.

The second method used is the introduction of a black resonance filter in the beam path. ^{113}Cd has a giant resonance below 1 eV with cross sections above 10^4 b and is consequently opaque for thermal neutrons. A 1 mm thick cadmium plate is, therefore, placed in front of the entrance of the last collimator tunnel to block the direct line of sight to the moderator. The result can be seen in Fig. 10 together with the cadmium cross section in red and a shot without the Cd filter, indicated by the blue line. The ratio between these two shots at 25 meV is (17±1):1 which is in good agreement to the first method with (16±1):1 as well as with the simulated result of (15±4):1.

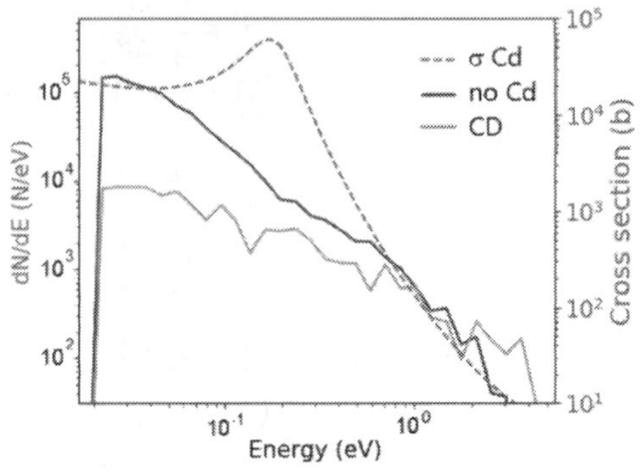

Fig. 10. Blue line: Moderated neutron spectrum without a sample. Pink line: Neutron spectrum with a 1 mm thick Cd sample in the beam path as black resonance filter. Red dashed line: The cross section of ^{113}Cd. With increasing Cd cross section, the blue and red curve start to deviate until cadmium is opaque and all contributions in the pink graph are from the background.

4.2 Characterization of the Moderated Neutron Spectrum

The moderated neutron spectrum consists of two parts. The first one is a Maxwellian distribution of thermalized neutrons with the temperature of the moderator. The second part is the result of stochastical collisions and energy transfer and it follows a $1/E^b$ distribution where E is the neutron kinetic energy and b is the leakage factor. b describes how many neutrons are either leaked out of the moderator or have been absorbed during the moderation process. An infinite moderator without absorption would have $b = 1$. In Fig. 11 the moderated neutron spectrum is compared to a simulated neutron spectrum.

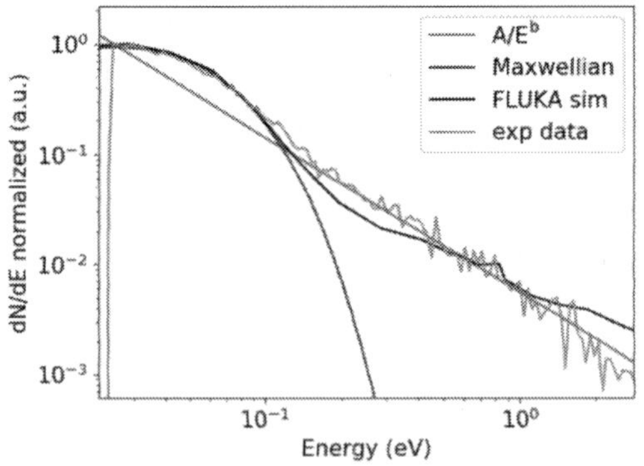

Fig. 11. Normalized experimental neutron spectrum (green) with a Maxwellian fit (blue) and A/E^b in red. The important fit parameters are $T = 291$ K and $b = 1.4$. The black line represents the corresponding FLUKA simulation with $T = 289$ K and $b = 0.92$.

For the experimental data, a Maxwellian distribution can be fitted to the thermal peak with a temperature $T = (291±15)$ K = (18±15)°C. This temperature is close to the room temperature of the PHELIX laser bay with 20°C. The leakage coefficient b on the other hand corresponds to a value of 1.4. This is not realistic since $b \leq 1$ has to be fulfilled for every moderator [15]. A FLUKA [16] simulation with a matched ion source spectrum for this experiment produces a leakage coefficient of $b = 0.92$. This discrepancy suggests that the used detector is less sensitive for fast neutrons than the $1/E^{0.5}$ scaling of the boron cross section would suggest. This is most likely caused by an oversaturation of the MCP during the γ-flash from the laser target interaction. For future applications, it is therefore advisable, to shield the direct line of sight between target and diagnostic with a high Z material.

5. Conclusion

The proposed laser neutron setup was developed and optimized to identify materials through the detection of nuclear resonances via NRS. The optimization was done with PHITS simulations for a setup at the PHELIX laser facility in Darmstadt. The setup was characterized with simulations and validated with an experiment, which was in agreement with the predictions. The catcher shielding, the moderator as well as the collimation system show large improvements to previous laser-driven neutron spectroscopy experiments and can be used to demonstrate the capability to detect various isotopes inside a sample.

6. Acknowledgement

The authors want to especially thank the PHELIX laser team at GSI for their work on the realization of this experimental campaign as well as the GSI Plasma Physics Department. This work is supported by the LOEWE excellence initiative of the state of Hessen.

7. References

1. R. Nelson, O. Ronald, et al. Journal of Imaging 4.2 : 45 (2018)
2. Z. Révay, et al., Handbook of nuclear chemistry: 1619 (2011)
3. P. Schillebeeckx, et al., Springer, **I/24.** 10-66. (2015)
4. Nuclear research and industry today. Nat Rev Phys 1, 171 (2019)
5. M. Roth, et al., Phys. rev. lett. **110.4** 044802 (2013)
6. B. Rus, et al., Vol. 8780. International Society for Optics and Photonics (2013)
7. E. Mauerhofer, et al., No. FZJ-2017-07855. Streumethoden, (2017)
8. S. F. Nowicki, et al., Physics Procedia 90: 374-380 (2017)
9. V. Bagnoud, et al., App. Phys. B **100.1** 137-150 (2010)
10. R. A. Snavely, et al., PRL, 85.14: 2945. (2000)
11. F. Wagner, et al., Phys. rev. let. **116.** 20 (2016)
12. A. Kleinschmidt, et al. PoP. **25.5** 053101 (2018)
13. A. J. Koning, et al., Nuc. data sheets 113.12: 2841-2934 (2012)
14. P. Tsai, et al., Nuc. Inst. and Meth. in Phys. 416: **16-29.** (2018)
15. I. Petrovic, P. Benoist, Adv. in Nuc. Sci. and Tec. Springer, Boston, MA, (2002)
16. G. Battistoni, et al., AIP Conference proceedings. Vol. 896. No. 1. Am. Inst. of Phys., (2007)

Neutrons production on the IPHI accelerator for the validation of the design of the compact neutron source SONATE.

Hoang Ngoc Tran[1], *Frédéric* Ott[2], *Jacques* Darpentigny[2], *Anthony* Marchix[1], *Alain* Letourneau[1], *Nicolas* Chauvin[1], *Fabien* Prunes[2], *Benoît* Homatter[2], *Burkhard* Annighöfer[2], *Alain* Menelle[2], *Jérôme* Schwindling[1]

[1] IRFU, CEA, Université Paris-Saclay, F-91191 Gif-sur-Yvette, France
[2] LLB, CEA-CNRS, Université Paris-Saclay, F-91191 Gif-sur-Yvette, France

Abstract. We aim at building a compact accelerator-based neutron source (CANS) which would provide a thermal neutron flux on the order of 4×10^{12} $n.s^{-1}.cm^{-2}.sr^{-1}$. Such a brilliance would put compact neutron sources on par with existing medium flux neutron research reactors for neutron scattering experiments. We performed the first neutron production tests on the IPHI proton accelerator at Saclay at a proton energy of 3 MeV. The thermal neutron flux were measured using gold foil activation and ^3He detectors. The measured flux were compared with GEANT4 Monte Carlo simulations (10.4) in which the whole experimental setup was modelled. There is a good agreement between the experimental measurements and the Monte-Carlo simulations. The available modelling tools will allow us to optimize the whole Target Moderator Reflector assembly together with the neutron scattering spectrometer geometries for the design of the neutron scattering facility SONATE.

1 Introduction

There is currently an interest in developing compact neutron sources based on low energy proton accelerators (10-100 MeV) [1]. Such sources could serve as neutron sources for neutron scattering to replace small ageing nuclear reactors [2] or in hospitals for Boron Neutron Capture Therapy [3-4]. There are already several such projects on-going around the world. They are generally referred to as CANS, Compact Accelerator-based Neutron Sources. The currently most advanced is LENS Low Energy Neutron Source at Indiana University [5]. The CANS operating or under construction have gathered into the UCANS, Union for Compact Accelerator-driven Neutron Sources [6].

At Saclay we are considering a similar possibility to replace the Orphée research reactor. Our final goal is to build a higher end facility SONATE which could serve a suite of about 10 neutron scattering instruments. The facility would operate at a proton energy around 20 MeV, in pulsed mode to use time-of-flight techniques with a typical power on the target of 80kW.

Our first aim has been to experimentally validate the neutron production and moderation obtained by Monte Carlo simulations using GEANT4. Once reliable simulation tools are available we shall be able to reliably estimate the performances of a CANS for neutron scattering experiments (from the source to the spectrometers) and compare its performances to existing facilities (reactor or spallation based).

As a starting point we are using the IPHI proton accelerator at Saclay which is able to produce high current proton beams (up to 90 mA) [7] with proton energies of 3 MeV.

The Monte Carlo simulation toolkit GEANT4 is a library for the simulation of the transport of particles through matter. This tool is widely used in various applications from high-energy to very low-energy physics, including astrophysics and space science, medical physics and radiation protection [8]. Within the neutron physics package of GEANT4, the *particlehp* (high precision) option has to be used for neutron energies below 20 MeV as it refers to the library G4NDL that is based on the evaluated nuclear data library ENDF/B-VII.0 [9]. Moreover, this latter includes the thermal scattering libraries that take into account the molecular effects occurring for neutron energies below 4 eV when considering moderator and reflector materials such as water, polyethylene, etc. Recently, the validation of the neutron transport treatment in GEANT4-10.3 for the thermal energy range was performed by inter-comparison with the standard Monte Carlo code MCNP-6.1 on microscopic and macroscopic observables [10]. It was pointed out some sampling issues for the data treatment of the thermal scattering libraries included in the G4NDL for the versions of GEANT4-10.4. These improvements concern the treatment of nuclear interactions involved during the propagation of neutrons in a complex geometry, in particular the moderation process. But, to ensure the reliability of simulation tools, the neutrons production has to be validated as the thermal neutron

* Corresponding author: frederic.ott@cea.fr

 https://doi.org/10.1051/epjconf/202023101007

distributions within the moderator are sensitive to the double differential cross sections in energy and angle of primary neutrons.

Within this framework, we have performed experiments in order to measure the primary neutron source characteristics associated to the reaction $^{9}Be(p,n)$, as well as the thermal neutron flux in various locations by using a simplified polyethylene (PE) moderator. This paper describes the experimental and simulation methods that was used to measure the thermal neutron flux inside and outside the moderator, including the thermal neutron spectra and the neutron beam divergences, as well as the fast-neutron angular distribution related to the reaction $^{9}Be(p,n)$. The method for calibrating the ^{3}He detector efficiency via wavelength is also discussed.

2 Experimental setup

2.1. Proton source

The experiments have been performed on the IPHI proton accelerator which is based at CEA Saclay, France. The accelerator consists of a proton source (SILHI) of energy 95 keV, a Low Energy Beam Transport Line and a Radio Frequency Quadrupole to accelerate the protons to an energy of 3 MeV. The accelerator is designed to operate in continuous mode with proton currents up to 100 mA which corresponds to a total power of 300 kW. For the current experiments dedicated to validations of neutron production and moderator Monte Carlo simulations we have operated the accelerator at a very low power of about 10 W, both to avoid any target damage and for radioprotection issues. The accelerator was operated in pulsed mode with proton pulses of length 100 µs and with a repetition rate of 1 Hz and a peak current of 30 mA. Rather narrow thermal neutron pulses are thus obtained so that precise time-of-flight measurements can be performed.

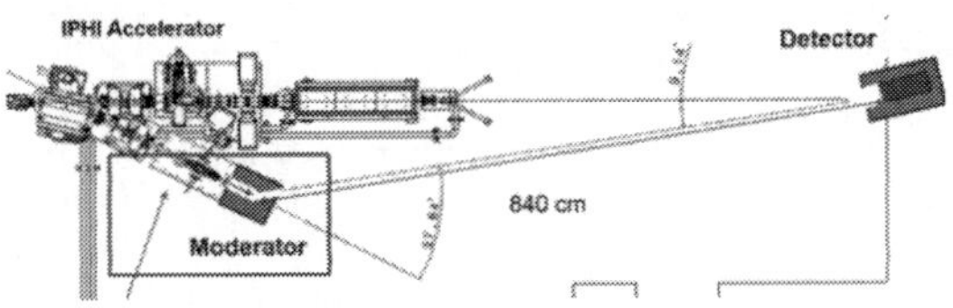

Fig. 1. Sketch of the experimental set-up. The target is installed on a deviator line and the neutron detectors are set at 8.4m from the target so as to be able to perform time-of-flight measurements.

2.2. Target and moderator assembly

In order to produce neutrons we opted for a beryllium target of thickness 0.5mm (99.0%) which stopped all incident protons. The target was attached on an aluminium support with titanium screws in order to minimize neutron activation. This support was air cooled. The proton beam size was limited to 16mm in diameter. The proton current incident on the target was continuously measured so as to have a precise value of the incident particle flux and be able to precisely estimate the neutron production. An electron repelling electrode set at a potential of 200 V was set in front of the beryllium target so as to prevent any bias in the target current measurement. The target was installed in a polyethylene (PE) moderator box (300x300x400 mm^{3}) so as to cool down the neutron to thermal energies (around 26meV). A 20 mm diameter hole was drilled through the moderator from the position where the thermal neutron density was expected to be the highest to the outside of the PE box (see Fig. 2).

In order to change the moderator geometry, it was possible to insert PE "plugs" inside the exit hole so as to fill more or less the neutron extraction channel.

The whole experimental setup (accelerator – target – moderator – detectors) was installed in a 2 m thick concrete casemate.

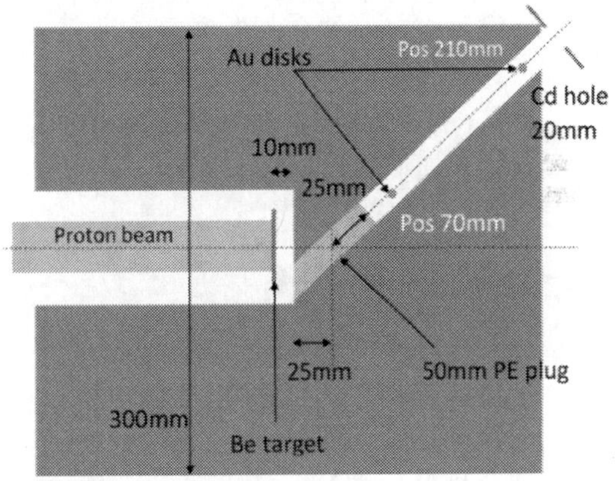

Fig. 2. Target and moderator assembly

2.3 Neutron detection

2.3.1 Fast neutron detection

A technical difficulty arises from the fact that since we operate in pulsed mode (100µs proton pulses width), the instantaneous neutron flux is very high. This prevented using a regular Bonner sphere. A dedicated acquisition chain was mounted consisting of a very small ^{3}He detector (1cm in diameter) encased in a cubic polyethylene setup. We made sure that even though we operated in pulse mode we did not have events pile-up in the ^{3}He detector. This led to acquisition times on the order of 5-10 minutes to obtain a good statistics in the measurement of the fast neutrons around the source.

2.3.2 Thermal neutron detection

The thermal neutron flux was measured at 8.4m from the source using time-of-flight mode and a ^{3}He detector. The ^{3}He detector was surrounded with a shielding made of polyethylene, as shown in
Fig. 3, and the detector was placed at 20 cm from the aperture of 6 cm in diameter. The solid angle covered by the detector was thus reduced to about 8° allowing to reduce the background noise from neutrons scattered and

moderated on the concrete walls of the casemate. Moreover, a layer of B_4C was installed around the ^{3}He detector to capture the thermal neutrons coming from the polyethylene shielding.
Even though the source was operated at a modest 10W power we were very close to the saturation limit of regular ^{3}He tubes (~20 kHz/tube) so that the detection area was limited to 20cm^2, corresponding to the front face of a single 2 inches ^{3}He detector tube (NH 50).
Thermal neutrons (1.8 Å) travel at a speed of 2200m/s so that the travel time over 8.4m is 3.8 ms. For a proton pulse width of 100 µs, the energy resolution is thus on the order 0.1/3.8 = 2.6% which is good enough to measure a rather precise neutron energy distribution. Note that besides the proton pulse length, the moderation process leads to an intrinsic neutron pulse broadening which reduces the energy resolution. Monte-Carlo simulations provide an estimate of a broadening on the order of 100-150 µs in a PE moderator [11-12].

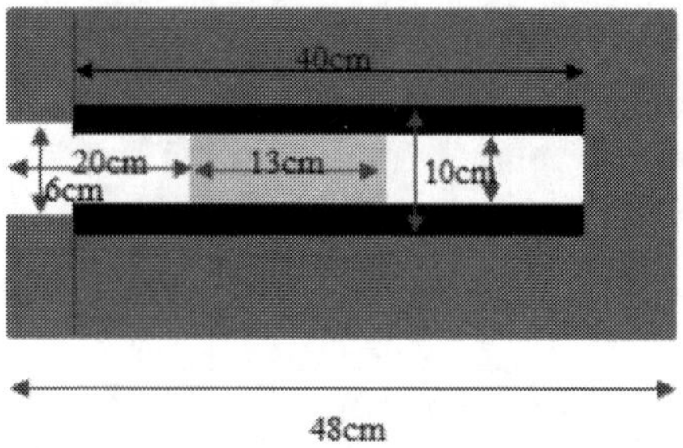

Fig. 3. Polyethylene (blue) moderator (left). ^{3}He detector system design (right).

The ^{3}He detector used in horizontal position suffered from an issue related to a dead zone at the front of the detector where neutrons are absorbed in the ^{3}He gas but not detected due to the absence of an amplification electric field at top end of the detector [13]. This dead zone bias the detection efficiency as a function of the wavelength. Short wavelengths penetrate deeply in the detector and are thus almost all detected while a large fraction of long wavelengths (>3Å) are absorbed in the front part of the ^{3}He gas but not detected. The "dead zone" of the detector was mapped with a pencil beam on the PRISM spectrometer at Orphée at a wavelength of 4Å (see Fig. 4a). The dead zone extends over 26mm from the top of the detector. The detector efficiency is deduced by integrating the ratio between the effective distribution and the ideal distribution to the detector length. Fig. 4b is provided the calculated detector efficiency as function of the neutron energy (via TOF). The measurement have been performed with a spectrometer of the Orphée reactor for a neutron beam of 4.3 ångström that confirms the calculated detector efficiency.

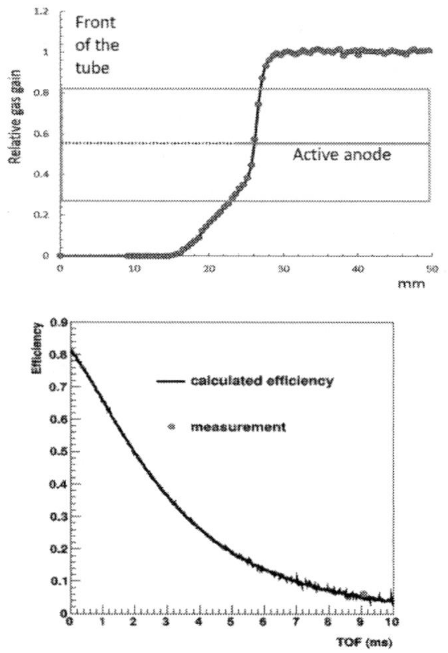

Fig. 4. (a) Detection efficiency along the length of the ^{3}He tube. The dead zone is approximately 26mm long; (b) Calculated detector efficiency with GEANT4 together with a superimposed experimental calibration point (red) measured on Orphée. The average detection efficiency of the detector was 30%.

2.3.2 Gold foil activation

In order to validate the Monte-Carlo simulations inside the moderators, gold disks were positioned at various positions inside the moderator. Disks with a diameter of 6 mm and thickness 200 µm were used. The weight of these disks was 112 mg so that even when operating at 10W, measurable activation of the gold was achieved within 15-30 minutes of operation. Bare disks and disks enclosed in Cadmium were measured so as to be able to estimate the thermal and fast neutron flux [14]. The Cd absorption cut-off edge is at around 500 meV (σ = 100 barn) so that the bare disks are activated by the whole neutron spectrum irradiation and the Cd covered disks are only activated by the fast neutrons. These gold activation measurements provide a simple and quantitative way of measuring the thermal neutron flux. However, while the measurement of the gold activation is quantitative and reliable, in order to calculate an absolute neutron flux, an a priori knowledge of the neutron energy distribution is necessary. Hence either an assumption of neutron energy distribution has to be made (Maxwellian distribution centered around 26meV for example) or the activation must be calculated using a simulated neutron energy distribution. Four gold disks with same size (with and without Cd casing) have been put into the moderator at positions 1 and 2 (see Fig. 2) at 70 mm and 210 mm from the intersection of the proton beam axis with the neutron extraction hole. Their activation has been calculated during the Monte Carlo simulations. The measured

 https://doi.org/10.1051/epjconf/202023101007

activation values have then been compared with the calculated activation values.

3 Experimental results

3.1 Fast neutron yield and angular distribution

In order to use a reliable model for the fast neutron term an accurate knowledge of the fast neutron emission rate and angular distribution is required. Moreover, no data are available for 3 MeV protons, while they are needed for validating the models implemented in GEANT4 in the framework of the prototype tests at 3 MeV protons. In consequence, we performed a new measurement of the angular distribution of the fast neutrons produced by a 3 MeV proton beam impinging on a thick beryllium target.

The angular distribution of neutrons related to the reaction $^{9}Be(p,n)$ was measured by using a PE-^{3}He detector system that consists of 10 cm thick PE in front of a ^{3}He counter (4.5 mm radius, 200 mm length). The aim of the polyethylene was to moderate part of incoming neutrons aiming to be captured in the ^{3}He detector. The PE-^{3}He detector was located at a distance of 2 m from the target. The detector was shielded from parasitic thermal neutron scattered from the concrete wall with a Cd sheet. The Proton beam intensity was limited to 2.2 μA to avoid pile-up effects in the ^{3}He counter. The fast neutron yield was measured for six angles around the target, from 0 to 150° by steps of 30°.

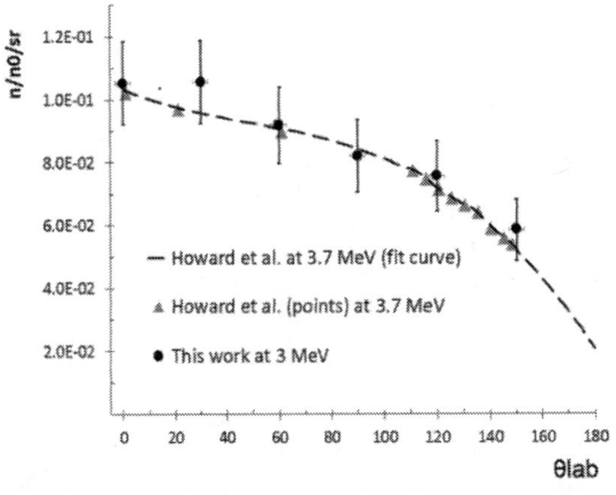

Fig. 5. Angular distributions of the neutrons from the ^{9}Be (p,n) reaction of 3 MeV proton beam compared with the results of Howard et al. [15] at 3.7 MeV. The angular resolution was ±2.8°.

Fig. 5 shows the measured angular distributions of neutrons produced in the reaction $^{9}Be(p,xn)$ for 3 MeV protons. Our new measurements confirms those from Howard et al. [15] performed with a proton beam of 3.7 MeV. The comparison is normalized to the neutron rate of each measurement (n0) which is adjusted by Howard's measurements [12]. The data were modelled with a fourth order polynomial to build the fast-neutron source used in the GEANT4 simulations.

3.2 Thermal neutron yield

3.1.1 Gold foil activation inside the moderator

Table 1 provides the data values obtained on the activity of the different gold foils. All the details have been fully described in Allen et al. [17]. The gold disks are irradiated for durations ranging from 15 to 22 minutes which lead to measurable gamma activity from ^{198}Au. It can be remarked that the Cd-layer strongly decreases the activity of each gold foils, especially for the furthest location (210 mm) that is below the detection limit. The measured activities of the gold foils have ~ 23% of incertitude due to determination of gold foils activation. Note that the measurements were performed by considering a 50 mm thick PE plug in the neutron extraction channel (see Fig. 2). The thermal neutron flux (showed in Table 1) associated to these measurements have been deduced based on the cross section of data library ENDF/B-VII.0.

Gold activation	Current (μC)	Irradiation time (s)	Activation (Bq/g)	Thermal neutron flux (< 0.4 eV) (n/cm²/μC)	Systematic errors
Foil 1 without Cd	2.9	900	2700	$1.21 x 10^{6}$	~23%
Foil 1 with Cd			533		
Foil 2 without Cd	2.6	1380	57	$1.70 x 10^{4}$	
Foil 2 with Cd			<9.7		

Table 1. The measured activation foils were used to reduce thermal neutron flux (less than 0.4 eV).

3.1.1 Thermal beam mapping

Outside the moderator, the thermal neutron flux was mapped with a ^{3}He detector by scanning the detector position in the X and Y directions to ensure we were aiming at the peak flux. Fig. 6 presents the thermal neutron beam divergences in both X and Y axis at the distance 8.4 m. We observe a fair agreement between the experimental data (red dots) and the calculation (dark line). The systematic underestimate between the experimental and calculated data may originate from secondary thermal neutron emission background in the experimental data.

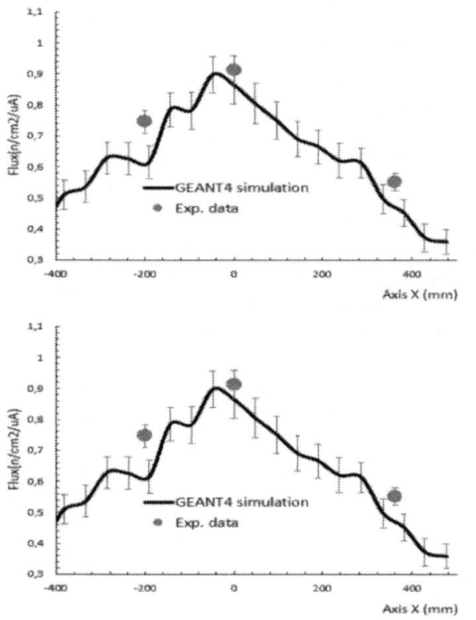

Fig. 6. Thermal neutron beam divergences at 8.4 m from the target. The positive direction of Y axis is in the proton beam's side. The origin is at 8.4 m.

Fig. 7 summarizes the thermal neutron flux measurements obtained from gold foil activation and time of flight measurement along the neutron flight path (red dots). The result of the simulation performed with GEANT4 is also provided. A very good agreement is obtained between the measurements and the simulation result. In this simplified moderator geometry, it is possible to obtain a thermal neutron flux of about 1000 $cm^{-2}.s^{-1}$ at the exit of the moderator.

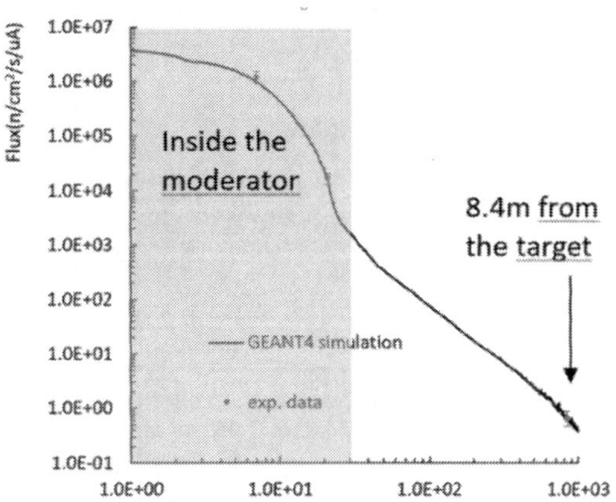

Fig. 7. Thermal neutron flux (less than 0.4 eV) as function of the distance between the target and 3He counter given by Geant4 simulation and compared with measurements. Measured flux inside the moderator using the gold foil activation technics and outside the moderator using ^3He counter.

3.1.1 Thermal neutron energy spectrum

The comparison of the neutron time-of-flight spectrum requires a high statistic for the simulations, and consequently a high computing time, as the number of neutron events reaching the ^3He detector at 8.4m is extremely low in the simulation. To reduce the computational time, the GEANT4 simulation has been split into two steps:

- The first step: 10^9 neutrons were generated at the target position according to the measured angular distribution discussed above. This simulation performs the neutron transport within the polyethylene moderator. The neutron spectrum is stored at the entrance of the ^3He detector, as well as the neutron angular distribution and the neutron position in the transversal plane.
- The second step: based on the neutron distributions in energy, in angle and in position obtained from the first step, we generated 10^7 neutrons from the surface in front of the ^3He detector. The $^3He(n,p)$ reactions are tracked in the detector volume and the energy distribution of incident neutrons is recorded.

Considering the time of the moderation processes (less than 2 % in average) is negligible in comparison with the TOF in the large distance from the moderator to the detector (8.4 m), we convert the neutron energy spectrums to the travel time. The amount of attenuation due to air is about ~8% for the distance of 8.4 m. Fig. 8. shows the comparison of the neutron TOF spectrum obtained for the 50 mm thick PE plug configuration (see the geometry in

Fig. 3a) between simulations and experimental data. A general agreement in the thermal Maxwell distribution with the measurement is showed in Fig. 7. An underestimation of simulation below 1 ms is due to the gamma background of the counter. Note that the calculation is corrected by the detector efficiency and the measurement is also corrected by the measured thermal neutron background in the casemate (figure 3).

The time-of-flight (ToF) measurements were performed by setting a single ^3He detector at a distance of 8.4m in the direction of the extraction hole. The incident neutron flux was measured with time channels of width 20 μs. The start of the acquisition was synchronized with the clock signal of the accelerator radio-frequency. Each ToF spectrum was acquired in 5 to 10 minutes. Figure 8 shows examples of ToF spectra measured in various conditions so that the contributions of various types of particles could be identified and quantified. The blue curve corresponds to the "reference signal" as measured with the box described on

Fig. 3. The peak in the region between 1 and 6 ms corresponds to the thermal neutrons. The thermal spectrum is peaked around t = 2.2ms corresponding to a neutron speed of 3800m/s, i.e. λ = 1.05Å. The neutron spectrum can be fitted with a Maxwellian distribution centred on a temperature of 350K. While this is higher than the actual moderator temperature, it is no surprise as neutron moderated with hydrogenated materials are always under-moderated contrary to thermal neutron produced in a D_2O reactor vessel where the interaction time is much longer.

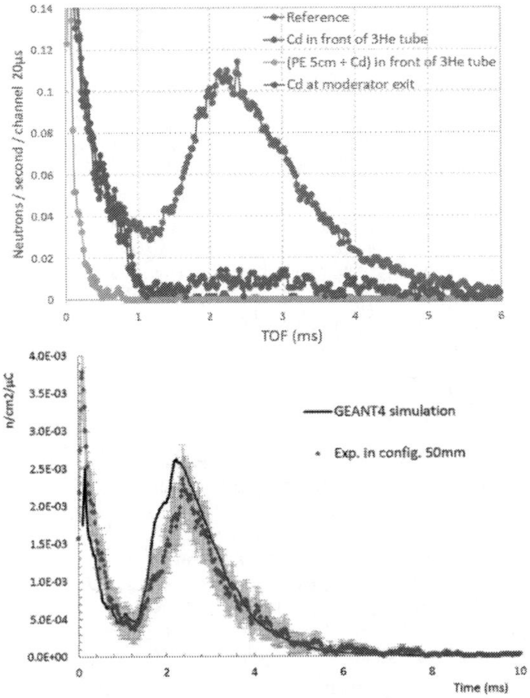

Fig. 8. The comparison of the ToF spectrum between experimental data and Geant4 calculation.

Extra contributions are observed below TOF<1ms. In order to identify these contributions, a Cadmium sheet was put in front of the detector shielding box (Figure 8a, red dots). Above t = 1ms, there is no residual signal. t = 1ms corresponds to a neutron energy of 0.5eV which is the neutron Cd cut-off. Hence it is demonstrated that the measured spectrum actually corresponds only to thermal neutrons. In a third measurement, thermal neutrons were prevented from exiting the moderator with a Cd sheet at the moderator exit; fast neutron could still escape the moderator. The measured signal (purple dots) shows that a residual neutron signal is then observed for t > 1ms. It corresponds to the fast neutrons moderated into the casemate walls and eventually arriving into the detector. This allows telling that 85% of the measured thermal neutrons are actually the ones coming from the moderator extraction hole while 15% of the neutron are secondary emission thermal neutron originating from the moderation of fats neutron in the casemate wall. In further measurements, the collimation system should be improved. In a 4th measurement the fast neutron contribution was stopped with a PE block followed by a Cd sheet in front of the detector shielding entrance. The data (green dots) shows that there is a remnant short time contribution of either epithermal neutrons or gammas. These different contributions can easily be discriminated using the ToF measurements. However, due to the finite proton pulse length (100µs), the finite moderation time (100µs) and the short flight path, no quantitative information can be retrieved for neutron energies above 30eV since above this energy, all ToF information is blurred in the first few time channels.

4 Conclusion

This paper describes the experimental and simulation methods to determine angular distribution of primary fast neutrons of $^{9}Be(p,n)$ and the thermal neutron flux inside and outside the PE moderator including thermal neutron energy spectra and neutron beam divergences based on 3 MeV proton beam of IPHI accelerator. The efficiency of the used ^{3}He detector for these measurements is estimated. Good agreements between GEANT4 simulation and measurement data are observed. This gives us confidence in the reliability of the GEANT4 for the quantitative description of neutron moderators.

References

1. C. Andreani, C-K. Loong and G. Prete, Eur. Phys. J. Plus **131** (2016) 217. DOI: 10.1140/epjp/i2016-16217-1
2. U. Rücker, T. Cronert, J. Voigt, et al, Eur. Phys. J. Plus **131** (2016) 19. DOI 10.1140/epjp/i2016-16019-5
3. A. J. Kreiner, M. Baldo, J.R. Bergueiro, D. Cartelli, et al., Appl. Radiat. Isot. **88** (2014) 185–189.
4. H. Kumada, A. Matsumura, H. Sakurai et al, Appl. Radiat. Isot. **88** (2014) 211–215.
5. C.M. Lavelle et al, NIM A **587** (2008) 324-341
6. UCANS, www.ucans.org/facilities.html
7. R. Gobin et al., Rev. Sci. Instrum. **73** (2002) pp 922.
8. J. Allison, et al., Nucl. Instr. and Meth. A, **835** (2016) 186–225.
9. M.B. Chadwick et al., Nucl. Data Sheets **107** (2006) 2931.
10. H.N. Tran et al., Nuclear Inst. and Methods in Physics Research A **893** (2018) 84–94
11. Lavelle CM et al, Nucl Instrum Methods A **587** (2008) 324–3410
12. B. Zhong, T.J. Liang, Q.X. Feng, G.L. Yu, K. Wang, C.-K. Loong, Physics Procedia **26** (2012) 25 – 30.
13. K. Kudo et al. Nuclear Instruments and Methods in Physics Research B **213** (2004) 305–309.
14. D.A. Allen et al., Nuclear Instruments and Methods in Physics Research A **353** (1994) 128-133.
15. W.B. Howard and J.C Yanch, Nuc. Sci. and Eng. **138** (2001) 145.

EPJ Web of Conferences **231**, 01008 (2020) https://doi.org/10.1051/epjconf/202023101008
UCANS-8

Short-Pulse Laser-Driven Moderated Neutron Source

Sven C. Vogel[1,*], *Juan C.* Fernandez[1], *D. Cort* Gautier[1], *Nikodem* Mitura[2], *Markus* Roth[2], and *Kurt F.* Schoenberg[1,2]

[1]Los Alamos National Laboratory, Los Alamos, NM 87545, U.S.A.
[2]TU Darmstadt, D- 64289 Darmstadt, Germany

Abstract. Neutron production with laser-driven neutron sources was demonstrated. We outline the basics of laser-driven neutron sources, highlight some fundamental advantages, and quantitatively compare the neutron production at the TRIDENT laser sources with the well-established LANSCE pulsed neutron spallation source. Ongoing efforts by our team to continue development of these sources, in particular the LANSCE-in-a-box instrument, are described. The promise of ultra-intense lasers as drivers for brilliant, compact, and highly efficient particle accelerators portends driving next-generation neutron sources, potentially replacing in some cases much larger conventional accelerators.

1 Introduction

Neutrons play an important role in material characterization. Neutron based characterization techniques include

- neutron radiography, covering cold, thermal, epithermal, and fast neutron radiography,
- neutron diffraction,
- inelastic neutron scattering,
- neutron reflectometry,
- small angle neutron scattering, and
- neutron cross-section measurements covering all neutron energies

For about half a century, research reactors enabled these applications. Major research reactor facilities include the Institute Laue-Langevin (ILL) in Grenoble, France, or the High Flux Isotope Reactor (HFIR) in Oak Ridge, USA. Reactors, however, have fundamental source brightness limitations that are set by constraints on reactor power density. Accelerator-driven sources on the other hand can surpass the intrinsic source brightness limitations of reactors and avoid the proliferation concerns and nuclear waste concerns arising from the use of highly enriched uranium, which often provides the fuel for these reactor sources..

Since the 1970s, spallation neutron sources became more and more prominent. Spallation neutron sources are driven by high-intensity proton beams and can achieve a source brightness that exceeds the performance of reactors. The most recent major research neutron sources, such as

- the Spallation Neutron Source (SNS) in Oak Ridge, TN, USA,
- the Japan Proton Accelerator Research Complex (J-PARC) at Tokai, Japan, or
- the Chinese Spallation Neutron Source (CSNS) at Dongguan, China,

are all spallation sources. Similarly, the planned European Spallation Source (ESS) in Lund, Sweden, will be a spallation neutron source.
The total facility cost for a high-power spallation neutron source exceeds $1B. Hence, there are few of these sources world-wide. This cost factor, along with the concurrent closure of many smaller research reactors, in particular in Europe, have decreased the ensemble of neutron sources available for research. The Union for Compact Accelerator-driven Neutron Sources (UCANS) has recognized this problem, pointing out that a reduction in available neutron sources will also adversely impact training of the next generation of neutron users at the larger facilities and ultimately jeopardizes neutron research as whole. This concern has driven projects, such as the Jülich high-brilliance neutron source project [1], which will pave the way for research neutron sources at a lower investment cost and with greater accessibility to the research community.

While neutron research plays an important role in the academic or scientific research communities, neutron-based sources for commercial (e.g. neutron radiography for quality control) or government applications (e.g. detection of nuclear contraband) are actively under development and commercially available. Typically, D-D or D-T neutron generators are utilized for these applications with attendant limitations such as the need to use tritium to reach ~10^{13} n/s versus the ~10^{12} n/s available with D-D sources. However, all accelerator-driven sources are ultimately flux-limited by charge-density considerations. However, most applications would greatly benefit from more intense sources, which could also enable entirely new applications such as neutron absorption resonance based characterizations of fission products in nuclear fuels [2] or bulk temperature measurements [3,4], at similar investment cost.

* Corresponding author: sven@lanl.gov

https://doi.org/10.1051/epjconf/202023101008

In this paper, we describe an alternative approach for the generation of short, intense neutron pulses based on high-power, short-pulse lasers. We describe the principle of this technique and lay out the advantages of laser-based neutron generation compared with conventional RF accelerator-based sources. We quantitatively compare the laser-driven, short-pulse neutron source with the spallation neutron source at the Los Alamos LANSCE facility. Examples of current applied research, as well as the prospects of such novel laser-driven compact neutron sources based on available or proposed laser systems, are discussed.

The promise of ultra-intense lasers as drivers for brilliant, compact, and highly efficient particle accelerators for electrons and ions (potentially replacing in some cases much larger conventional accelerators) also portends driving next-generation neutron sources. Laser-driven sources, at present, cannot compete with accelerator-driven sources or high flux reactors in terms of average neutron flux (where average flux is the number of particles per second regardless of temporal substructure). However, for medium size neutron sources as well as for high peak flux neutron sources, the performance of laser sources is evolving to where they are competitive.

2 Principles of Laser-Driven Neutron Sources

Neutrons are produced by nuclear reactions. These reactions can be driven by the interaction of electromagnetic radiation, energetic electrons, or ions with atomic nuclei. For example, a high-velocity ion can break up in the Coulomb field of another atom releasing one or more neutrons. In addition, neutrons can be released by fission reactions, fusion reactions, or by proton-induced spallation of heavy target nuclei.

Los Alamos National Laboratory (LANL) and collaborators have pioneered a novel, short duration yet extremely intense neutron source using the short-pulse laser at LANL's TRIDENT laser facility [5-9]. The TRIDENT laser facility was, until it's closure in 2016, one of the most intense and powerful short-pulse lasers in the world, providing a very high contrast laser pulse of ~600 fs in duration, ~80 J energy, and with a wavelength of 1053 nm. The laser beam is concentrated to a peak intensity up to 10^{21} W/cm^2. That intensity, when interacting with a sub-micron ultrathin CD (or CD_2) foil target, drives a high-energy (peaked around 15-30 MeV, see [5] for details) deuteron beam. The intense deuteron beam then interacts with a beryllium target ("catcher") to produce neutrons.

The TRIDENT laser-driven neutron source featured high intensity and directionality, with approximately 10^{10} fast neutrons per shot in a ~1 steradian cone and with an extremely short neutron pulse of <1 ns duration [5,6]. The source has been used to develop a new generation of active interrogation concepts to detect clandestine nuclear material [9]. In order to enable other applications, such as nuclear resonance spectroscopy for isotopic identification of (irradiated) nuclear fuel [10], temperature measurement in shock–driven dynamic material experiments [3,4], and pulsed neutron diffraction and scattering, we developed a bright, moderated short-pulse laser-driven neutron source [6]. Beryllium and high-density polyethylene were used to achieve this aim. The single pulse, moderated neutron source was fully characterized with respect to its energy spectral components by a set of neutron time-of-flight detectors. The energy analysis included both the fast neutron component (~1 MeV to ~50 MeV, see [5] for details) coming from the neutron production, as well as the epithermal, thermal, and cold (few meV) components from the neutron moderation.

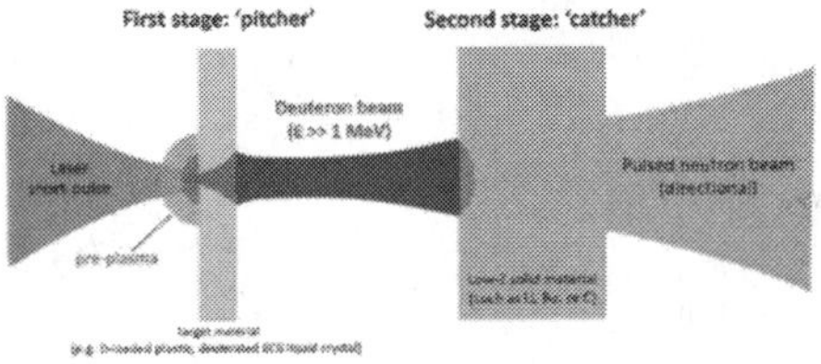

Fig. 1. Schematic of the principle of laser-driven neutron sources, see text for details.

Fig. 1 shows a schematic of a laser-driven neutron source using the ion-pitcher-catcher scenario. Two basic processes dominate: isotropic emission from nuclear reactions, where the incoming proton thermalizes its energy in the convertor and a neutron is emitted, and a directed component that generates neutrons while conserving the incoming ion momentum. The directed neutron generation component is dependent on two processes. The first is the so-called pre-equilibrium (or pre-compound) reaction, where the energy and momentum of the incoming proton or deuteron is directly transferred to a single nucleon in the convertor, resulting in the emission of an energetic neutron that conserves the initial momentum [11]. The second reaction is a stripping or breakup reaction. Here the incident deuterium is split or fissioned in the convertor and the resulting neutron conserves its initial momentum [12].

Results of our team and more detailed descriptions of the technique can be found in [5-7]. The reader is referred to these articles for more information.

3 Qualitative Comparison with Conventional Pulsed Neutron Sources

For compact neutron sources, p and D accelerator driven sources are well-developed. Such sources provide cost efficient access to thermal and fast neutrons, enabling applications in neutron radiography or neutron diffraction. The RANS source at the Riken accelerator complex [13] is a good example of such a source. While time-of-flight techniques were demonstrated for

https://doi.org/10.1051/epjconf/202023101008

diffraction [13], the source, in pulsed mode, generates particle pulses that are at least 10μs long. As shown in Fig. 2, the moderation time for neutrons in the epithermal energy region (E>0.4eV) is 1μs or shorter. Therefore, this type of accelerator driven source cannot be used for energy-resolved epithermal neutron applications, e.g., to resolve nuclear resonances.

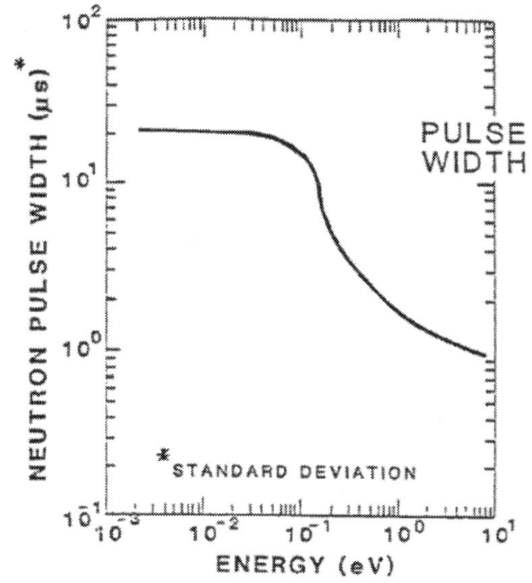

Fig. 2. Neutron pulse width emitted from a LANSCE thermal moderator as a function of neutron energy. While thermal neutron applications (E<0.4 eV) are possible with initial p/d pulse widths of ~10μs, for epithermal applications pulses of this length would be much longer than the moderation time and therefore do not allow such applications (From: Gary Russell et al. ICANS-VIII Proceedings (1985) [14]).

The pulse structure in conventional RF accelerator-driven sources is strongly influenced by beam space-charge. Short pulse spallation sources typically use proton LINACS, coupled with storage rings, or proton synchrotrons to produce intense, sub-microsecond neutron pulses. There is, however, a limit on the number of protons or ions that can be stored in a conventional storage ring due to space-charge tune-depression [15]. Limits on the number of protons per pulse directly imply limits on the pulse density of neutrons. For example, the proton storage ring (PSR) at LANSCE compresses a 750 ms LINAC macro-pulse of 4×10^{13} protons into a 270 ns pulse at 800 MeV (the number of neutrons produced per proton by spallation in a tungsten target at 800 MeV proton energy is approximately 20). Further reductions in compressed pulse length result in a concomitant reduction in protons per pulse, due principally to space charge effects. For example, pulse lengths of 1.5 ns, 5 ns, and 10 ns in the LANSCE PSR were projected to store 1×10^{11}, 3×10^{11}, and 6×10^{11} protons, respectively [16]. Finally, the cost for conventional accelerator pulse compressors or storage rings are significant and far exceed the investment cost for a compact neutron source such as RANS. Furthermore, space-charge limits the number of ions in a pulse, which constrains applications where high instantaneous flux with short temporal resolution is required, like in dynamic radiography or temperature measurements in dynamic events using neutron absorption resonance spectroscopy [3,4].

Laser-driven neutron sources utilize photons to first accelerate electrons to high energy. The energetic electrons then couple to and accelerate ions by generating electrostatic fields due to charge separation. Charge separation, driven by the transport of energetic electrons, generates electric fields of TV/m. Ions are accelerated over typical distances of $L_{ac} \approx 10\mu m$ up to energies of approximately 94 MeV for protons [17] and up to a few MeV/u for heavier ions [18] using thin foil targets. Typical beam angular divergence, which is energy-dependent, is between 20°- 50°. The expansion of the ion beam results in an exponential energy distribution. The ion beam is highly directed with an emittance well below ion beams driven by conventional RF accelerators [19-21].

In a conventional accelerator-driven source, significant effort is required to keep the ion pulses short and prevent the space-charge repulsion from spreading them in space. In contrast, in a laser-driven particle accelerator, the ions are accelerated in sub-picosecond timescales and propagate essentially with co-moving electrons. Thus space-charge driven beam divergence is small and beam emittance is preserved from the point of acceleration to the neutron-producing target. *This fundamental difference between conventional accelerators and laser-driven particle accelerators allows lasers to achieve ultra-short pulse durations at intensities far greater than conventional accelerators.* For example, at TRIDENT 10^{11} deuterons above 14 MeV were produced in 500 fs [5], equivalent to an instantaneous current of 10^{11} deuterons $\times$ $1.6\cdot10^{-19}$ C/d / 500 fs = 32 kA. Furthermore, there is a direct scaling between laser intensity and neutron pulse intensity. That is, more powerful lasers translate to higher neutron intensities per pulse depending on parameters such as the laser-matter interaction accelerating the deuterons or the energy-dependent cross-section of the deuteron/target interaction. The details of the relationship between laser power and neutron output are subject of ongoing research and therefore beyond the scope of this paper.

4 Quantitative Comparison with the LANSCE Pulsed Sources

This section provides a quantitative comparison between a pulse produced at the TRIDENT laser and at the LANSCE spallation source.

At TRIDENT, ~10^{10} n/pulse were achieved during campaigns in March and July 2016, utilizing 70J output energy deposited in a 600 fs long pulse (0.1 PW) of ultra-high contrast [6]. This pulse produced deuterons with kinetic energies ~20 MeV interacting with the beryllium catcher target. Since the neutrons were pre-dominantly produced by the deuteron break-up reaction, the momentum of the incident deuteron was preserved and the produced neutrons travelled pre-dominantly in the direction of the laser pulse. Therefore, the majority of the neutrons were moderated by the polyethylene moderator in which the beryllium target was embedded (target/moderator coupling efficiency of ~1). For the sake of this estimation, we assume that all ~10^{10} neutrons originating from a sub-nanosecond pre-moderation neutron pulse width were moderated (i.e. neglecting the small fraction of neutrons emitted into directions not covered by the moderator, see [5] for directional dependence of the neutrons emitted from the target).

LANSCE, on the other hand, is typically operating at ~100μA proton current at 20 Hz with protons having an energy of 800 MeV. This converts to ~3×10^{13} protons per pulse hitting the tungsten spallation target. Assuming a conversion rate of 20 neutrons per proton, this results in ~6×10^{14} neutrons per pulse. However, these neutrons are emitted isotropically from the ~10 cm diameter, 20 cm tall upper spallation target [22]. To avoid exposure of the beam lines with the intense γ flash from the spallation pulse, the moderators serving the beam lines are below this target and above a lower target in a so-called split target configuration [22]. Estimating that 2% [23] of the produced neutrons cross the moderator surface of one of the four moderators (because of the moderators viewing a void rather than target material), ~1×10^{13} moderated neutrons are emitted per proton pulse with a ~270 ns proton pulse length.

A beam line viewing the moderator at TRIDENT therefore receives about three orders of magnitude less neutrons than a beamline viewing a LANSCE moderator. However, concepts to improve the deuteron flux over that produced by simple deuterated plastic targets as well as optimizations of the thermal and epithermal neutrons produced for a given application over that of the simple polyethylene block used as a moderator at TRIDENT are predicted to allow for a gain of about an order of magnitude. Furthermore, lasers providing higher energies than the ~70J provided by the TRIDENT system are available today, e.g. kJ systems (albeit not yet for repetition rates suitable for a pulsed neutron source). A higher laser energy would produce both more accelerated deuterons as well as a higher kinetic energy of the deuterons, translating to a gain in neutron pulse intensity. Laser systems such as the Scalable High-average-power Advanced Radiography Capability (SHARC) from Livermore National Lab provide 150J at 10 Hz [24]. Systems that are even more powerful and deliver laser pulses at ~10Hz are under development. However, no laser system was designed yet *specifically* as a driver for a laser-driven neutron source. The latter is likely to offer further improvements as current lasers for laser facilities are designed for a broad range of laser parameters, serving many applications ranging from plasma physics to laser-driven dynamic compression [25,26].

It is worth pointing out that the cross-section of the deuteron break-up reaction increases neutron yield better than linear with deuteron energy, which correlates linearly with laser energy. In the foreseeable future, therefore, laser-driven neutron sources that are about an order of magnitude weaker than LANSCE appear feasible. For spallation sources the neutrons produced by the intranuclear cascade process reach energies up to the proton energy (e.g. ~1 GeV) [27]. However, it is this small fraction of neutrons which drive the shielding demands. Modern spallation sources have their closest sample position outside the bulk shield at ~15-20m, e.g. the high pressure beamline SNAP at the SNS has a moderator to sample distance of ~15m [28]. At LANSCE, the bulk shield ends at 4.5m with the closest possible sample position around 6 m. Laser-driven neutron sources, similar to the RANS source [13], require significantly less shielding of ~1m since neutrons are not produced by spallation. For some applications, such as epithermal energy-resolved neutron imaging, allowing to measure isotope densities [10], the resolution available with a shorter flight path is acceptable. The concurrent flux increase from the $1/L^2$ ratio of the LANSCE FP5/ERNI beamline [29], where these experiments are presently conducted at around 10 m sample to detector distance, would therefore enable some applications at neutron count times similar to what is required at LANSCE.

Compact neutron sources typically serve only a single or a few rather than several tens of beamlines as is the case for large-scale user facilities. All design studies and optimizations conducted by the UCANS or projects such as the Jülich high-brilliance neutron source project [1] for compact accelerator-driven neutron sources are immediately applicable for laser-driven neutron sources.

4 Ongoing Efforts

Concepts and efforts for laser deuteron catcher target systems were summarized in a report by our team [7], including a discussion of alternative solutions for the solid targets, left with a hole after each laser shot, such as gas or liquid cryo-jet target systems. Our team collaborates with laser facilities and will utilize them to produce neutrons as they become available after required upgrades, e.g. to improve laser contrast or enable higher repetition rates. In order to demonstrate the usability of these laser sources for applications, a multi-modality neutron instrument called "LANSCE-in-a-box" was designed and is in commissioning at the FP5 beam line at LANSCE.

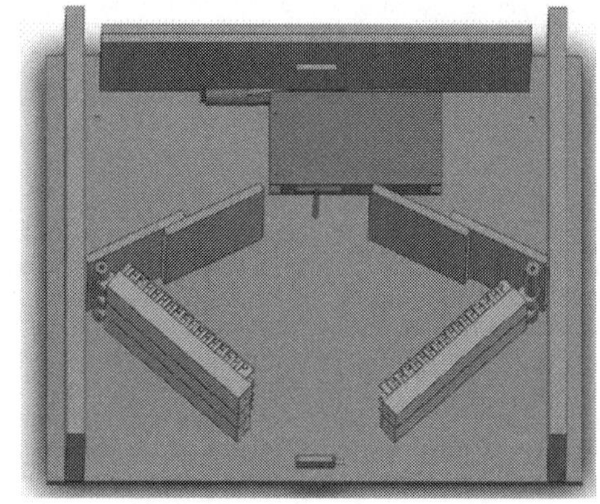

Fig. 3. Schematic of the LANSCE-in-a-box multi-modal portable neutron instrument. The dimensions of the square base-plate are ~1 m. From bottom to top, the detectors are a fission chamber (green), 96 ^{3}He tubes on three panels (yellow), a light-box for scintillator-based imaging with a CMOS camera (magenta), and a dual-photomultiplier neutron transmission detector viewing a ^{6}Li glass scintillator. The sample is along the centreline of the instrument in front of the imaging lightbox. The green walls are borated polyethylene.

Fig. 3 shows a schematic of the setup. The instrument is designed to be portable and to require only ~1 m^2 in area to allow deploying in the target area of laser sources not necessarily designed for time-of-flight neutron instruments. Collimation and shielding will be part of the setup. After calibration at LANSCE, the diffraction detectors will only require the source to sample distance to be adapted at a different facility before the histograms from the 96 ^{3}He tubes can be integrated into a single histogram of high statistical quality, removing the requirement to calibrate sample-detector distances at a weak source. A sample such as an additively manufactured tungsten object, approximately of cylindrical shape, will allow to demonstrate simultaneous diffraction with the ^{3}He detectors, imaging with an imaging detector, and neutron resonance spectroscopy from the intense 18.6 eV ^{182}W resonance with the dual-photomultiplier coincidence setup viewing a ^{6}Li glass scintillator. Data collected at LANSCE with single, ten, one hundred and one thousand LANSCE neutron pulses will allow comparison of the per-pulse neutron intensity produced at a laser source. Fig. 4 shows a photograph of the first experiments with the setup in September 2019 at the FP5 beamline at LANSCE.

Fig. 4. Photograph of the first tests of the LANSCE-in-a-box multi-modal portable neutron instrument. The ^{3}He detector panels as well as the multi-channel plate imaging detector (behind green borated polytehtylene walls) are visible. A 6 mm diameter vanadium can with a sample is visible in front of the imaging detector. The black borated plastic, rolled into a pipe, reduced air scatter.

5 Conclusions

The promise of ultra-intense lasers as drivers for brilliant, compact, and highly efficient particle accelerators portends driving next-generation neutron sources, potentially replacing in some cases much larger conventional accelerators. For medium size neutron sources, as well as for high-peak-flux neutron sources, the performance of laser sources is evolving to where they are both cost and performance competitive with conventional accelerators. Furthermore, laser sources are relatively compact (similar to the RANS source, i.e. several hundred m^2 including power supplies and diagnostic systems, fitting in a modest lab) and can be easily attached to secondary drivers such as accelerators, other lasers, or pulsed power machines. The fundamental difference is that the acceleration potential for particle acceleration in short pulse lasers is TV/m compared to the tens of MV/m in conventional RF accelerators. Laser plug to beam efficiencies today are slightly below that of RF accelerators with rapid improvements occurring. The precedence of high-powered lasers in rugged environments for military or industrial (e.g. welding) applications may even provide a pathway to mobile compact short-pulse neutron sources. The Lyncean Compact Light Source [30], replacing an undulator with a laser to produce synchrotron X-rays in-house, may be transformative for the synchrotron community. The neutron community may experience a transformation enabled by lasers, too.

We wish to thank an anonymous reviewer for thoughtful comments greatly improving the manuscript. MR and NM acknowledge financial support for visits to LANSCE by the Institute for Materials Science at Los Alamos via an IMS Rapid Response grant. SCV acknowledges support by the DOE/NE Fuel Cycle R&D research program. This work has benefitted from the use of the Los Alamos Neutron Science Center (LANSCE) at Los Alamos National Laboratory. Los Alamos National Laboratory is operated by Triad National Security, LLC, for the National Nuclear Security Administration of the U.S. Department of Energy under contract number 89233218NCA000001.

References

1. U. Rücker, T. Cronert, J. Voigt, J. P. Dabruck, P-E. Doege, J. Ulrich, R. Nabbi et al., European Physical Journal Plus **131** 19 (2016).
2. H. G.Priesmeyer, and U. Harz, Atomkernenergie **25.2** 109-113 (1975).
3. V.W. Yuan, J.D. Bowman, D.J. Funk, G.L. Morgan, R.L. Rabie, C.E. Ragan, J.P. Quintana, and H.L. Stacy, Phys. Rev. Lett. **94** 125504 (2005).
4. J.C. Fernández, C.W. Barnes, M.J. Mocko, L. Zavorka, Rev. Sci. Instr. **90** 094901 (2019).
5. M. Roth et al., Phys. Rev. Lett. **110** 044802 (2013).
6. J.C. Fernández, D. Cort Gautier, C. Huang, S. Palaniyappan, B.J. Albright, W. Bang, G. Dyer, A. Favalli, J.F. Hunter, J. Mendez, S. C. Vogel, M. Roth et al., Phys. Plasmas, **24** 056702 (2017).
7. M. Roth, S.C. Vogel, M.A.M. Bourke, J.C. Fernandez, M.J. Mocko, S. Glenzer, W. Leemans, C. Siders, C. Haefner, *Assessment of Laser-Driven Pulsed Neutron Sources for Poolside Neutron-based Advanced NDE–A Pathway to LANSCE-like Characterization at INL*, Los Alamos National Lab. Los Alamos, NM, U.S.A., report No. LA-UR-17-23190 (2016).
8. S.C. Vogel, M. Roth, J.C. Fernandez, D.C. Gautier, K.F. Schoenberg, Neutron News, **29** 32-36 (2018).
9. A. Favalli, N. Guler, D. Henzlova, S. Croft, K. Falk, D.C. Gautier, K.D. Ianakiev, M. Iliev, S. Palaniyappan, M. Roth, and J.C. Fernandez, Scientific Reports, **9** 2004 (2019).
10. A.S. Tremsin, S. C. Vogel, M. Mocko, M. A. M. Bourke, V. Yuan, R. O. Nelson, D. W. Brown, W. B. Feller, J. Nucl. Mat. **440** 633-646 (2013).
11. G.D. Harp, J.M. Miller, Phys, Rev. C, **3(5)**, 1847 (1971).
12. Y. Watanabe, T. Ye, K. Ogata, *In*: EPJ Web of Conferences, **2**, 11003 (2010).
13. Y. Otake, Plasma and Fusion Research **13** 2401017-2401017 (2018).
14. G.J. Russell, C. D. Bowman, E. R. Whitaker, H. Robinson, M. M. Meier. *In:* ICANS-VIII: Proceedings of the Eighth Meeting of the International Collaboration on Advanced Neutron Sources, Rutherford-Appleton Laboratory, Report RAL-85-110, pp. 272-293 (1985).
15. A.G. Ruggiero, AIP Conference Proceedings, **468** 244-257 (1999).
16. R Garnett, Los Alamos National Laboratory, private communication (2019).
17. A. Higginson, R. J. Gray, M. King, R. J. Dance, S. D. R. Williamson, N. M. H. Butler, R. Wilson et al. Nature Comm. **9** 724 (2018).
18. M. Hegelich, S. Karsch, G. Pretzler, D. Habs, K. Witte, W. Guenther, M. Allen et al., PRL **89** 085002 (2002).
19. T.E. Cowan, J. Fuchs, H. Ruhl, A. Kemp, P. Audebert, M. Roth, R. Stephens et al., PRL **92** 204801 (2004).
20. M. Schollmeier, M. Roth, A. Blazevic, E. Brambrink, J. A. Cobble, J. C. Fernandez, K. A. Flippo et al., NIMA **577** 186-190 (2007).
21. J.A. Cobble, R. P. Johnson, T. E. Cowan, N. Renard-Le Galloudec, M. Allen, J. Appl. Phys. **92** 1775-1779 (2002).
22. M. Mocko, G. Muhrer, NIMA **704** 27-35 (2013).
23. M. Mocko, personal communication.
24. E. Sistrunk, D. A. Alessi, A. Bayramian, K. Chesnut, A. Erlandson, T. C. Galvin, D. Gibson et al., *In:* Short-pulse High-energy Lasers and Ultrafast Optical Technologies, **11034**, *1103407*. International Society for Optics and Photonics (2019).
25. G. Mourou, T. Toshiki, Optics and Photonics News **22** 47-51 (2011).
26. National Academies of Sciences, Engineering, and Medicine. *Opportunities in Intense Ultrafast Lasers: Reaching for the Brightest Light*. National Academies Press (,2018).
27. G.H. Russell, Proceedings CANS-XI : International Collaboration on Advanced Neutron Sources, KEK. Tsukuba. October 22-26,1990, 291-299 (1990).
28. A.S. Tremsin, J. B. McPhate, J. V. Vallerga, O. H. W. Siegmund, W. B. Feller, H. Z. Bilheux, J. J. Molaison et al., J. Phys.: Conference Series, **251** 012069 (2010).
29. R. Nelson, Ronald, S.C. Vogel, J.F. Hunter, E.B. Watkins, AS. Losko, As. Tremsin, N.P. Borges et al., J. Imaging **4** 45 (2018).
30. E. Eggl, M. Dierolf, K. Achterhold, C. Jud, B. Günther, E. Braig, B. Gleich, F.Pfeiffer, J. Synchrotron Rad. **23** 1137-1142 (2016).

Development of the radio frequency quadrupole proton linac for ESS-Bilbao

Juan Luis Muñoz[1]*, *Ibon* Bustinduy[1], *Igor* Rueda[1], and *David* de Cos[1]**

[1]ESS-Bilbao, Parque Tecnológico de Bizkaia, Laida Bidea, Edificio 207-B 48160 Derio, Bizkaia

Abstract. The Radio Frequency Quadrupole (RFQ) linear accelerator for ESS-Bilbao is described. This device will complete ESS-Bilbao injection chain after the ion source and the LEBT. The design was finished in 2015 and machining of the accelerator cavity started in 2016. The RFQ is a 4-vane structure, aimed to accelerate protons from 45 keV to 3.0 MeV and operating at 352.2 MHz in pulsed mode with a duty cycle up to 10%. Total length is about 3.1 m divided in 4 segments. Each segment is itself assembled from four components, named vanes, by using polymeric vacuum gaskets with no brazing among them. Notable aspects of the design are the constant mean aperture R_0, vane radius ρ and thus ρ/R_0 ratio and also uniform inter vane voltage. Novel procedures for the design of the modulation and integrated beam dynamics and electromagnetic design have been developed for this task. In this paper, the complete design procedure and its results are presented, including beam dynamics, RF cavity design, field flatness and frequency tuning, cooling and thermo-mechanical design.

1 Introduction

ESS-Bilbao[1] is an Spanish public Consortium of the Central and Basque Governments. It is the institution selected to supply Spanish in-kind contribution to the European Spallation Source ERIC, ESS [2]. The contributions are focused in different areas: accelerator (the complete MEBT (Medium Energy Beam Transport) section and the RF systems of the warm section), target and neutron instruments (MIRACLES). In addition to these contributions, local projects are also under development. The ESS-Bilbao injector is a proton injector that consist on an ECR (Electron Cyclotron Resonance) ion source and a LEBT (Low Energy Beam Transport) system, already in operation[3]. This injector will be completed with the RFQ linac presented in this communication. The 3 MeV can then be used as the linac of a compact accelerator-driven neutron source.

The ESS-Bilbao RFQ design was carried out by a local team. The ISIS-FETS[4] and Linac4[5] (also ESS[6], based on IPHI[7]) RFQs were taken as references and initial state-of-the-art models. However, the frozen design incorporates mixed selection of characteristics. The rounded lobe shape (the so-called Montgolfier shape), external geometric characteristics and, more importantly, the assembly procedure without brazing were adapted from ISIS-FETS. The tuning approach (based on movable tuners) and the cooling strategy of the FETS RFQ were replaced by more standard cooling based on longitudinal drilled channels in the vane, and fine tuning by cooling water temperature. But due to the external characteristics (two minor and two major vanes with vacuum pump grids) there are no longitudinal cooling channels in the body of the cavity, so the approach to tuning is different than in other RFQs. The complete design process has been collected in a Technical Design Report (TDR)[8]. This paper discusses the different approaches and the design route chosen and the corresponding tools that have been used. The conclusions can be useful to other groups attempting to design an RFQ linac for use in a CANS facility.

The non-conventional approach to the design of the modulation and the resonant cavity needed the design of home-made computer tools or to adapt existing ones. The software used for each design step will be described in the corresponding section. The RFQ is currently under fabrication, and first tests are expected to start during 2019.

The RFQ is divided in four segments of around 800 mm in length. Each segment is itself an assembly of four components, two major vanes and two minor vanes, as depicted in Fig. 1. There are no coupling structures between segments. The vanes are assembled by means of polymeric vacuum gaskets (3D O-rings), using no brazing or welding. Finger strips are also added in all copper contact surfaces to improve RF contact. This strategy was adapted from the ISIS-FETS RFQ[4], and has been also proposed for other projects[9]. The vacuum ports location (in the upper and lower major vanes) is also taken from FETS RFQ. Cooling scheme is different from this RFQ though, as drilled channels instead of cooling-pockets, are used for the vacuum grid and along the vane tips.

Cavity tuning and field stabilization will be provided by static plunger tuners, while cooling water temperature will be used to fine tune the RFQ operation. Movable plunger tuners can, nonetheless, be used if needed. Cooling channels are drilled only along the vane tips (and not in the structure body). This was decided by mechanical and

*e-mail: jlmunoz@essbilbao.org
**Now at Departamento de Física Aplicada II, Facultad de Farmacia, Universidad del País Vasco, Paseo de la Universidad 7, 01006 Vitoria-Gasteiz, Spain

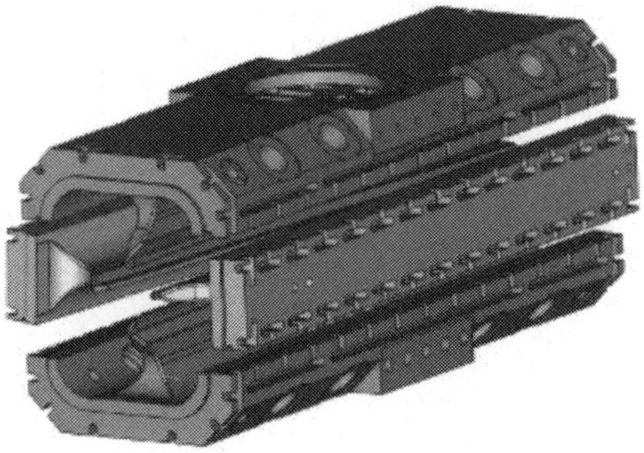

Figure 1. Exploded view of first segment of the RFQ, showing the so-called major and minor vanes.

Table 1. ESS-Bilbao RFQ Main Characteristics.

	ESS-Bilbao RFQ
Type	4 vane
Particle	Protons
RF frequency	352.2 MHz
Innervate Voltage	85 kV (uniform)
Energy	45 keV ⟶ 3.0 MeV
Design current	60 mA
Input emittance (rms norm)	0.25 π mm mrad
Duty cycle	Up to 10%
Kilpatrick factor	1.85
Number of cells	273
R_0	3.44 mm
ρ/R_0	0.85
Input/Output matcher	16.674 mm / 14 mm
Total length	3.12 m (3.66 λ)
Number of segments	4 (about 800 mm each)
Method of assembly	Polymeric gaskets + RF seals
Plunger tuner ports	16 per segment

fabrication criteria. This means that the frequency control by the cooling water temperature will operate only on vane-tip channels, contrary to other RFQs where two water circuits (vane tip and body) are available for tuning ([5–7, 10, 11]). Extensive thermo-mechanical and electromagnetic simulation studies have been done to proof that the operational control can be performed this way. Tests with the first segment will verify this approach.

The RFQ is designed to accelerate protons from 45 keV to 3.0 MeV. It is a pulsed machine, operating at 352.2 MHz with an expected duty cycle in operation of 5% (designed up to 10%). Notable aspects of the design are the constant mean aperture R_0, vane radius ρ and thus ρ/R_0 ratio. Inter-vane voltage is also uniform, with a value of 85 kV. Main characteristics of the RFQ can be found in Table 1.

In each segment there are 16 tuner ports (⌀37 mm) that can be used for static plunger tuners. The power coupler flange also fits in these ports, so that they can be inserted in any position. Additionally, 8 ⌀16 mm ports are built in order to be used for pick-ups or other sensors needed at any moment. Vacuum grid ports are designed for standard ⌀210 mm flanges. All ports are machined in the major vanes.

2 MODULATION AND BEAM DYNAMICS

2.1 Modulation design

The ESS-Bilbao RFQ modulation is the result of an optimization process. The modulation is designed for an inter-vane voltage of 85 kV, uniform throughout the entire length. Vane radius (ρ) is also constant, so to obtain a uniform local frequency and field flatness the mean aperture R_0 should also be constant. Modulation follows the shape of a 2-term expansion of the inter-vane voltage[12] and has been designed using a modified version of RFQSIM code[13].

The aim of the modulation design was to obtain an RFQ shorter than $4\lambda = 3.4$ m to simplify tuning operations. Also, copper blocks 800 mm long were already available, so an additional target was to fit the total length of the RFQ in four segments, instead of five, reducing inter segments joints and the risks associated. A final length limit of 3.1 m was selected.

Another figure of merit target for optimization was the Kilpatrick factor (ratio of designed maximum electric field value with respect to the theoretical maximum for copper at the operating frequency), that was chosen to be below 1.85.

2.1.1 Modulation optimization process

The optimization procedure starts by obtaining a modulation geometry, shorter than the maximum length allowed, that fulfils the beam dynamics criteria (transmision and beam losses). The main tool for this task is a modified version of RFQSIM code that uses a 2-term voltage expansion of the vanes geometry. Electromagnetic calculations for surface electric field and RF electromagnetic field for particle tracking were then performed. These computations used the actual geometry of the vane tips in FEM codes. The initial parameters were then modified according to the FEM results, and a new optimization run starts. This procedure continues until an optimal solution is obtained.

The input parameters that define a particular modulation are (using the standard parameter naming convention [14]): synchronous phases and particle energy at the end of each section (Shaper (ϕ_{sh}, W_{sh}), Gentle Buncher (ϕ_{gb}, W_{gb}) and Accelerator (ϕ_f, W_f)), clear aperture (a_{gb}) and modulation (m_{gb})) parameters, and input and output matcher sections radius.

In order to obtain a short modulation the range of parameters during optimization was oriented towards this target, with preference over other goals such as maximizing the transmission or maintaining a conservative Kilpatrick factor. In this sense, we concentrated the parameter space search on areas with lower clear aperture a_{gb}

 https://doi.org/10.1051/epjconf/202023102001

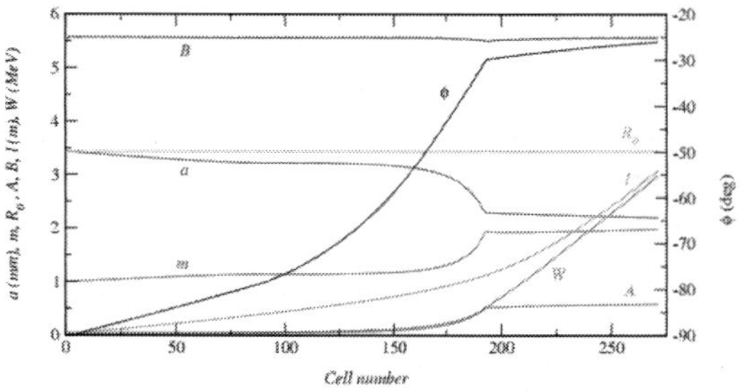

Figure 2. Evolution of several modulation parameters as a function of the cell number.

and energy W_{gb}, and less negative synchronous phase ϕ_{gb} (all referred to the end of the Gentle Buncher). Particle tracking results obtained with RFQSIM presented transmissions below 90 % for modulations found in this region of the parameter space. However, later simulations with other codes presented much better results. This strategy also produced higher surface fields, with a peak at the end of the Shaper typically above 1.8 times the Kilpatrick limit.

Also, a progressive reduction of the aperture in the Acceleration section was implemented in order to shorten the modulation. RFQSIM originally built the 2-term based Acceleration section using a constant aperture and modulation factor, following the rules proposed by Kapchinskiy-Tepliakov[12]. This produces slowly decaying fields along z, due to the increased cell length. We modified the part of RFQSIM that creates the 2-term modulation to incorporate custom aperture reduction strategies in the Acceleration section, as well as the ability to set a goal total length for the modulation.

The evolution of the main parameters as a function of cell number is shown in Fig. 2, while the final modulation shape and the corresponding accelerating field are shown in Fig. 3.

Concerning the high energy end of the modulation, it has been designed with the aim of optimizing the transmission through the designed MEBT and DTL. After the last regular cell of the accelerator section, a transition cell is included so both X and Y vanes end with the same aperture. A circular output matcher ends the modulation. The transition cell has a length of 17 mm and the output matcher has a radius of 14 mm. The last cells are shown in Fig. 4.

2.1.2 Surface field calculations

Although RFQ design software packages provide a calculation of Kilpatrick factor (surface electric field) we preferred to perform these calculations in an external finite element software, in order to gain control and accuracy over the process. The modulation description provided by RFQSIM consist on the modulation amplitude at the beginning and end of each cell (of length $\beta\lambda/2$). The actual shape of the modulation spine is built cell by cell by performing the 2-term interpolation in a Matlab script. The output of this script is a pair of curves $V_x(z)$ and $V_y(z)$ for horizontal and vertical vanes, respectively. The cell 3D geometry vane region is then built as a parametric surface for each vane in COMSOL Multiphysics FEM software. For example, the geometry of the profile of an horizontal vane cell between z coordinates z_0 and z_1 is defined as a parametric surface dependent on parameters $u \in [0, 1]$ and $v \in [-\pi/2, \pi/2]$ as defined in Eq. 1. Vertical vanes are defined in a similar way.

$$\begin{cases} x = V_x(z(u)) + \rho(1 - sin(v)) \\ y = \rho\, sin(v) \\ z = z_0 + u(z_1 - z_0) \end{cases} \tag{1}$$

The surface field is computed running an electrostatics simulation. To avoid border effects the model is built for three consecutive cells, but only the results for the central one are considered. The process is fully automated using Matlab/COMSOL scripts. The result for a particular modulation is the curve $E_{s,max}(z)$, that is the maximum value of the surface field for the cell starting at coordinate z. The overall maximum value of this curve is taken as the figure of merit characterizing the particular modulation. Using this approach we managed to scan a huge and fine set of parameter ranges in an automatic, brute-force approach. An example of such electrostatics model is shown in Fig. 5.

The results obtained for the final modulation are shown in Fig. 6. The peak that can be seen at around $z = 0.5$ m is caused by the transition between Shaper and Gentle Buncher sections, while the gradual increase from $z = 1.3$ m is a consequence of designing for higher accelerating fields in order to reduce modulation length.

Surface field calculations have also been used to determine the shape and position of gaps between the segments. The modulation is described as a continuous function from the beginning of the input matcher to the end of the output matcher. But the segmentation of the RFQ structure forces cuts in the modulation at certain z coordinates corresponding to the end of each segment. In order to reduce the perturbation that these cuts cause on the electric field, extensive simulation studies were made[8]. In these studies the position of the end-beginning of segments were considered between the cell-end position and the Lloyd position [15]. In the first case the longitudinal accelerating field is zero for all particles in the gap, but not the transverse components. In the second case, the field vanishes when the bunch center crosses the gap. Also, the rounding of the vane tips in the cut was studied, finally selecting an elliptical shape instead of a circular one. These studies included surface field calculation and also beam dynamics simulations. The overall best results, considering figures of merit and segment mechanical length conclusion was to use gaps 200 µm long placed at the end-cell position; the end of the vanes are rounded in an elliptical shape with long semi-axes of 2 mm length and short semi-axes having 0.75 mm (see Fig. 7).

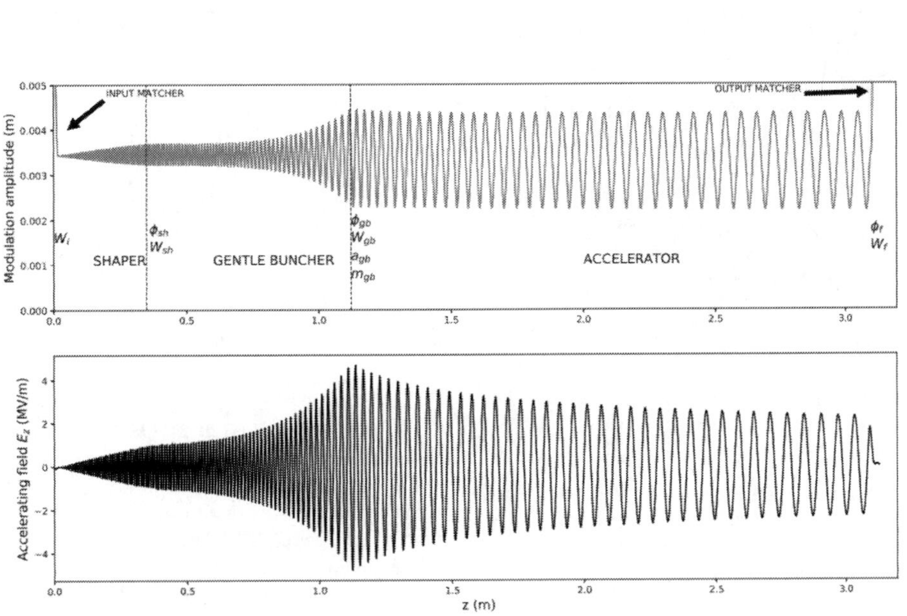

Figure 3. Modulation amplitude and accelerating field for ESS-Bilbao RFQ.

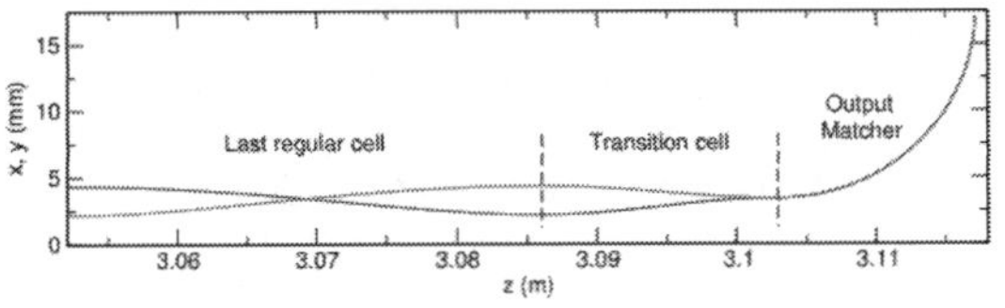

Figure 4. Characteristics of the last cells of the RFQ.

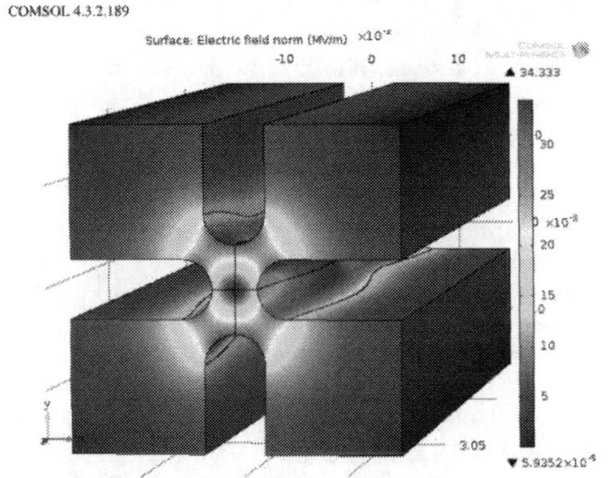

Figure 5. Surface electric field in an electrostatics model of three cells computed to extract the maximum surface field per cell. Only the vane region is considered for the calculation.

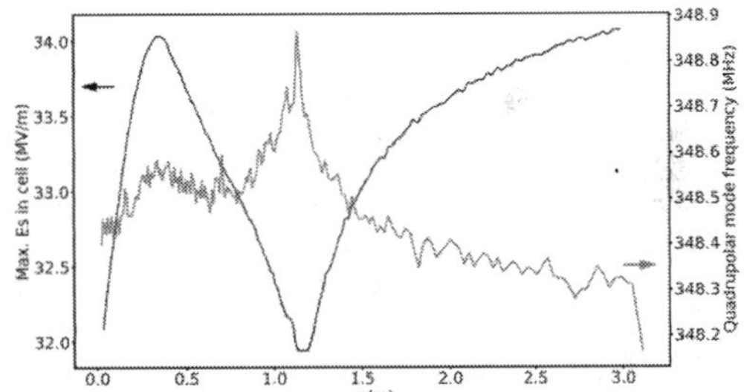

Figure 6. Surface maximum electric field per cell (left) and local frequency (right) as a function of cell starting coordinate z. Frequency is calculated in a slice of RFQ with one cell length and no tuners.

2.2 Beam dynamics results

From the modulation description obtained by using the optimization procedure, the full-vane 3D geometry of the RFQ was built in COMSOL Multiphysics software. Electrostatic simulations were run and the electric field was then exported in an external file. Particle tracking analysis using GPT[16] code were then run with the precise field map as input. Several optimization runs were made, exploring different regions of the parameter space until the final modulation was selected. Results were then cross-checked using different codes (Toutatis[17], PARMTEQ

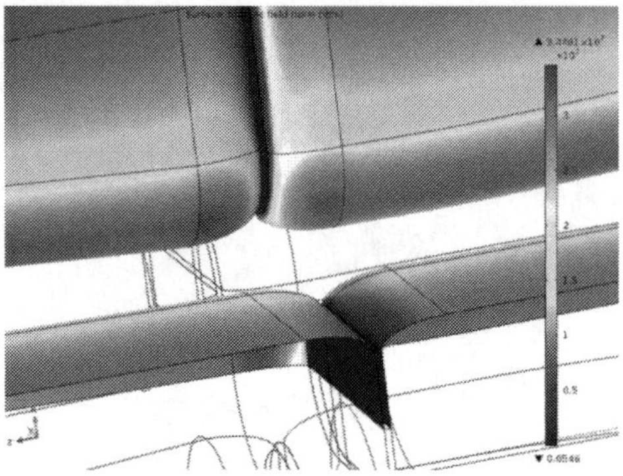

Figure 7. Electric field in the surface of vane tips around a gap between segments. The shape of the end of the vanes is elliptical.

Table 2. Simulation results at the output of the RFQ for the four methods used. The figures of merit are: full transmission (%), accelerated transmission (%), transverse normalized rms emittance (π mm mrad) calculated as the average between the horizontal and the vertical values, longitudinal rms emittance ($\pi\, deg$ MeV) and maximum surface electric field (MV/m).

Code	T_{all}	T_{good}	ϵ_t	ϵ_z	E_s
RFQSIM	90.7	86.8	0.243	0.120	30.4
TOUTATIS	94.6	93.6	0.266	0.131	33.4
GPT + COMSOL	95.2	94.1	0.264	—	34.1
RFQGen	96.2	94.1	0.241	0.125	33.7

/ RFQGen). Additional characteristics of the modulation can also be found in Table 1.

Beam dynamics simulations have been performed using different codes for comparison. All the simulations were performed using the same input beam characteristics: 45 kV input beam energy, 60 mA current intensity and 0.25 π mm mrad transverse emittance. The C-S (Courant-Snyder[14]) of the input beam varied slightly from code to code, depending on the optimal match calculated in each case, with values of alpha typically a little above 1, and beta about 0.03 m/rad.

The codes used for comparison of results were: GPT + COMSOL, RFQSIM, TOUTATIS and RFQGen (an improved fork of PARMTEQ). The implementation of our design in RFQGen is not completely accurate, since this code did not include an option to import a cell-by-cell description of the modulation at the time. So in consequence, the current RFQGen results must be considered as approximate.

Table 2 presents the main results of the particle tracking simulations. The accelerated ($\sim$ 3 MeV) beam transmissions of TOUTATIS, GPT + COMSOL and RFQGen are almost identical ($\sim$ 94 %). The emittance results are very positive, with virtually null increase of the transverse emittance, and a value of 0.12–0.13 deg MeV for the longitudinal emittance.The surface fields, while not identical (they are calculated by different approaches), lie within 1.81–1.85 times the Kilpatrick limit (18.47 MV m^{-1} for 352.2 MHz).

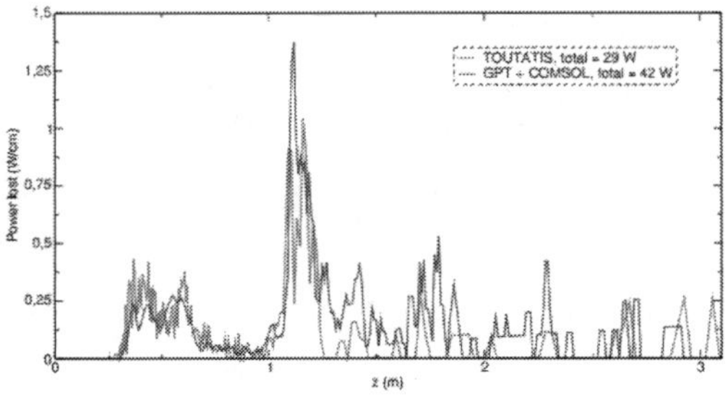

Figure 8. Beam power loss per centimeter of RFQ length, simulated with two different methods.

Results obtained with RFQSIM, however, have a poorer transmission than the other three codes, which is mainly due to a predicted transverse loss of almost 10 % of the beam. This is due to the method of field calculation in each code [18]. The other three codes represent the actual transmission of the RFQ more accurately, given that TOUTATIS and RFQGen (PARMTEQ) are the most widely used codes in the field of RFQ beam dynamic and that the field maps provided by COMSOL comes from a finite element simulation with accurate representation of 3D vane geometry, and the tracking is performed with GPT, a code extensively used beam dynamics simulations of other accelerator elements.

Fig. 8 shows the transverse losses (due to particles impacting with the vanes) along the RFQ, by depicting the beam power lost per centimeter of length. The curves are very similar in the Shaper and Gentle Buncher regions (first 1.2 m), with slightly higher power loss predicted by GPT + COMSOL in the Acceleration section. The total power lost (integral of the curves) remains within reasonable values in both cases, since most of the lost particles impact the vanes with energies in the hundreds of keV.

In summary, the results presented above prove the validity of the proposed vane modulation in terms of beam dynamics, as tested with different particle tracking and field map calculating methods. Although some of the new design constraints (especially the ~20 % length reduction) were initially expected to reflect negatively on the beam transport performance, this is only apparent in the RFQSIM results, with the other three codes presenting transmissions of ~94 %.

2.3 Local frequency profile

The vane modulation changes the local capacitance per unit length of the ideal quadrupole resonator. As a consequence there is a perturbation on the voltage profile $V(z)$, that no longer is constant. This perturbation can be modeled by considering a local resonant frequency computed in a cell-by-cell approach. Considering a modulation cell that is located between z_i and z_{i+1}, the local frequency $f_q(z_i)$ is the frequency of the resonator built from a slice of RFQ volume between z_i and z_{i+1}, considering perpendicular magnetic field boundary conditions at both ends and perfect electric boundary condition at the rest of the surfaces. This frequency profile should be as flat as possible because voltage profile in the real cavity will be dependent on this perturbation[14]. For the first (2013) design of ESS-Bilbao RFQ a sinusoidal (instead of 2-term) vane profile was used. This resulted in a very perturbed frequency profile[8]. For the final design a 2-term modulation shape (meaning that the geometry mimics the profile of an expansion of voltage truncated to two terms) was used. This resulted in the rather flat profile shown in Fig. 6. Deviation of local frequency around the 2D ideal design frequency is of the order of 0.1 MHz, while with a sinusoidal shape it was of several MHz (see the TDR[8] for details).

3 CAVITY DESIGN

3.1 Cavity cross-section design

The cross section of the cavity is based on the circular lobe approach used by ISIS-FETS[4], but using straight (vertical or horizontal) segments in the vanes, to aid the alignment verification operations later on. The machining of the rounded shape of the lobes was tested by fabricating different models in copper and aluminum[19]. The cross section is uniform for all the length of the RFQ. Although a variable lobe diameter along length is common in the current state-of-the-art, the uniform approach was selected as a conservative decision concerning mechanical engineering. For future developments, a variable approach would be probably preferred.

The optimization of the geometry was made by parametric simulations. The optimum values were selected by minimization of the power losses (or maximization of the quality factor, Q_0). The range of parameters available during the optimization procedure was restricted by mechanical constraints, like the total width of copper blocks, cooling channels diameter and corresponding wall thickness. In the optimization process, one of the parameters was always free, so it could be adjusted later to obtain the right frequency. The design frequency for the 2D models was chosen to be 348.6 MHz (several MHz below operating frequency) to avoid problems due to machining. The frequency will be raised to the operational one by means of plunger tuners. The average modulation aperture (3.438 mm) was used as vane tip aperture in the 2D models. A sketch of the cross-section is shown in Fig. 9, while in Fig. 10 electric and magnetic field maps for whole cross section in the quadrupolar mode are shown.

3.2 3D cavity design

The 3D body of the cavity is constructed by extruding the cross-section, adding the vane modulation solids, the input and output regions and other details like tuner and vacuum ports. Again, the shape has been optimized by finite element simulations aiming at reducing power deposition. An overall longitudinal cross section is shown in Fig. 11. The whole structure is an assembly of the four segments.

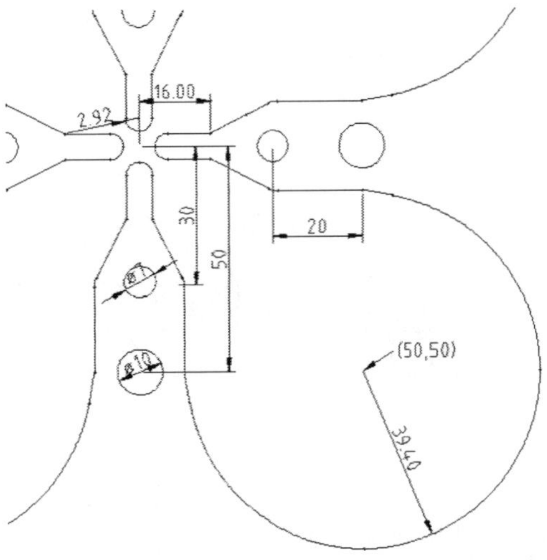

Figure 9. Sketch of a portion of the RFQ cross section, showing some dimensions (all in mm). The vane tips geometry comes from the average values of the modulation and the cooling channels position and diameter, that restrict vane thickness due to impose a minimum wall thickness, are fixed values. All the other parameters were subject of optimization.

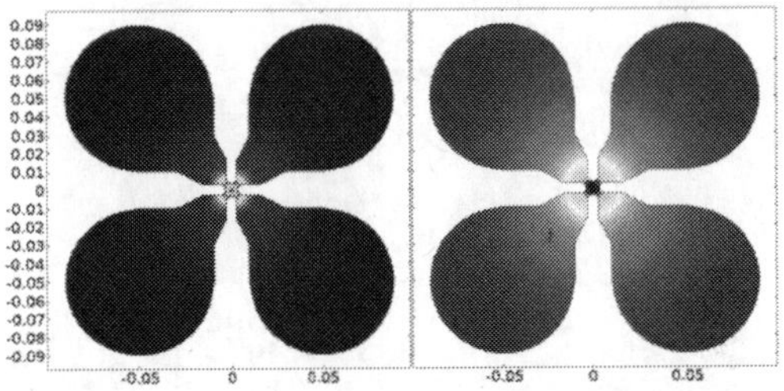

Figure 10. Cross section of the RFQ, showing normalized electric (left) and magnetic (right) field distribution for the quadrupolar mode.

Each segment has its own set of channels for cooling of the vanes and for cooling of the vacuum grid region.

The input and output undercut sections were designed so they have the same frequency as the bulk of the RFQ, contributing in this way to maintain the field flatness. The optimization looked for the optimum set of parameters that reduced power losses and kept electric voltage flatness. An schematics of the input section with the optimization parameters is shown in Fig. 12. Parameter values can be found in Table 3.

The field flatness was verified by 3D simulations of the whole length of the RFQ (not only the input or output undercut regions), and by mathematical transmission line models of the structure. One of the parameters is the distance between the vane tips and input (or output) lid (see Fig. 12). This distance can be easily modified after fabrication and testing of the RFQ, as it only involves changes in the lid, and not in the main body of the cavity.

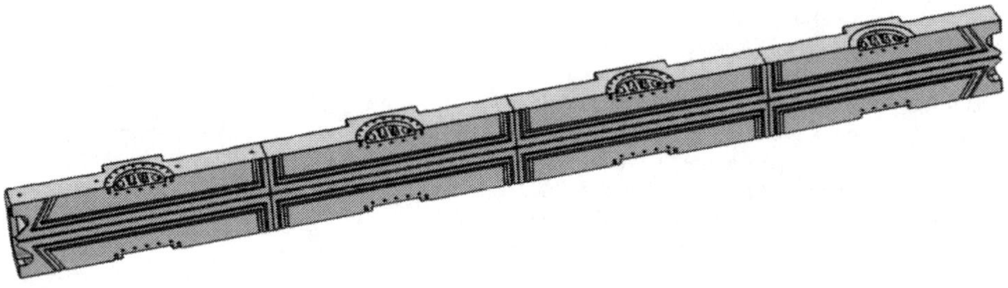

Figure 11. Longitudinal cross section of the RFQ structure. The four segments can be seen, showing the cooling channels and the vacuum ports.

Table 3. Parameters for input and output sections of the RFQ. Refer to Fig. 12 for description of parameters.

Parameter	Input section	Output section
R_{matcher} [mm]	16.674	14.0
z_{cover} [mm]	7.5	7.5
xy_{lobe} [mm]	88	88
r_{matcher} [mm]	25	25
z_{matcher} [mm]	22.60	21.0

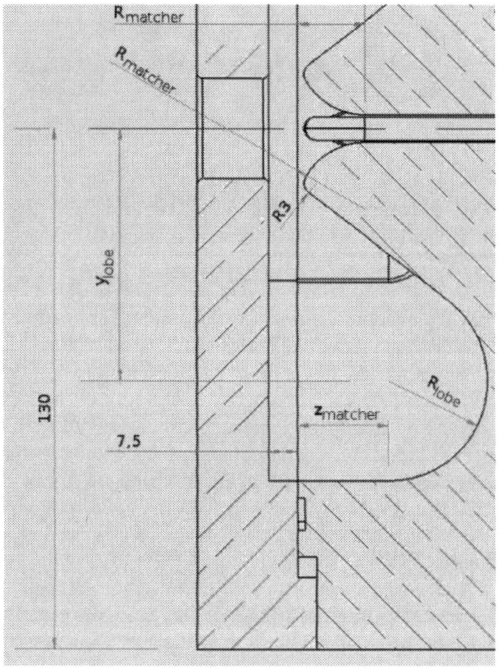

Figure 12. Schematics of the input undercut section of the RFQ, showing the parameters chosen for optimization.

The vacuum port grid also changes slightly the local frequency of the cavity. In order to compensate this detuning, the solid part of the grid penetrates the cavity as ridges, that have been also designed to provide the right frequency. These ridges are cooled by the vacuum port cooling channels. The ridges have been rounded to facilitate the machining process.

4 3D ELECTROMAGNETIC SIMULATIONS AND TUNING

Electromagnetic FEM simulations of the whole length of the RFQ including the details of the modulation and undercuts have been performed to verify the design. These simulations require a lot of resources in preparation of the models and computation time. For this purpose, models with a simplified geometries were used whenever possible. Simulations have been solved efficiently using COMSOL Multiphysics eigenvalue solver. A more detailed description can be found elsewhere [20].

The vane tip region must be defined very finely in order to catch all the details of the modulation. The geometric construction of the vanes themselves is challenging, and we couldn't do it in a reliable way using the CAD packages that we had available, so a home made software that makes use of OpenCascade 3D technology[21] was coded and used for this. The tool allows to import the file describing the modulation for the whole length with a resolution of 50 points per cell. The vane modulation CAD files produced this way were also used for the final solids supplied to the manufacturer.

The electrostatics (*ES*) simulations with constant voltage boundary condition were used during the design of the RFQ modulation. *ES* simulations were also used to scale the electromagnetic fields obtained in an eigenvalue simulation (that have no power input reference). This procedure is detailed in [20]. An example of the results of the 3D simulations is shown in figure 13, where power losses

 https://doi.org/10.1051/epjconf/202023102001

Table 4. Frequency spectrum of the RFQ (in MHz), computed from FEM simulations of the whole structure.

Mode index	Q	D1	D2
0	351,9651	339,9509	339,9814
1	355,2243	346,2343	346,309
2	364,8562	356,3018	356,423

computed from the fields in the surface, and adequately scaled to an innervate voltage of 85 kV, are depicted.

A mathematical model based on a transmission line description of the RFQ cavity[14] has also been developed. The model has been adjusted using the FEM calculation results and implemented in a computer tool to assist in the tuning operations. A comparison between the innervate voltage computed with the transmission line model and extracted from the FEM models is shown in Fig. 14.

4.1 Mode spectrum

From the 3D computer simulations of the whole length the frequency spectrum of the RFQ cavity can be obtained. Previously, this was estimated by extrapolation from only first segment simulations. Computed modes frequencies are grouped in table 4. Modes near the first quadrupolar one are well separated in frequency. This will contribute to an easy tuning and stabilization of the RFQ.

4.2 Static tuning

Static tuning of the RFQ will be provided by a set of plunger tuners. For each 800 mm segment there are 4 sets of 4 tuner ports, so a maximum of 64 tuners could be installed in the whole length (two ports will be used by the power couplers). The static tuners will increase the cavity frequency from the 348.6 MHz of the design to a value close to the operational frequency of 352.21 MHz. The voltage profile (field flatness) should be kept as uniform as possible, and this will probably require that the tuners penetrate the cavity in a non-uniform way. The combination of FEM and mathematical models will be used for this once the actual voltage profile is measured in the final cavity. An automatic procedure will calculate the tuners penetration that correct the measured profile and result in a flat one. This has been implemented in a software tool and tested with numerical examples, as shown in Fig. 15.

5 COOLING DESIGN AND DYNAMIC TUNING

The RFQ is water cooled. The cooling removes the excess heat and also is used to fine tune the RFQ cavity during operation by controlling the thermal expansion driven frequency changes. For each segment, there are cooling channels near the vanes and also in the vacuum grid area.

5.1 Heat load

The heat load coming from the RF standing wave excited to the RFQ cavity walls is computed from the electromagnetic fields at the surface [14, 20], using a copper surface resistance of $R_s = 0.0052\,\Omega$. The eigenfrequency problem is solved and the fields scaled as previously described. A power loss of about 17 kW is estimated at 5 % duty cycle.

5.2 Cooling channels and thermo-mechanical results

There are two types of cooling channels in the structure. Longitudinal channels run along the vane tips (with inlets and outlets in the external surface), while transversal channels cool the vacuum pump grid region. These channels can be seen in Fig. 16.

Thermo-mechanical simulations have been done using a simplified geometry, in order to increase the speed particularly for transient simulations. Simulations are done also using COMSOL Multiphysics, taking the power losses density maps as input heat flux in the internal surfaces. Coupled heat transfer, CFD and thermo-mechanical models of the cavity are then solved. The deformation of the vacuum region of the cavity allows to compute again the resonant frequency. In this way, the detuning caused by the thermo-mechanical deformations can be studied in any condition: steady state or transient simulations, with different RF duty cycles as power inputs and different water input temperatures and flows.

An example of the temperature distribution in the solid and the mechanical deformation can be seen in Fig. 16. For a duty cycle of the 5 % and the temperature of the input water fixed at 25 °C, the temperature of the copper is below 31 °C. In Fig. 17 the dependence of cavity frequency (first segment) with cooling water temperature is shown for steady state simulations. The detuning is very small (about 19 kHz/°C. This is due to the thickness of the walls. A similar detuning of 9.3 kHz/duty is obtained by changing the duty cycle percentage. These results point out the stability of the operation when an adequate water flow is used.

Transient conditions have also studied this way, in order to study the fine control of frequency during operation of the cavity. As an example of these calculations, in Fig. 18 the dynamic change of frequency during a power on step (and the opposite power off step) is shown. These results are used to tune a control model of the cavity frequency, where a mixture of cold and warm water can stabilize the frequency of the RFQ easily. These results are beyond the scope of this paper and will be duly presented elsewhere. Additionally, movable plunger tuners can be used at certain lengths if needed. One plunger tuner per quadrant will provide a tuning range of 20 kHz/mm, enough for tuning the cavity in a LLRF loop. This solution is a back-up plan.

6 FABRICATION

The fabrication of the first segment of ESS-Bilbao RFQ started in 2016. As mentioned before, the RFQ is a 4-vane structure. It has a total length of about 3.1 meters, divided in 4 segments of about 800 mm length each. The segment length is determined by the machining equipment

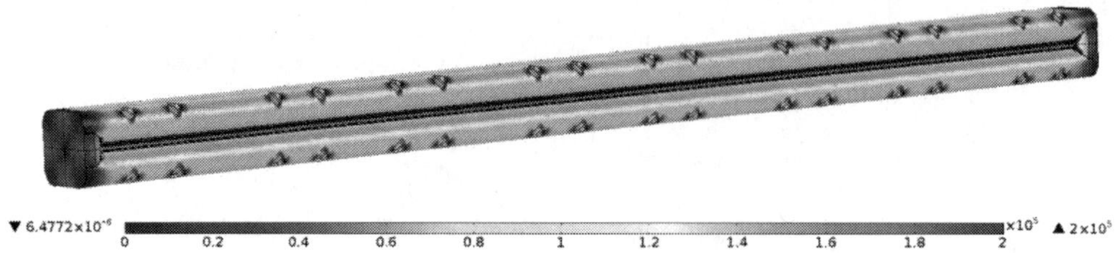

Figure 13. Peak power losses (W/m^2) in the surface of the RFQ cavity, with all the tuners in uniform penetration. The model shown is a simplified model with no pumping pick-up ports included, but with modulation in full detail.

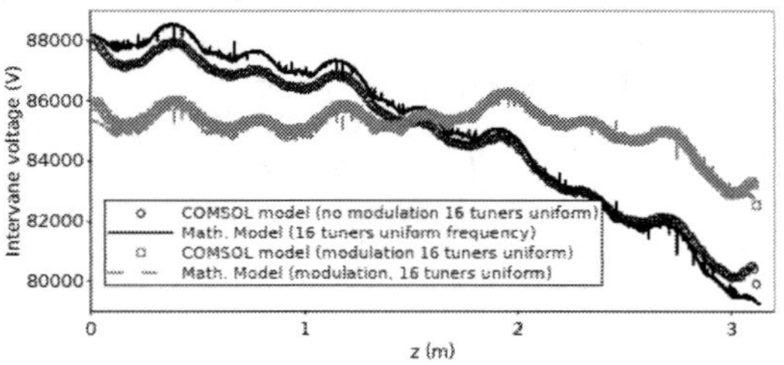

Figure 14. Comparison of the innervate voltage computed using the transmission-line model (full lines) and the voltages extracted from the FEM model of the whole length of the RFQ (open markers). In one situation the vane modulation is not considered and in the other the full details are included. The overlapping of the FEM and the mathematical model curves is very good.

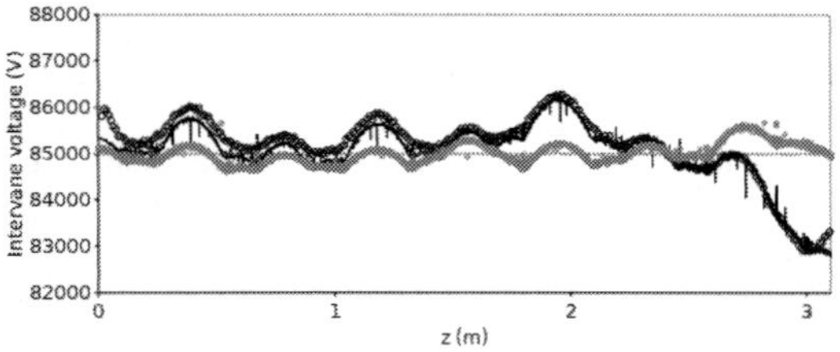

Figure 15. An optimized voltage profile (in red) as compared to the original one without optimization (black).

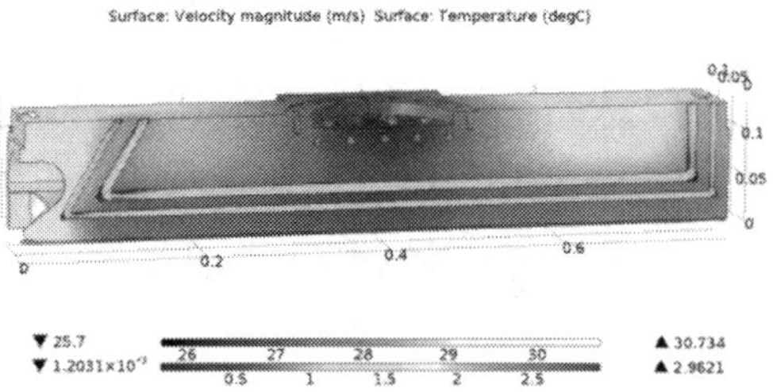

Figure 16. Temperature distribution in the solid and water velocity in the cooling channels, obtained from the thermo-mechanical model of the first segment. Input power corresponds to a duty cycle of 5%. Water enters all the channels at 25 °C.

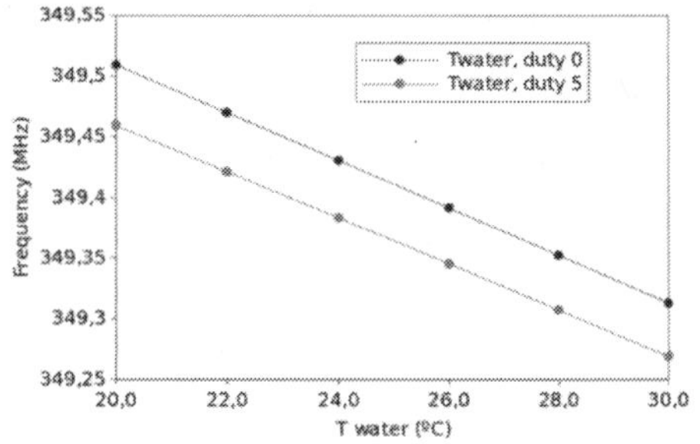

Figure 17. Change in cavity frequency with cooling water temperature, for two different RF power conditions. The results come from steady-state simulations.

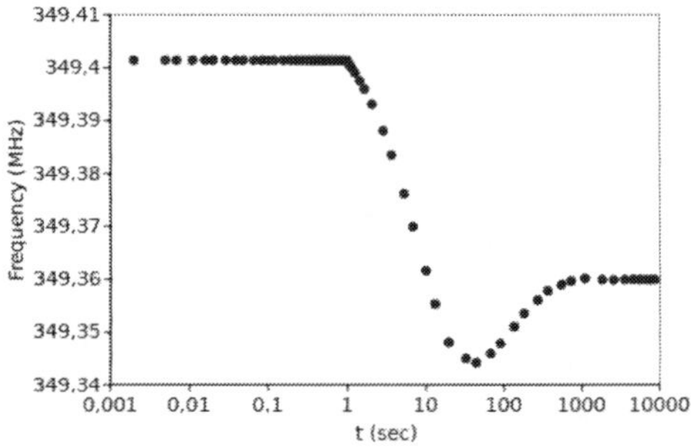

Figure 18. Frequency of the RFQ cavity (first segment) during a transient steps, from zero RF power to operation power at duty cycle 5 % (power up). Cavity frequency is computed at all times of the simulation.

available for fabrication of the vanes. Each segments is an assembly of four elements, 2 major and 2 minor vanes, assembled together by using polymeric vacuum gaskets instead of brazing or other welding system. Material is copper OHFC (Cu 10100) quality.

6.1 Mechanical model

Starting from the electromagnetic design of the RFQ cavity a CAD model of the structure was built. The ports for tuners, vacuum grid and all other mechanical features were

 https://doi.org/10.1051/epjconf/202023102001

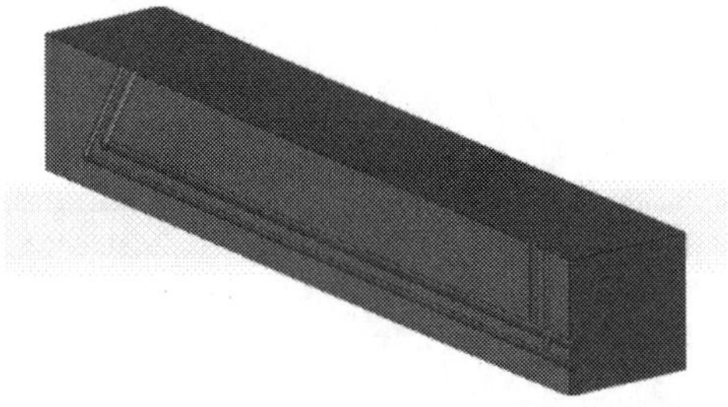

Figure 19. Model showing the deep drilling of the longitudinal cooling channels.

implemented in the model. RF and thermo-mechanical simulations were then ran again to validate the design. As described in a previous section, the vane modulations solids were built up using a home made CAD tool to avoid certain issues with the interpolation of the modulation curve[22].

6.2 Fabrication steps

6.2.1 Raw material

The material for the fabrication of the RFQ was supplied in blocks of two different sizes: 270x140 mm for major vanes and 115x140 mm for the minor ones. Length of all blocks is 830 mm. The copper grade selected is Cu OFE C10100.

6.2.2 Squaring and deep drilling

The first step in the fabrication process is to evenly square the copper blocks, to assure that each face is parallel to the opposing one and perpendicular to the others. Marks to fix the position of the cooling channels are machined in the corresponding faces.

A first rough machining is done, leaving an excess of about 2 mm, followed by a stress-relieve annealing. The deep drilling of the longitudinal cooling channels is then done. This is made during the first steps of the fabrication process, and channels will then serve as reference for the rest of the geometry. In this way, the effects of a possible deformation due to the machining is minimized as major machining is done afterwards. A model showing the channels is shown in Fig. 19. Fig. 20 shows a detail of the first fabricated vane after this stage.

At this stage, the two channels are connected and open in the back side (higher energy) of the RFQ segment. They will be sealed in a later step.

6.2.3 EBW of the cooling channels plugs

Cooling channels are then sealed by inserting copper plugs and welding them. This is carried out by our staff at ESS-Bilbao's Advanced Welding Facility by means of electron beam welding (EBW). Three plugs are inserted and welded, one to separate the two channels and the other two to seal them from the external face (Fig. 21). In the final steps of the process this face will be machined and no external evidence of the welding of the plugs will be visible.

Figure 20. A major vane after rough machining. The holes of the cooling channels can be seen in the vane.

6.2.4 Fine machining

The last fabrication step for a vane is the fine machining, where all the final details are included. Particularly, the vane modulation needs a careful process in a temperature controlled machine to avoid over-heating that could give rise to deformations. Fig. 22 shows a major vane in the milling machine.

The milling of the modulation is made using a CAM controlled, 3-axes, HERMLE C800V machining center. Many displacements of the tool on the copper surface are used, removing a very thin layer each step. This increases milling time but also provide excellent surface quality (roughness 0.8 Ra and mechanical tolerances around 0.005 mm.

6.2.5 Metrology

The final step is the metrology of the fabricated piece (Fig. 23). Special care is taken in the measurement of the vane modulation profile. After control of the first major vane, a deviation of the modulation measured with respect to the design one was detected, despite the attention paid to the machining process. To correct this, a second machining of the modulation profile was performed, lowering the height from the bottom face by 100 um. The contact faces between major vanes and the two minor vanes were also machined removing the same height of material, so the four vane tips have the correct distribution after assembly.

Apart from this issue, the metrology results of the first vane of the first segment is very satisfactory (Fig. 24).

Figure 21. A major vane after cooling channels plugs have been inserted and welded by EBW.

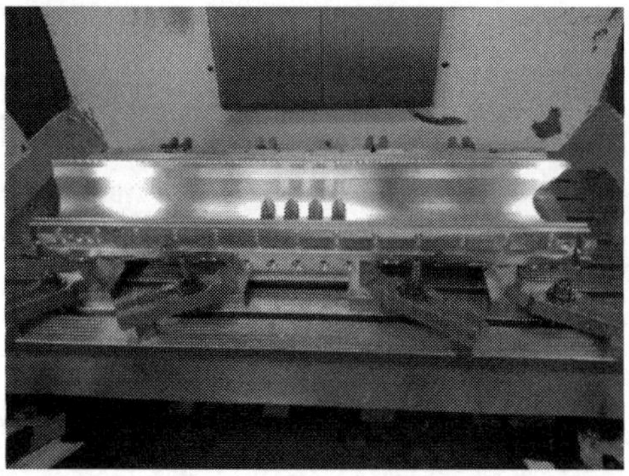

Figure 22. A major vane of the RFQ after fine machining in the milling machine.

6.3 Assembly and vacuum strategy

The vacuum strategy for the cavity is based on the use of polymeric vacuum gaskets at the unions between major-minor vanes, and also on the contact faces between the assembly of four vanes and the cover or inter-segment ring. RF contact seals (so-called finger strips) are considered for all contacts between surfaces. Once the four vanes are assembled (as in Fig. 1) and the alignment verified, the groove for the O-ring in the front side will be machined. This approach will allow assembling or disassembling the RFQ in case of misalignment or other problems. This strategy will be thoroughly tested with the first segment, in terms of vacuum levels and other issues. If results are satisfactory, the same procedure will be used for the rest of the segments. Otherwise, the strategy will be revisited and brazing of vanes will be considered. The results of these tests (expected for the last quarter of 2019) will be presented in future UCANS meetings.

7 CONCLUSIONS

ESS-Bilbao RFQ design has been summarized, including details on the beam dynamics, cavity electromagnetic design, cooling and thermo-mechanical studies. Fabrication of the first segment of the RFQ is ongoing. The whole segment will be received, and tests performed on it, before the end of 2019.

References

[1] ESS-Bilbao, *ESS-Bilbao* (2017), `http://www.essbilbao.org`

[2] ESS, *European Spallation Source ERIC* (2017), `https://europeanspallationsource.se/`

[3] Z. Izaola, I. Bustinduy, J. Corres, D. de Cos, C. de la Cruz, G. Harper, R. Miracoli, J. Muñoz, I. Rueda, A. Vizcaino et al., *Advances in the Development of the ESS-Bilbao Proton Injector*, in *Proceedings of HB2016* (Malmo, Sweden, 2016), TUPMLY01, `http://accelconf.web.cern.ch/AccelConf/hb2016/papers/tupmly01.pdf`

[4] A. Letchford et al., *Status of the RAL front end test stand*, in *Proceedings of IPAC 2015* (Richmond, VA, USA, 2015), THPF105, `http://accelconf.web.cern.ch/AccelConf/IPAC2015/papers/thpf105.pdf`

[5] C. Rossi et al., *The Radiofrequency Quadrupole Accelerator for the LINAC4*, in *Proceedings of LINAC'08* (Victoria, BC, Canada, 2008), MOP040, `http://accelconf.web.cern.ch/accelconf/linac08/papers/mop040.pdf`

[6] N. Misiara et al., *Recent RF and mechanical developments for the ESS RFQ*, in *Proceedings of the LINAC2016 Conference, THPLR054* (2016), `https://accelconf.web.cern.ch/accelconf/IPAC2016/papers/thplr054.pdf`

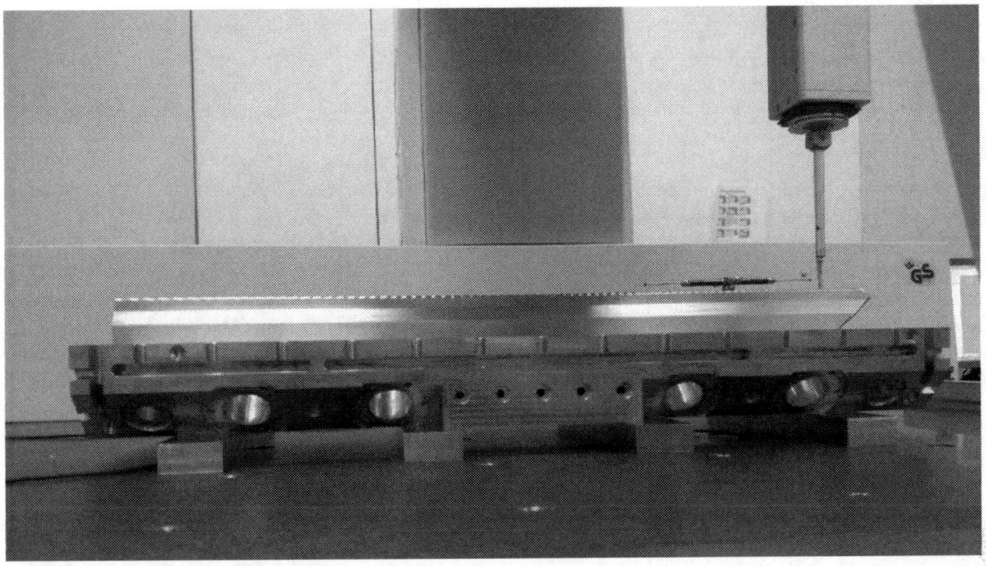

Figure 23. Major vane in the metrology station.

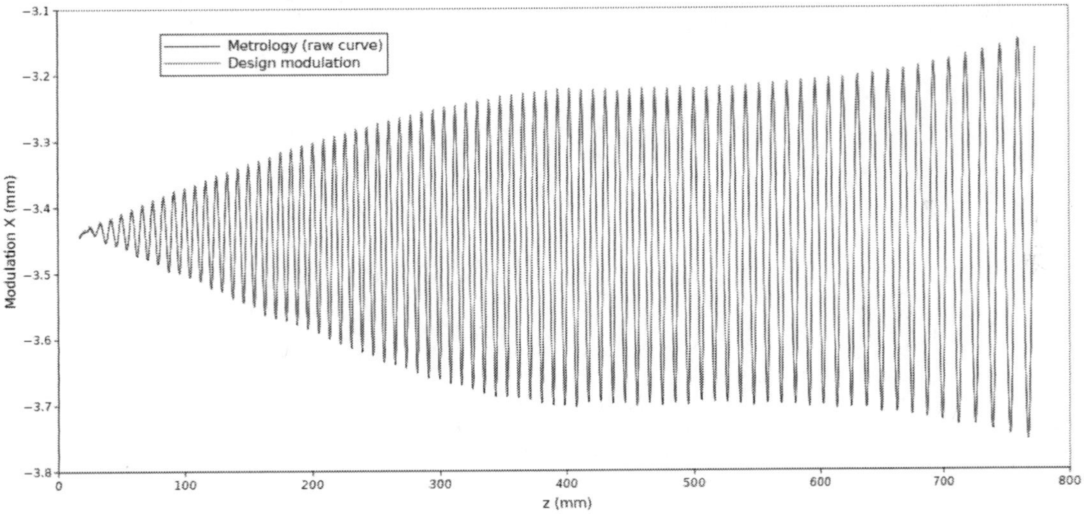

Figure 24. Comparison of measured and designed modulations.

[7] R. Ferdinand et al., *Status Report on the 5 Mev Iphi RFQ*, in *Proceedings of the XX International Linac Conference* (Monterey, California, 2000), http://accelconf.web.cern.ch/AccelConf/l00/papers/TUD04.pdf

[8] I. Bustinduy, J.L. Muñoz, eds., *Technical Design Report: ESS-Bilbao RFQ* (2015), ISBN 978-84-616-5445-1, https://www.essbilbao.org/st_docs/technical-design-report-ess-bilbao-rfq

[9] J. Staples, S. Abbott, R. Caylor, R. Gough, D. Howard, R. MacGill, *The Radiofrequency Quadrupole Accelerator for the LINAC4*, in *Proceedings of 1988 Linear Accelerator Conference* (Williamsburg, Virginia, USA, 1988), http://accelconf.web.cern.ch/accelconf/linac08/papers/mop040.pdf

[10] O. Delfierrea, M. Desmons, A. France, R. Ferdinand, *A new RF tuning method for the end regions of the IPHI 4-vane RFQ*, in *Proceedings of the EPAC 2006, MOPCH105* (Edinburgh, Scotland, 2006), https://accelconf.web.cern.ch/

https://doi.org/10.1051/epjconf/202023102001

accelconf/EPAC2006/papers/mopch105.pdf

[11] L. Young, L. Rybarcyk, *Tuning the LEDA RFQ 6.7 MeV accelerator*, in *Proceedings of the International LINAC Conference, 1998, MO4088* (Chicago, Illinois, 1998), accelconf.web.cern.ch/AccelConf/l98/PAPERS/MO4088.PDF

[12] I.M. Kapchinsky, V.A. Tepliakov, Tech. rep., Stanford Linear Accelerator Center (1969), sLAC-TRANS-0099

[13] A. Letchford, A. Schempp, *A comparison of 4-rod and 4-vane RFQ fields*, in *Proceedings of EPAC 98 Conference* (Stockholm, Sweden, 1998)

[14] T. Wangler, *RF Linear Accelerators*, Physics textbook (Wiley, 2008), ISBN 9783527406807, http://books.google.es/books?id=5EO4_nyndUMC

[15] M.J. Browman, L. Young, *Coupled RadioFrequency Quadrupoles as Compensated Structures*, in *Proceedings of LINAC1990* (Albuquerque, USA, 1990), p. 70, a-12004-C

[16] *General Particle Tracer*, Pulsar Physics (2011), www.pulsar.nl/gpt

[17] R. Duperrier, Ph.D. thesis, University of Orsay, Orsay (2000)

[18] R. Ferdinand, R. Duperrier, J.M. Lagniel, P. Mattei, S. Nath, *Tuning the LEDA RFQ 6.7 MeV accelerator*, in *Proceedings of the International LINAC Conference, 1998, MO4029* (Chicago, Illinois, 1998), accelconf.web.cern.ch/AccelConf/l98/PAPERS/MO4029.PDF

[19] I. Bustinduy, J.M. et al., *Design and Fabrication of the ESS-Bilbao RFQ Prototype Models*, in *Proceedings of IPAC 2012, THEPPB001* (New Orleans, Luisiana, USA, 2012), https://accelconf.web.cern.ch/accelconf/IPAC2012/papers/theppb001.pdf

[20] J. Muñoz, D. de Cos, I. Madariaga, I. Bustinduy, *Design of ESS-Bilbao RFQ Linear Accelerator*, in *Proceedings of COMSOL Conference* (2015)

[21] Open Cascade, *Open Cascade Technology* (2017), https://www.opencascade.com/

[22] J. Muñoz, I. Bustinduy, J. Martin, A. Ortega, I. Rueda, A. Zugazaga, M. Carrera, A. Garbayo, *Mechanical fabrication of ESS-Bilbao RFQ*, in *Proceedings of IPAC 2017, THPIK083* (Copenhagen, Denmark, 2017), https://accelconf.web.cern.ch/accelconf/IPAC2017/papers/thpik083.pdf

https://doi.org/10.1051/epjconf/202023102002

Proton Beam Multiplexer Developments for Multi-Target Operation at the High-Brilliance Neutron Source HBS

Marius Rimmler[1,*], *Johannes* Baggemann[2], *Paul* Doege[2], *Olaf* Felden[1], *Eric* Mauerhofer[2], *Ulrich* Rücker[2], *Helmut* Soltner[3], *Raimund* Tölle[1], *Paul* Zakalek[2], *Ralf* Gebel[1], *Thomas* Gutberlet[2], and *Thomas* Brückel[2]

[1]Institut für Kernphysik (IKP-4), Forschungszentrum Jülich GmbH, 52425 Jülich, Germany
[2]Jülich Centre for Neutron Science (JCNS-2), Forschungszentrum Jülich GmbH, 52425 Jülich, Germany
[3]Zentralinstitut für Engineering, Elektronik und Analytik (ZEA-1), Forschungszentrum Jülich GmbH, 52425 Jülich, Germany

Abstract. The High-Brilliance neutron Source project (HBS) aims at developing a medium-flux accelerator-driven neutron source based on a 70 MeV, 100 mA proton accelerator. The concept intends to optimize the facility such that it provides high-brilliance neutron beams for instruments operating at different time structures. This can be realized by three different target stations irradiated with different proton pulse sequences. The appropriate proton pulses will be distributed by a Multiplexer unit. In the following, we present the integration of this Multiplexer with the HBS beam optics as well as ongoing developments of this unit and several components of the Multiplexer at the COSY facility in Jülich.

1 Introduction

Within the last decades, neutron scattering has proven to be a powerful tool for the observation of complex phenomena in condensed and soft matter science with impacts on innovations in our everyday life. Compact Accelerator-driven Neutron Sources (CANS) represent a promising new type of neutron sources to replace fading out reactors and to support future facilities as the European Spallation Source (ESS) [1]. In CANS, a primary proton or deuteron beam in the sub-100-MeV energy range impinges on a metal target generating neutrons via nuclear reactions. The High-Brilliance neutron Source project (HBS) aims to develop a pulsed medium-flux accelerator-driven neutron source facility based on a high current 70 MeV linear proton accelerator, optimized to deliver high brilliance neutron beams to a large variety of neutron instruments [2]. This requires a thorough adaption of the HBS layout to provide ideally suited neutron beam properties, e.g. energy resolution, energy spectrum and bandwidth for individual instruments simultaneously. Thus, it was decided to operate different target stations in parallel, each featuring a different group of neutron instruments with dedicated time structures. Corresponding adjustments of the primary proton beam pulsing scheme permit to efficiently generate neutrons serving the needs of the instruments.

The correlation of proton and neutron pulse parameters is schematically shown in figure 1. The proton beam pulse length t_p and frame length T_p are correlated by a fixed duty cycle $c = t_p/T_p$.

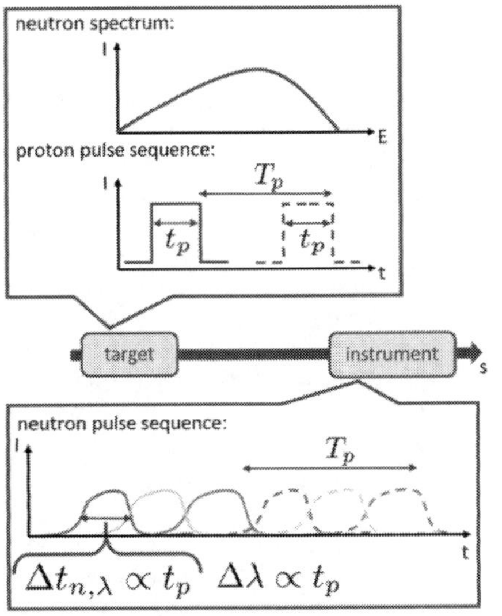

Figure 1. Schematic representation of the interplay of incoming proton beam pulse sequence with variable pulse length t_p and frame length T_p on the target with the observed neutron pulse sequence at the instrument. The colors in the lower plot indicate neutrons with short, intermediate and long wavelengths as red, gray and blue respectively. For simplicity the neutron pulses with different wavelength were drawn separately, note however that a real neutron pulse consists of continuously distributed energies, as indicated by the neutron spectrum in the top graph. The time interval within which neutrons of the same energy arrive at the instruments $\Delta t_{n,\lambda}$ and the absolute wavelength uncertainty $\Delta\lambda$ of neutrons at the instruments scales with the proton beam pulse length t_p.

*e-mail: m.rimmler@fz-juelich.de

Table 1. Comparative proton pulse sequences for three target stations at HBS.

Target	Merit	t_p	$f_p = 1/T_p$
1	high resolution	short	high
2	compromise	medium	medium
3	large bandwidth	long	low

Table 2. Proton pulse sequences for three target stations at HBS adjusted according to 2 % duty cycle.

Merit	t_p /µs	f_p /Hz
high resolution	52	384
compromise	208	96
large bandwidth	833	24

At the instrument, neutrons of the same energy or wavelength λ arrive in a certain time interval $\Delta t_{n,\lambda}$ which is proportional to the primary proton pulse length t_p. Thus the absolute wavelength uncertainty $\Delta\lambda$ of the neutrons is directly proportional to the proton beam pulse length.

For certain instruments one aims for a high resolution, i.e. minimum relative wavelength uncertainty $\Delta\lambda/\lambda$. For this reason it is favorable to use a short proton pulse length t_p. With a fixed duty cycle, this corresponds to a short frame length T_p, i.e. high frequency $f = 1/T_p$. On the other hand, instruments that use a large bandwidth require a large frame length T_p enabling neutrons with long wavelengths to arrive at the instrument prior to neutrons with short wavelengths generated from the subsequent proton pulse. A trade-off between high resolution and bandwidth can be achieved by maintaining a target station that operates at intermediate proton pulse and frame length. The compositions for the three target stations are summarized in table 1.

At the HBS such operation will be realized by introducing three different pulse sequences to the primary proton beam prior to acceleration. The different pulses will then be distributed to the according target stations by a device called *Multiplexer*.

2 Integration with HBS framework

The HBS will be operated with three different pulse sequences as presented in table 2. The different frequencies are multiples of a base frequency (factor 4) such that a total pulsing scheme can be created where lower frequency pulses will be always centered between two subsequent higher frequency pulses as shown in figure 2. In order to distribute the different pulses to the appropriate target station, it is planned to use a magnetic kicker - septum combination to pick the individual pulses and to send them into a different beamline. Hereby, a fast kicker magnet deflects the lower frequency components (green and blue components in figure 2) in between two subsequent higher frequency components (red in figure 2) oppositely. As a result, these deflected pulse components enter a certain region of the septum magnet which generates further deflection of these pulse components. At the same time, the higher frequency proton pulses pass this Multiplexer unit unperturbed. This method directly defines the minimum settling time of the kicker magnetic field at the beam position to be the time between the shortest and the longest pulses, i.e. 859 µs as shown in figure 2. In order to derive geometric constraints on the Multiplexer, i.e. kicker-septum combination, in terms of distances, deflection angles and gap heights of the magnets and thus necessary magnetic flux densities one has to analyze the beam dynamics of the HBS facility. This can be done by dividing the beam transport line of the HBS facility from the accelerator to the targets into three separate segments as shown in figure 3. It should be mentioned that this beam transport segments distribute over a ground and a top floor in order to irradiate the targets from the top. This is required to operate a target cooling concept based on liquid metal as thermal interface material.

The facility is based on a 70 MeV, 100 mA proton LINAC with an approximate normalized rms emittance of $\epsilon_{x(y),\mathrm{norm,rms}} = 1$ mm mrad, a momentum spread $\Delta p/p = 1\%$ and beam Twiss parameters $\beta_{x(y)} = 1$ m, $\alpha_{x(y)} = 0$ at the matching point of the beam transport line. The proton beam is transported to a top floor at roughly 10 m height by the first segment which will feature two double-bend achromat units as shown in figure 6 deflecting into the x-y plane and into the x-z plane successively. The corresponding beam dynamics of these units, each consisting of two quadrupole doublets in the beginning and in the end in order to control the beam Twiss parameters, a vertically focusing quadrupole to compensate the dispersion and two 45° sector bending magnets, are shown in figure 4. Since we employ bending of the beam in both horizontal and vertical degrees of freedom over the whole beam transport line, we aim for dispersion-free fully achromatic systems in order to avoid dispersion mixing. This works for the first segment according to figure 4 in combination with a point-to-point imaging of the beam size. Note that it was in general important to keep the betatron amplitude $\beta_{x(y)}$ smaller than 40 m in order to utilize an inner beam tube diameter of 60 mm.

After beam transport to the top floor, segment two takes over, which comprises the Multiplexer device. Hence, this segment separates into three beam lines during propagation. Thereby the angular separation, which is governed by the septum magnet, is chosen such that the horizontal spacing allows the positioning of quadrupole magnets in the deflected beam lines after about 1 m drift length in order to treat dispersive effects emerging from the septum magnet.

Therefore, the deflection angle of the septum magnet was chosen to be 300 mrad. Together with an effective length of the septum of 1 m, this enables a horizontal separation of neighboring beamlines of more than 450 mm after 1 m drift behind the septum. This allows the positioning of a quadrupole at this distance in order to realize achromatic optics as shown in the bottom graph of figure 6.

The floor plan shows a slightly asymmetric position of the centered quadrupole Q23 in a double-bend achromat composition with four additional quadrupoles, the septum magnet and a 45° sector bending magnet identical to the ones used already in segment one. The corresponding

https://doi.org/10.1051/epjconf/202023102002

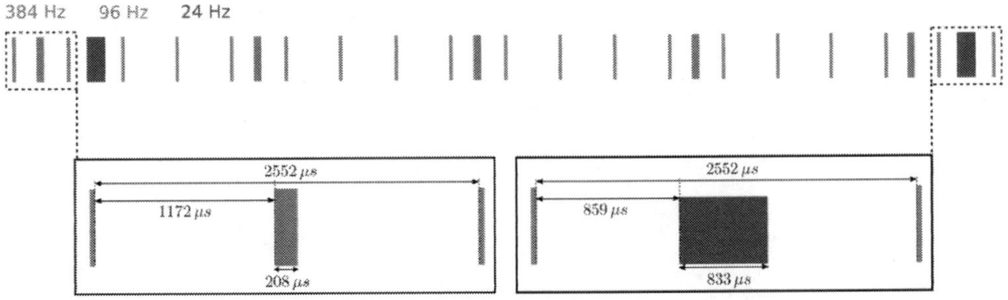

Figure 2. Pulsing scheme of the primary proton beam of HBS. Red, green and blue pulses will be sent to the different neutron target stations. The lower part of the figure shows the time between subsequent pulses.

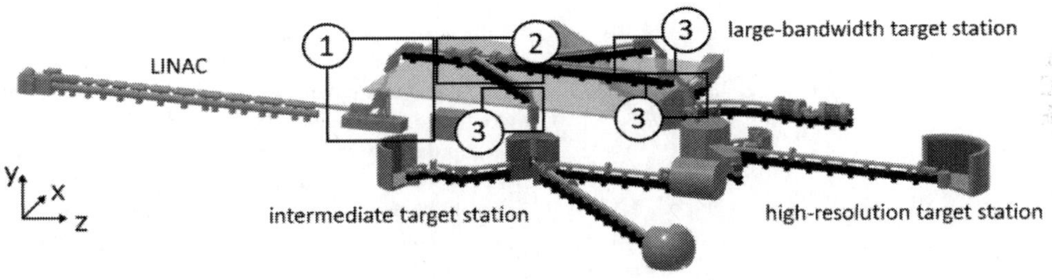

Figure 3. Layout of the HBS facility with three target stations. The neutron production will be driven by a 70 MeV, 100 mA proton LINAC as shown in the left part of the figure. The target stations are classified according to their proton pulse structures. The beam transport from the accelerator to the target station can be subdivided into three segments as shown in the figure. Segment one represents a achromatic system to transport the beam to the top floor. Segment two comprises the Multiplexer unit and segment three uses units from segment one in order to bend the beam towards the target station without dispersion.

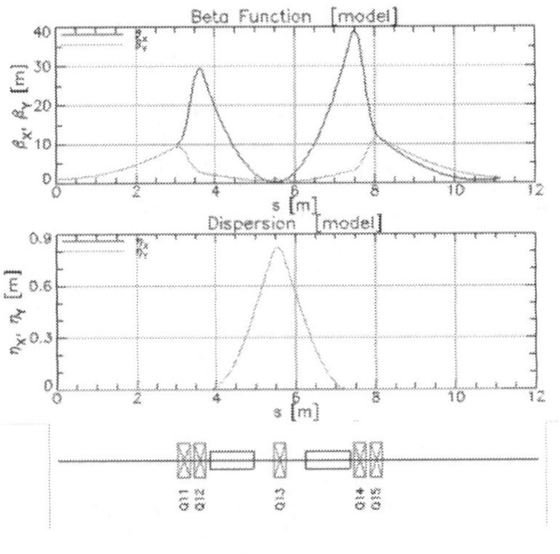

Figure 4. Double-bend achromat systems as used in the first and third segment according to figure 3. This system will be used in three different orientations. One for bending up into the vertical plane and another one for bending back into the horizontal plane in segment one. Another one for bending down into the vertical plane and towards the target station in segment three.

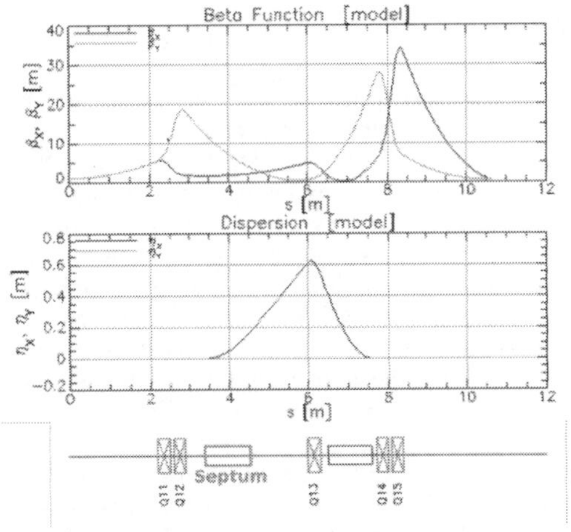

Figure 5. Second segment according to figure 3. This graph shows the deflected beam of the Multiplexer. The beam which passes through the Multiplexer without perturbation will only see the first quadrupole doublet of the displayed beam line.

 https://doi.org/10.1051/epjconf/202023102002

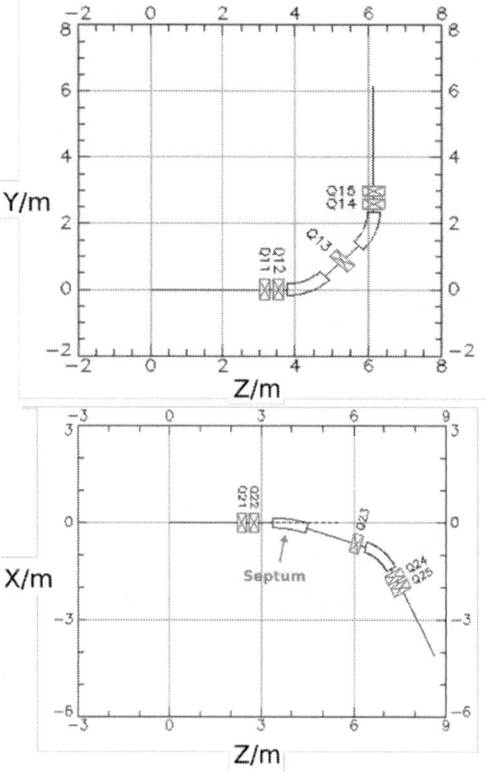

Figure 6. Floor plan of double-bend achromat unit used in segment one and three (top) and the second segment (bottom) according to beam dynamics shown in figure 4 and figure 5, respectively, and as introduced in figure 3. The dashed line in the lower graph indicates the beamline for unperturbed proton pulses.

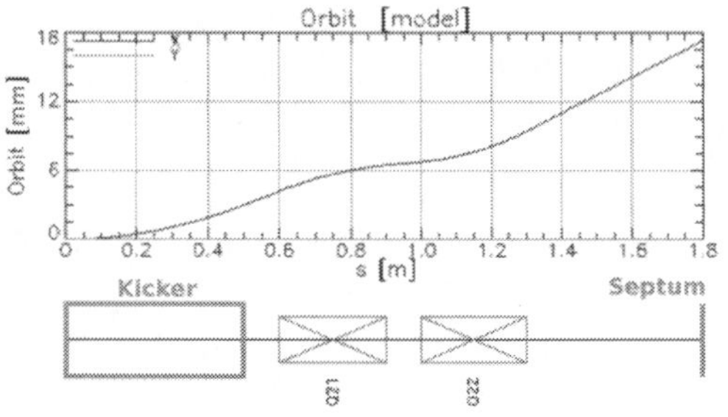

Figure 7. Deflected orbit during propagation from kicker magnet to septum magnet of Multiplexer setup. 12 mrad angular deflection at the kicker result in 18 mm offset at the septum position. Note the effect of feed-down dipolar kicks from the quadrupole magnets.

beam dynamics calculations as shown in figure 5 prove a dispersionless point-to-point imaging for these optics. These beam optics will be used identically for the opposite deflection in the Multiplexer setup. The straight beamline transporting the unperturbed pulse sequences will only see the quadrupole magnets Q21 and Q22 of segment two. Finally, after segment two, the three beamlines introduced by the Multiplexer will obtain a angular separation of more than 60° amongst each other. The spacing between the target stations is therefore controlled by the length of the beamline from segment two to segment three (after each meter, one gains approximately 1 m additional horizontal separation), which can be either realized by a set of quadrupole triplets, doublets or FODO depending on the desired beam quality at the target.

At last, segment three reuses the double-bend achromat configuration utilized in segment one in all of its three beamlines to bend the beam vertically into the x-y plane in order to irradiate the targets from the top. By using quadrupole triplet configurations in the beamline prior to segment three, one would obtain a beam spot on the target with a diameter of 10 mm (as extracted from the LINAC) which can then be scanned by a kicker system within one pulse of minimum 52 µs over a target area of 100 cm^2.

The quadrupole magnets as planned are operated below 9 T/m gradients with 0.3 m effective length while the sector dipoles operate below 0.9 T on 1 m effective length. All dipoles can be connected in series while segment one and three use three quadrupole power converters together and segment two uses another three quadrupole power converters.

The lattice of the HBS beam transport from accelerator to target now defines the design parameters for the Multiplexer. The beam size at the septum magnet is $6 \cdot \sigma_x = 12$ mm and $6 \cdot \sigma_y = 35$ mm according to figure 5. Aiming for a horizontal separation of the beam spots at the septum of 18 mm provides 6 mm distance between the beams in the beginning of the septum. This will be realized by a kicker magnet generating a 12 mrad kick, which is positioned 1.3 m in front of the septum, i.e. 0.1 m in front of the first horizontally focusing quadrupole of segment two as shown in figure 7. The close positioning of the kicker with respect to this quadrupole helps reduce feed-down dipolar kicks from this quadrupole magnet on the deflected beam. For the given geometry, it is planned to use a kicker magnet with 0.5 m effective length and a magnetic flux density of 30 mT.

The deflecting sectors of the three-field-septum magnet need to provide a magnetic flux density of 400 mT with an effective length of 1 m. The developments towards these magnets will be described in the following section.

The beam dynamics calculations in the present section have been carried out using the *Bmad* library [7]. Space charge effects were not taken into account at this stage of development. The overall beam size was considered to be six rms beam sizes $6 \cdot \sigma_{x(y)}$ which takes into account about 99.5 % of the total beam intensity.

3 Developments

In order to efficiently drive the developments of an HBS Multiplexer unit, we plan to operate a Multiplexer test setup at the cyclotron injector of the COSY (Cooler Synchrotron) facility, i.e. JULIC (Juelich light ion cyclotron). It provides 45 MeV H^- [3] and 76 MeV D^- [4] beams with a maximum current of 10 μA in DC or pulsed mode. This accelerator serves as a test bench for multiple HBS related developments, e.g. target material selection and moderator performance tests. Furthermore, it is intended to build a CANS facility based on this accelerator which comprises the Multiplexer test unit during routine operation providing a second target station for fast neutrons [8].

The Multiplexer test unit will be scalable to HBS requirements which mainly manifests in an upgrade of the achievable magnetic flux densities of the magnets (kicker and three-field-septum magnet) due to an energy increase from 45 MeV to 70 MeV, corresponding to a factor of $(B\rho)_{70\,\mathrm{MeV}}/(B\rho)_{45\,\mathrm{MeV}} = 1.26$, and appropriate material selection for enhanced radiation hardness of the magnets. Furthermore the power supply of the kicker magnet has to be chosen such that it provides the demanding HBS pulsing scheme.

3.1 Evaluation of kicker magnet rise time for HBS Multiplexer

For further developments towards HBS Multiplexer operation, one proceeds with the derived magnetic flux densities in section 2, i.e. 30 mT for the kicker magnet and 400 mT for the septum magnet. It is planned to use an array of permanent magnets for the septum as described in the following subsection. For the kicker magnet, however, a high-performance power supply will be required to provide short rise times in order to distribute the pulses. Therefore, one needs to know precisely the specifications of such a power supply governed by the rise time of the current pulse delivered to the magnet. For this purpose, a simulation of eddy currents induced effects in the conducting walls of the beam pipe which predominantly limit the achievable rise times of the magnetic field seen by the beam was necessary.

Studies on the effect of eddy-current induced delays from the stainless steel beam tube during current ramping are displayed in figure 8. From section 2 and figure 2 we know that the required rise time of the magnetic field seen by the beam must be less than 800 μs. By taking into account a safety margin of roughly 50 μs, the rise time of the kicker magnet itself must be below 600 μs according to figure 8 in order to properly distribute the pulses to the target stations.

With the knowledge of the required magnetic excitation rise time of the magnet and the proposed beam tube diameter of d_{ID} = 60 mm one can now start to construct a magnet and power supply to provide 30 mT over on effective length of 0.5 m.

3.2 Developments of Multiplexer test setup

The main features of the test setup will be a kicker magnet, previously employed at the COSY facility for injection,

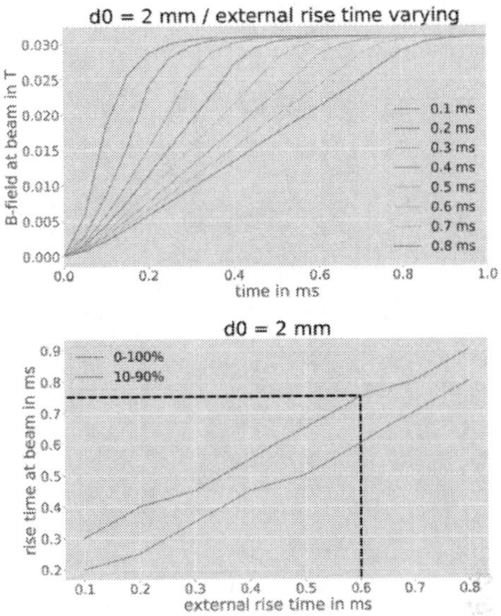

Figure 8. Evaluation of magnetic flux density rise time at beam position in the kicker magnet of the Multiplexer setup governed by eddy-current induced delays from the stainless steel beam tube with d_0 = 2 mm wall thickness and inner diameter d_{ID} = 60 mm. The top plot shows the magnetic flux density (y-component) time dependence at the beam position during ramping of the current in the magnet with different magnet excitation rise times. The lower plot shows the rise time of the magnetic flux density at the beam position (0-100 % and 10-90 % rise time) versus different magnet excitation rise times as derived from the top plot. The dashed line indicates the magnet excitation rise time that yields 750 μs rise time at the beam position. The magnet is ramped up to 30 mT as described in section 2.

and an in-house developed septum magnet with three field regions as it may also be used in the HBS setup presented in section 2. The two magnets will be separated by 1.5 m along the direction of the beam line given by the spatial constraints of the building. In order to make further considerations on the design of the septum magnet, one has to determine the maximum deflection that can be provided by the kicker magnet.

The reused COSY injection kicker magnet is routinely operated with a 250 A power supply offering pulse rise, break and fall times matched to the COSY injection procedure, i.e. 1 ms, 15 ms and 10 ms, respectively. The pulse parameters imply a slightly relaxed pulse sequence with respect to the HBS pulse sequence as shown in figure 2, which is however sufficient for a proof of principle. For future and more ambitious operation of the Multiplexer one could plan an upgrade or replacement of the power supply to offer appropriate pulse parameters as requested by the instruments at this HBS test facility.

The geometry of the Multiplexer test setup and especially the design of the septum magnet generally requires an upgrade of the kicker magnet to larger deflection angles to realize a larger horizontal separation of the beam

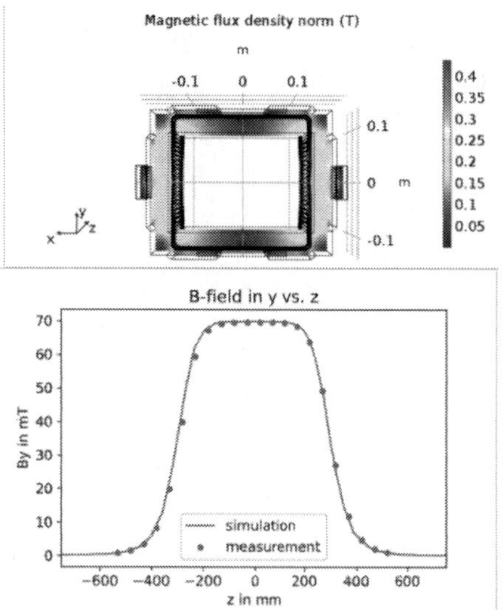

Figure 9. FEM simulation and measurement of COSY kicker magnet model powered with 500 A. The upper figure shows the magnetic flux density distribution in the iron yoke of the magnet. The colorplot scale indicates that the magnetic flux density in the iron does not exceed 0.45 T which is well below saturation. The lower graph shows the simulation and the measurement of the magnetic flux density y-component in the centre of the magnet along the z-axis according to the coordinate system shown in the top figure.

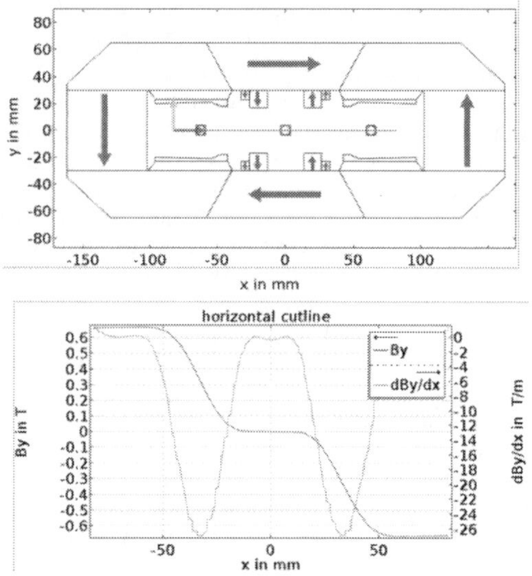

Figure 10. The top graph shows the 2D geometry of the new three-field-septum magnet used in the FEM simulation. The beam is normal to the image plane. The areas comprising gray arrows indicate the permanent magnets and their orientation of magnetization. All other enclosed areas which are mounted on the permanent magnets represent the iron yoke, i.e. carbon steel. The large centered area shows unoccupied space for the vacuum chamber including the three beam spots. The measured beam size at this position is 40 mm. The red line shows the segment which was used for the data in the lower graph. This plot shows the magnetic flux density y-component and its gradient versus x coordinate.

spots at the septum. So far one would obtain 40 mm horizontal separation of the beam spots with 40 mm beam diameter, which turned out to be very challenging concerning the design of the septum magnet. Thus, it is planned to power the kicker magnet with an additional identical power supply in order to reach 500 A total current. To ensure the absence of saturation in the iron yoke of the magnet, an FEM simulation of the COSY injection kicker magnet model as shown in figure 9 was performed. From these simulations one sees that saturation during 500 A operation of the kicker magnet does not occur since the magnetic flux density in the yoke is well below 2 T. Based on this simulation, it was proven experimentally that the integrated vertical magnetic field of the kicker is raised to 42 mT · m. This enables a maximal horizontal separation of the beam spots at the position of the septum magnet of 77 mm facilitating the design of this magnet as described in the following paragraph.

Figure 10 shows a cross-sectional view of the design of the septum magnet as it would be used in the Multiplexer test setup. The test setup magnet will be based on NdFeB permanent magnets. The direction of magnetization is shown in figure 10 by grey arrows. The remanent magnetic flux density for the calculations was set to be 1.35 T.

The motivation to build a permanent-magnet based septum originated from the HBS proton beam parameters. Since the facility will be driven at a fixed design energy of 70 MeV, there is no necessity to tune the magnets which makes it economically favourable to use permanent magnets rather than a design based on electromagnets [5]. Additionally it is aimed to built a so-called massless septum, i.e. no destructive elements in the aperture of the magnet, e.g. return conductors in current-driven septa. This lowers the risk of radiation damage of the magnet due to a malfunction of the kicker magnet.

The dimensions of the three-field-septum magnet are presented in figure 10. They were adjusted to fit the beam size of 40 mm diameter at the future position of the septum in the beamline. The separation of the outer beam spots to the centered unperturbed beam is 62.5 mm which can be achieved with 80 % of the maximum power of the upgraded kicker magnet.

The lower plot in figure 10 shows the vertical component of the magnetic flux density and its derivative with respect to horizontal position x versus horizontal position. This plot represents the figure of merit during the design of the septum magnet. It was primarily important to realize three regions where the magnetic flux density attains a plateau over the whole beam size with no gradient. This allows to generate a left region which further deflects the negative charged particles to left, vice versa for the right region and a zero-field region in the center. By design it-

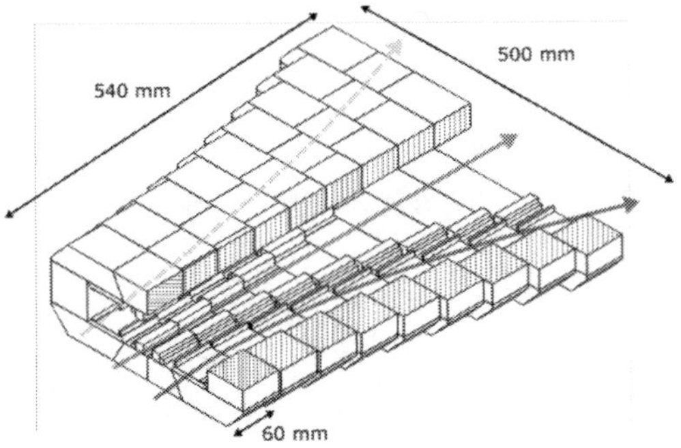

Figure 11. 3D layout of the septum magnet consisting of nine consecutive layers with 60 mm thickness, each. The horizontal apertures of these layers increase according to the trajectory of the deflected beams. The total length of the septum will be 540 mm with 500 mm width at the end. The colored arrows indicate the trajectories of proton beams with different pulse structures as presented in figure 2.

erations including the positioning and shape of the permanent magnets and the pole face structure appropriate results as shown in figure 10 were achieved.

The septum will be extruded by 0.54 m for the Multiplexer test unit. A potential 3D layout of the septum magnet is shown in figure 11. It consists of nine units with 60 mm depth appended onto each other. The horizontal size of this magnet increases in order to accommodate the trajectory of the beam. This design uses identical parts in each section and additional individual iron bricks which will be inserted in the center of the magnet units.

4 Conclusion

The HBS Multiplexer unit will be based on a kicker magnet and a three-field-septum magnet, in which the kicker magnet deflects different pulses into different directions such that the highest frequency pulse components will remain unperturbed. This requires a rise and fall time of the magnetic field at the beam position of less than 800 µs. The break time varies between 208 µs and 833 µs depending on the pulse component that is going to be deflected. Taking into account eddy-current induced delays in the vacuum tube this calls for a rise time of the magnet which is below 600 µs at 30 mT flat top field. The Multiplexer will be integrated with the HBS beam transport line such that dispersion is suppressed. As to the conceptual design of the HBS Multiplexer unit it is planned to built a new septum magnet which features three different magnetic field regions - two regions of finite magnetic dipole field with opposite polarity in order to deflect the proton beam and one region which is supposed to remain field free during operation. Such a magnet will be developed at the COSY facility and integrated with a Multiplexer test unit. This magnet will be adapted to a 45 MeV H^- beam and a significantly larger beam size at the position of the septum due to the larger emittance at this facility compared to the HBS ion beam parameters. After successful operation of this test unit we will scale this magnet to smaller horizontal dimensions to match the smaller HBS ion beam size and elongate it longitudinally to 1 m length by increasing the number of consecutive layers to tackle the higher magnetic rigidity at 70 MeV. Furthermore, it is planned to select radiation hard permanent magnets of SmCo [6] with slightly lower remanent magnetic flux density for the HBS setup which requires additional lengthening of the magnet.

References

[1] ESFRI Physical Sciences and Engineering Strategy Working Group, Neutron Landscape Group, ESFRI Scripta **I** (2016)

[2] U. Rücker et al.,The Jülich high-brilliance neutron source project, The European Physical Journal Plus **131**, 19 (2016)

[3] W. Bräutigam et al., H- operation of the cyclotron JULIC as injector for the cooler synchrotron COSY-Jülich, IOP, 654 (1999)

[4] W. Bräutigam et al., Extraction of D- beams from the cyclotron JULIC for injection into the cooler synchrotron COSY, AIP **600**, 123 (2001)

[5] K. Halbach et al., Application of permanent magnets in accelerators and electron storage rings, Journal of Applied Physics **57**, 3605 (1985)

[6] C.H. Chen, J. Talnagi, J. Liu, et al., The effect of neutron irradiation on Nd-Fe-B and Sm2Co17-Based high-temperature magnets, IEEE Trans. Magn. **41**, 10 (2005)

[7] https://www.classe.cornell.edu/bmad/

[8] O. Felden et al., Recent Extensions of JULIC for HBS Investigations, Proc. Cyclotrons'19, 196–199 (2019) doi:10.18429/JACoW-Cyclotrons2019-TUP019

EPJ Web of Conferences **231**, 03001 (2020) https://doi.org/10.1051/epjconf/202023103001
UCANS-8

Diffusion bonded Be neutron target using 8MeV proton beam

Toshikazu Kurihara, *Hitoshi* Kobayashi

High Energy Accelerator Research Organization (KEK), Tsukuba, Ibaraki 305-0801, Japan

Abstract. The development of the high intensity compact neutron source is mainly conducted by using accelerators for medical purposes. Recently, a lot of compact neutron sources have been developed, and most of them are for boron neutron capture therapy (BNCT). Compared with the common accelerators used for industries, accelerators for BNCT have to accelerate a 100 times larger current of charged particles because of the low conversion efficiency of neutron moderators. To attain a reliable target for the BNCT neutron source, two obstacles have to be overcome; radiation damage (blistering) and heat issue. We introduce diffusion bonding method aiming to make defect free Be neutron target and 1 MPa latent heat water-cooling system. We also install beam expansion optics using quadrupole and octupole magnets to reduce the charge density. Our facility has been operating for more than three years from the commissioning period and we report the recent situation as following. 1) Diffusion bonding method was applied to the solid neutron target of Be with latent heat cooling system. 2) Amount of integrated coulomb value of protons implanted to the Be target is nearly 3000 C now 3) Amount of neutron dose rate was measured. 4) A laser reflection microscope (LRM) method has been developed to observe the neutron target condition (especially radiation damage) through viewport with a few meters working distance. 5) Direct observation of the surface of the Be target was done during the period of improving nearly target vacuum and residual radiation conditions. From these evidences, we conclude our neutron target already produces neutrons capable to treat nearly 500 patients of malignant melanoma.

1 Introduction

After the 3.11, 2011 earthquake and tsunami in Japan, new demand to construct compact accelerator driven neutron source occurred. One comes from universities and one comes from hospitals. In 1980's, Prof. Fukumoto and his group joined together with University of Tsukuba group to establish proton therapy using the KEK booster synchrotron [1]. The upstream of this accelerator was used to study initial phenomena of radiation damages (blisterings) of neutron target related materials in the later years [2].
Compared with the common accelerators used for industry, accelerators for boron neutron capture therapy (BNCT) have to accelerate a 100 times larger current of charged particles because of the low conversion efficiency of neutron moderators. To attain a reliable target for the BNCT neutron source, two obstacles have to be overcome; the heat issue and defects control (blistering). And we have solutions. Cooling with latent heat and bonding with diffusion. First of all, existing neutron source is presented. Design concept of our neutron target is as following: Separate function target. Top layer consists of neutron production materials. Middle layer consists of mitigation materials for hydrogen embrittlement (blistering tolerant materials). Bottom layer consists of heat sink with water cooling. Usually these layers consist of different materials. That requires joining dissimilar metals. To satisfy this goal, we try to introduce diffusion bonding method with hot isostatic pressing (HIP) in care of defect control. Direct observation of a blistering phenomenona under irradiation of incident proton beam reveals us an initial stage of blisterings [2]. The combination of blistering tolerant materials does not always enables us to make blistering tolerant neutron target. It is necessary to take much care about the newly coming reaction layers with diffusion bonding method. Selection of blistering tolerant materials and estimation of the quality of newly emerged reaction layer is important.

2 Neutron target

For the compact accelerator driven neutron source, a lithium target or a beryllium target is generally used with proton beams. We selected beryllium for the neutron target with 8 MeV incident proton beams and 10 mA average beam current [3]. Reasons are stable chemical characteristics including melting temperature (lithium vs. beryllium) and residual radio-activities (8 MeV vs. 30 MeV). Considering the complexity of handling of hydrodynamics and phase change of target material related to the ambient atmosphere or temperature, we abandoned the liquid lithium target.

* Corresponding author: toshikazu.kurihara@kek.jp

Design concept is simple. Separate function target is our keyword. Our target component composed of three parts. The nuclear reaction part (beryllium), the proton implantation part (palladium) and the heat removal part (copper with water for cooling). For each part combination of suitable materials and methods are applied. Figure 1 shows a rough sketch of our neutron target.

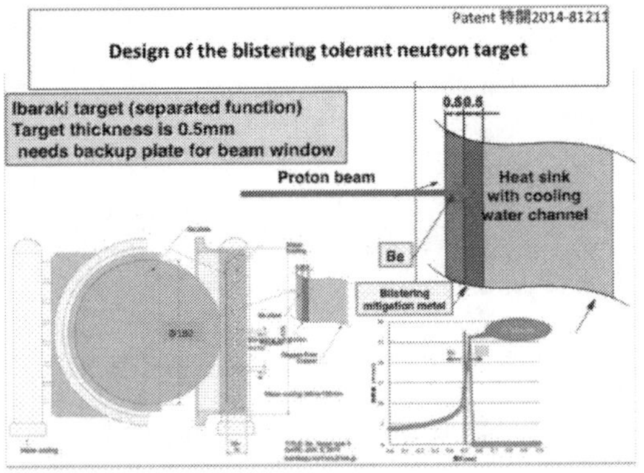

Fig. 1. Design of the separate function three-tier target.

Three tier of different metals are used. The Bragg curve of 8 MeV proton in beryllium has its peak at the path length of 0.54 mm. It is proposed to use 0.5 mm beryllium for neutron production and stop proton beam in palladium for blistering tolerance. Thickness of 0.5 mm beryllium is not suitable mechanically to sustain high-pressure differences.

2.1. Cooling with latent heat

Cooling with latent heat is one of the established heat management methods. One can see a lot of example in the field of modern power plants and fusion reactors [4]. Critical heat flux and minimum heat flux of Nukiyama curve are key words for modern heat exchange [5]. Three parts exists in its curve, nucleate boiling, transition boiling and film boiling. The heat will be transported in the domain of nucleate boiling. In other word latent heat cooling [6] will be applied to the neutron target. Designed beam power at our neutron target will be 4.5 MW/m^2. Available beam current density on the target material will be 0.02 ~ 0.05 mA/cm^2, special method to reduce beam current density is necessary. Two methods exist. One is to scan beams and the other is to expand beam diameter. Beam expansion optics [7] is applied for this purpose. It is difficult to obtain average low current density, spatial density minimum was selected in this case.

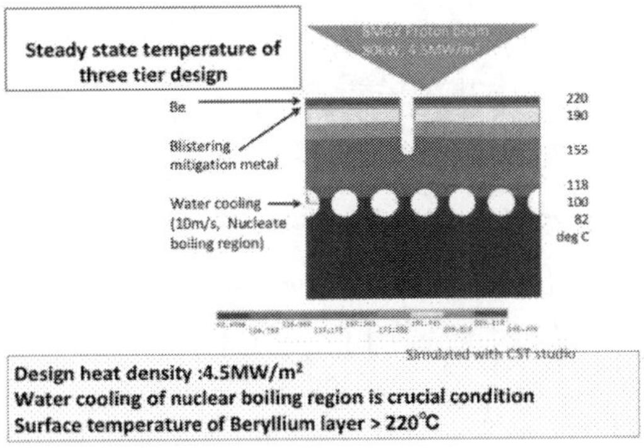

Fig. 2. Cooling with latent heat - high pressure, high speed, microchannel water flow.

Simulation results by CST is presented (Fig. 2). 10m/s fast speed water flow together with high pressure will enable to attain latent heat cooling.

2.2. Defects control and diffusion

8MeV proton beam injected into three tiers neutron target will stop in the blistering tolerant materials (Bragg peak at palladium, not inside of beryllium). Injected proton at non-equilibrium state will diffuse owing to the potential barrier. Hydrogen embrittlement is a phenomenon where hydrogen atoms gather together around the vacancies and/or vacancy clusters and form hydrogen molecules. During this process, vacancies play an important role to form two-dimensional or three dimensional vacancy complex. These become site of a source of hydrogen molecules. Around these defects, interfaces scatter electrons and phonons. It is possible to detect these defects through electrical resistivity measurement, heat transfer coefficient, ultrasonic testing (UT, or ultrasonic inspection) and so on. Figure 3 shows the potential for electrons or protons in a metallic lattice. Left part shows a potential near vacancy. Lack of ion core makes potential difference. Practical material in use contains these lattice defects and vacancy complex. We have to select proper materials and also proper methods to obtain blistering tolerant neutron target.

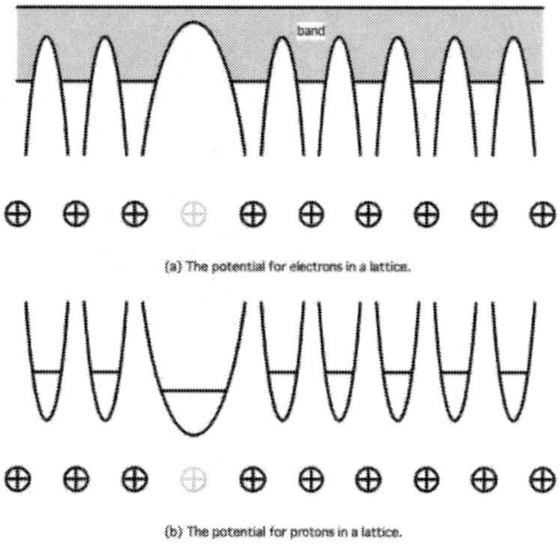

Fig. 3. Potential for proton in a lattice near a vacancy.

Let us change the viewpoint from nanometer scale to micrometer scale. Proton or hydrogen atoms diffuse in the bulk of metals. Figure 4 shows the state of diffusion of impurity atom. Three stage of diffusion process (G:grain, GB:grain boundaries, S:surface) is represented. Diffusion of impurity atom in a grain is slow, that at a grain boundary is faster and diffusion on the surface is very fast. The difference between the grain boundary diffusion and the grain diffusion is not so different at high temperature. This is a generally accepted opinion. It is important to point out that the diffusion of hydrogen is different from that of impurity metal or atom. For example diffusion of hydrogen in metal is the fastest of all the atomic diffusion in solids. Diffusion in grain boundaries and dislocations for large atom is easier with vacancy mechanism. On the contrary, the diffusion of hydrogen is slower owing to the trapping mechanism of hydrogen [9].

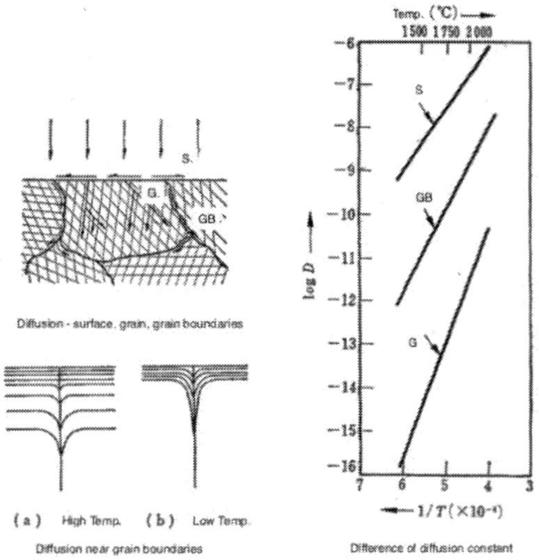

Fig. 4. State of diffusion of impurity atom [8].

Experimental results with electrons, positrons, muons and also protons will introduce new perspectives to leptons and hadrons.

2.3. Diffusion bonding with defect control

Joining dissimilar metals by HIP aims for few defects in the intermediate boundary region. Figure 5 shows the schematic diffusion bonding process. Five marked regions reveal the vacant space between two dissimilar metals. After the suitable time and pressure of HIP process, these vacant space shrink and vanish.

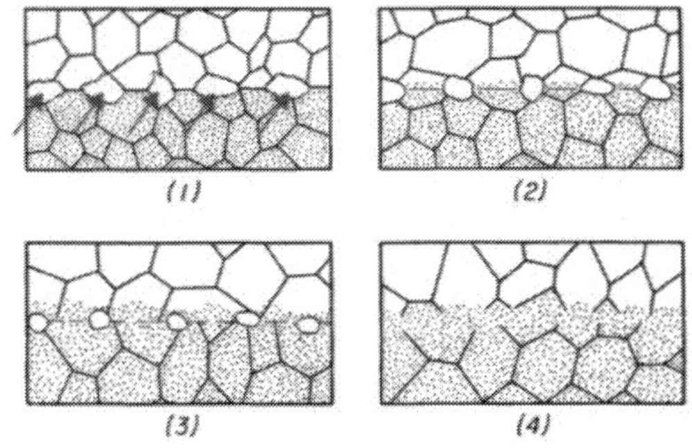

Fig. 5. Bonding interface tends to disappear over time. [9].

Joining dissimilar metals by HIP is possible by this process. Dissimilar metals can be joined together. But it does not mean the completion of the blistering tolerant neutron target. Again, let us change the viewpoint from micrometer scale to nanometer scale. Are there any other defects exist at the interface? After the HIP process, thin newly emerged layers were observed (Fig.6). We call them reaction layers. Testing and estimation for neutron target related materials were done to select the suitable combination of materials. Optical microscope, Scanning Electron Microscope, UT, Laser Flash Measurement, Tensile Stress Measurement and other experiments were performed [3]. Backscattered electron image of scanning microscope was shown (Fig.7). After these observations and measurements, combination of optimal three dissimilar metals (beryllium, palladium and copper) and parameters of the HIP process were fixed.

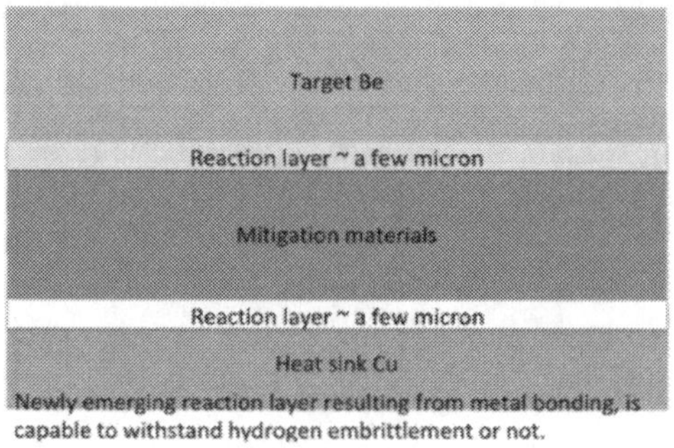

Fig. 6. Newly emerged reaction layer.

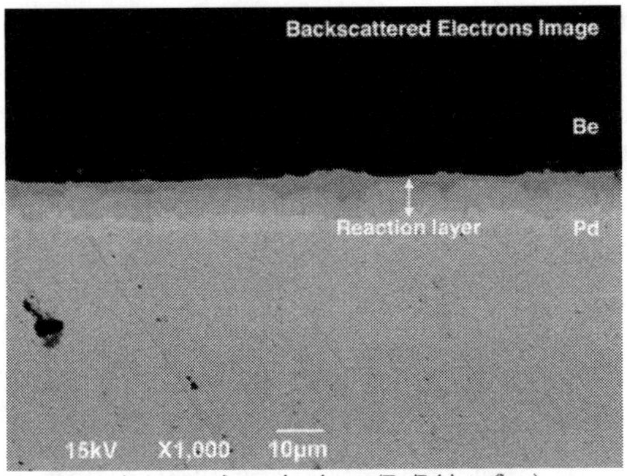

Fig. 7. Newly emerged reaction layer (Be/Pd interface).

Tantalum, niobium, and other candidate materials were also tested and observed. For those samples, the quality of the interfaces were inferior to palladium. There were voids, intermetallic compounds, precipitations.

2.4. Beam expansion optics

Beam expansion optics is composed of two quadrupole (QM) magnets and two octupole(OM) magnets (Fig.8). The area of the typical normal distribution beam upper stream is 200mm^2. Owing to this optics, we can get 140 mm x 140mm square beam. The ratio of the beam intensity of expanded beam is 1/100 compared with that of normal distribution beam. 20kW beam power on the neutron target was established on Feb. 2019.

Fig. 8. Beam expansion optics.

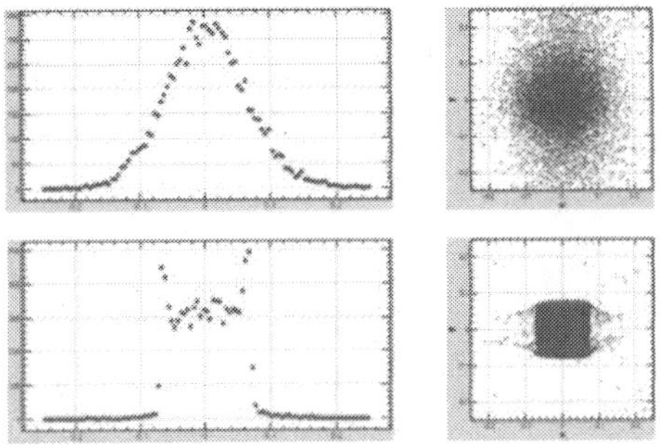
Fig. 9. Calculated beam expansion optics.

Figure 9 shows the calculated beam profile on the neutron target. Actual profile was monitored by FLIR thermo camera and temperature measurement with thermo couples. Due to a lack of spatial resolutions for this method, it is difficult to compare with experimental results.

3 Performance and Conclusions

Recent performance of iBNCT is shown (Fig.10). Average beam power on the target is 21.7 kW.

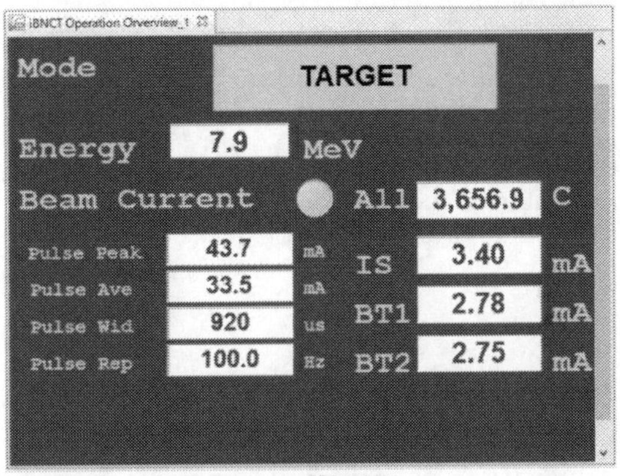

Fig. 10. Operation parameter of iBNCT.

Integrated charge from ion source and on neutron target, average current and normalized neutron flux are presented (Fig.11). Recorded time started from Jan. 01, 2017 to April 2019. The integrated charge (Coulomb), the average current for Ion source and Be target (mA) and the normalized neutron flux (n/cm^2/s/mA) are presented.

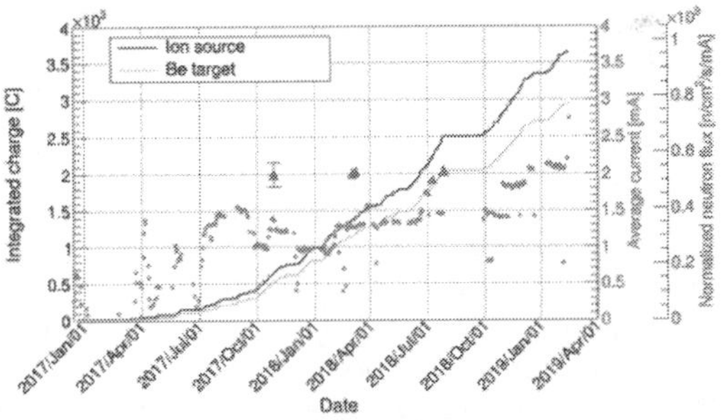

Fig. 11. Operation parameter of iBNCT.

Our three-tier beryllium neutron target has sustained integrated proton fluences on the order of 3000C during 2years and there is no degradation of the neutron flux or the neutron yield.

The compact accelerator driven neutron source for BNCT is operational. The 20 kW operation is stable over extended periods of time. We are heading to 40 kW or 80 kW.

4 Summary

Three tier diffusion bonding HIP beryllium target, its performance and its basic strategy are presented. The target has sustained integrated proton fluences on the order of 3000C during 2years and there is no degradation of the neutron flux. The author wants to write down that palladium is selected for blistering tolerant materials, but it is not the only one candidate. Please refer Supplement. Surface potential, nano structure, electronic structure and relativistic effects from the reconstruction of inside wall of void surface might play an important role.

Acknowledgments The author wish to thank H. Kumada, F. Naito, M. Sato, T. Sugimura and other members of the Ibaraki iBNCT group. This work was supported by the JSPS Grant-in-Aid for Scientific Research (B) Grant Number 23340080 and Scientific Research (B) Grant Number 26293136.

References

1. S. Fukumoto, CYCLOTRON VERSUS SYNCHROTRON FOR PROTON BEAM THERAPY *Proceedings of the 14th International Conference on Cyclotrons and their Applications, Cape Town, South Africa* , pp.533-536 (1995)
2. T. Kurihara, H. Kobayashi, H. Matsumoto, M. Yoshioka, J. Radioanal. Nucl. Chem. **305**, 3 (2015)
3. H. Kumada, T. Kurihara, M. Yoshioka, H. Kobayashi, H. Matsumoto, T. Sugano, H. Sakurai, T. Sakae, A. Matsumura, Appl. Radiat. Isot. **106**, 78 (2015)
4. T. D. Marchall, D. L. Youchison, L. C. Cadwallader, Modeling the Nukiyama curve for water-cooled fusion divertor channels, *Proceedings of the Fourteenth Tropical Meeting on the Technology of Fusion Energy. October 15-19, 2000. Park City, Utah*, pp.849-855 (2001)
5. S. Nukiyama, Maximum and Minimum Values of Heat Transmitted from Metal to Boiling Water under Atomspheric Pressure, Journal of the Japanese Society of Mechanical Engineers, **37**, 367 (1934)
6. S. Yu. Taskaev, Physics of Particles and Nuclei, **46**, 956 (2015)
7. Y. Yuri, T. Ishizaka, T. Yuyama, I. Ishibori, S. Okumura, K. Yoshida, Nucl. Instr. and Meth. Phys. Res. Sect. A, **642**, 10 (2011)
8. S. Kouda , *Introduction to Metal Physics*, CORONA PUBLISHING CO., LTD. TOKYO (1964)
9. T. Mutsschele, R. Kirchheim, Scripta Metall., **21**, 135 (1987)
10. I. Kawakatsu, S. Kitayama, J. Japan Inst. Metals, **40**, 96 (1976)

Supplement

The author wants to underline that palladium is selected for blistering tolerant materials, but it is not the only one candidate. Prof. Baxter pointed out the low diffusion speed of hydrogen in the bulk of beryllium in this meeting UCANS8. Time spectrum of the hydrogen desorption from our neutron target were measured (not presented this time). It looks like desorption spectrum of room temperature. At a glance, it violate Fick's low. But I think it is a matter of surface barrier. Surface barrier of Be-H might be stronger than normal M-H (if the specific conditions supplied or satisfied). For example Ogorodnikova and their group's work [Ogorodnikova, JNM277(2000)130]. Surface potential, nano structure [Mutschele, Scripta Metall.21(1987)135] and relativistic effects from reconstruction of inside wall of void surface play an important role. Selection of palladium itself is not essential. Surface potential, Fermi energy of palladium is important. You will find this is the issue of surface phenomenon or boundary between gas phase and solid phase and also the surface physics and chemistry. Recently, the valence band structures of face-centered-cubic Ag-Rh alloy nanoparticles, which are known to have excellent hydrogen-storage properties, were investigated using bulk-sensitive hard x-ray photoelectron spectroscopy [for example Yang, Appl. Phys. Lett. 105, 153109 (2014) Fig.5].

CoolGAL: a Galinstan bathed Be fast neutron production target at the NEPIR facility

Rogelio Alfonso Barrera[1], *Dario* Bisello[2], *Juan* Esposito[1], *Pierfrancesco* Mastinu[1,*], *Gianfranco* Prete[1], *Luca* Silvestrin[3] and *Jeffery* Wyss[2,4]

[1]INFN Laboratori Nazionali di Legnaro, vale dell'Università 2, I-35020 Legnaro (Padua), Italy
[2]INFN Padova, via Marzolo 8, I-35131 Padova, Italy
[3]University of Padova, Physics and Astronomy Department, via Marzolo 8, I-35131 Padova, Italy
[4]DICeM, University of Cassino, via Di Biasio 43, I-03043 Cassino (Frosinone), Italy

Abstract. In this contribution, we present the CoolGAL fast neutron production target system, to be used in the initial phase of the NEPIR irradiation beamline at the SPES facility, that will be operational in 2022. Initially, NEPIR will be used for shielding studies against fast neutrons for space applications and to investigate neutron-induced single event effects in microelectronic devices and systems. In CoolGAL, the neutron production component, a thick Be cylinder, is immersed in a static bath of Galinstan, a liquid alloy of Ga, In and Sn, contained by an outer water cooled copper cladding. MCNPX calculations indicate that, by using a 1 μA current of 70 MeV protons, it can produce a fast neutron energy spectrum that is somewhat flat the 30-65 MeV energy range and with a sharp cut-off at the beam energy. At the standard test point, located 2.6 m downstream from the source, the beam spot diameter, defined by the peculiar collimation scheme of the initial phase of NEPIR, is 10 cm and the integral fast neutron flux is $\Phi_n(1< E_n<65\ \text{MeV}) \sim 3\times10^6\ \text{n cm}^{-2}\text{s}^{-1}$. Using proton beams with two different energies, one can calculate, by subtraction, the effects due to the neutrons in the energy interval defined by the two cut-off values. Preliminary results of ANSYS calculations, for a 1 μA proton current of 70 MeV protons (70 W), show a limited regime temperature (28 °C) of the Be component, capable of ensuring the exceptional safety level required for the operation at SPES. The thermal reliability of CoolGAL is very promising for future developments that require higher proton currents.

1 Introduction: NEPIR

NEPIR (NEutron and Proton IRradiation facility) is a multidisciplinary project at the new 30-70 MeV cyclotron of SPES [1] at the INFN National Laboratories of Legnaro (LNL). NEPIR will be built in stages, according to available funds. In the final configuration, it will deliver complementary fast neutron beams: quasi mono-energetic neutron (QMN) beams in the 30-70 MeV energy range and an intense neutron beam, with a continuous energy distribution in the 1-65 MeV range, similar to that produced by cosmic rays in the atmosphere found at flight altitudes and at sea level. Direct variable energy proton irradiations will be also available to study the effects by protons (e.g. solar protons) too. By using proton-beam energy degraders, the lowest energy of both the QMN and direct protons will be lowered below 30 MeV down to 20 MeV, possibly lower.

NEPIR was originally conceived to study the effects of neutrons and protons on electronic components, an important field of both scientific and technological research. The capability of electronic circuits to survive exposure to radiation is essential for many scientific and technological applications in radiation hostile environments: outer space, surface of planets, high-energy physics experiments, nuclear reactors, medical and industrial applications. But the issue radiation-assurance of state-of-the-art electronics used in everyday life applications is a growing concern.

The trend towards extreme miniaturization of commercial electronics (not specifically rad-hard) has brought an increased sensitivity to certain kinds of effects induced by neutrons present at sea level. The industrial concern for the effects of these invasive and ubiquitous atmospheric neutrons in vital electronic instrumentation has always been and continues to be found in avionics, but it is an ever growing concern for the fast growing high-reliability ground-based information-technology electronics industry too. At sea-level the fast (E > 1 MeV) neutron flux on earth is 21 n $\text{cm}^{-2}\ \text{hr}^{-1}$; it is much lower than that at flight altitudes (~300-400 times), but the number of ground-based

* Corresponding author: pierfrancesco.mastinu@lnl.infn.it

digital electronic devices used in ever-increasing and diverse applications is enormously larger [2][3][4][5][6].

Big European and Japanese industrial groups have shown interest in the NEPIR project at LNL and are waiting for a chance to use the facility. NEPIR will be a precious applied research and test facility also for small/medium high-tech industries, typical of the local and national economy, that develop high-reliability electronic devices and systems for various applications; e.g. information technology, industrial automation, telecommunication, automotive, biomedical, security and military.

NEPIR will also be used to study shielding against fast neutrons and protons. For this purpose, the facility has been partially funded by the MIUR within the SPARE (Space Radiation Shielding) project: a joint INFN, ASI and Centro Fermi collaboration that deals with health risks due to cosmic radiation: a major showstopper for safe space exploration and colonization, where shielding techniques are the only practical countermeasure against radiation. The main goal of the SPARE project is to perform experimental tests, with neutrons and protons, of physical (active and passive shielding materials) and biomedical (hibernation) countermeasures for human space exploration. For novel shielding materials, based on composites and nanomaterials (proprietary, with undisclosed exact composition), or made of complex in-situ available planetary resources, code predictions have high uncertainties or may be completely lacking. Accelerator-based measurements are in those cases an essential tool to characterize shielding capabilities and to tune models. Compared to the surface of the earth, the fast neutron flux on the surface of the Moon is two orders of magnitude higher, as on the surface of Mars where the spectrum is even harder [7]. The SPARE project includes a test campaign of active and passive shielding materials for fast neutrons at LNL and for protons at the accelerator facility of TIFPA [8].

With the limited SPARE funds, we are building NEPIR-0: a simplified phase-0 fast neutron irradiation facility, adequate for the SPARE project, but still attractive for academic and electronics industrial users. NEPIR-0 is completely funded and could be completed and commissioned by 2022 and then made available to users.

Indeed the goal of NEPIR-0 is to catalyze further support in order to obtain the necessary funds to upgrade to the final configuration, that would make NEPIR a unique multidisciplinary facility in Europe, very useful for research: applied, industrial and basic.

2 NEPIR-0

The floor plan of NEPIR-0 is shown in Figure 1. Its neutron production target is based on a ~30 mm thick proton stopping Be target, that produces a continuous energy neutron beam. NEPIR-0 has no independent shielding: it uses instead the shield wall of the SPES facility by placing the Be target inside the conduit that connects the cyclotron hall to the experimental one (section 2.1 and Figure 2). For this reason, the installation of the beam optics for QMN is not possible in this phase of the project.

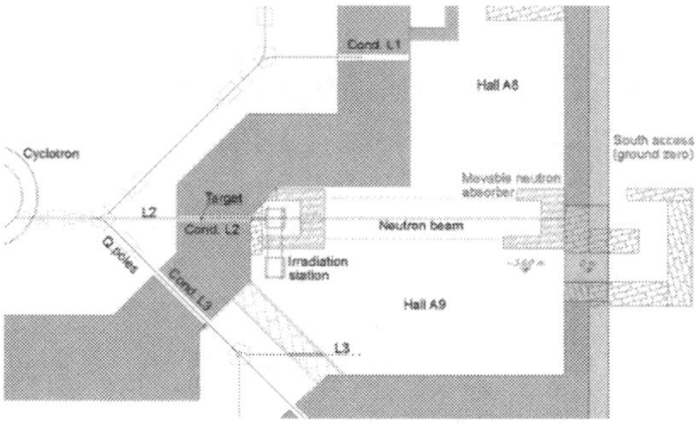

Fig. 1 Floorplan of the L shaped NEPIR experimental area (Halls A9 and A8) and part of the cyclotron hall. The cyclotron is visible on the left. The position of the NEPIR-0 neutron production target system is placed inside the conduit through the 3m thick shield wall that separates the cyclotron hall from the experimental hall A9. A movable neutron beam absorber is schematically shown in both open and closed positions.

Because there is not well defined bunker, a low pressure containment system of the activated air of the experimental hall will not be available. For this reason the proton current on the Be target is at most 1 μA, to ensure that the average activation, diluted by the large ventilated volume of the experimental hall, remains below 1 Bq/gr.

A movable neutron beam stopper, shown schematically in Figure 1, will reinforce the shielding of the back wall; in the closed position, it will allow users to access the experimental hall A9 while the cyclotron delivers beams to other users; it will also shield against the radiation of the activated beamline materials.

2.1 The target station and collimation system

The placement of the Be-based neutron production target inside the conduit is shown in Figure 2.

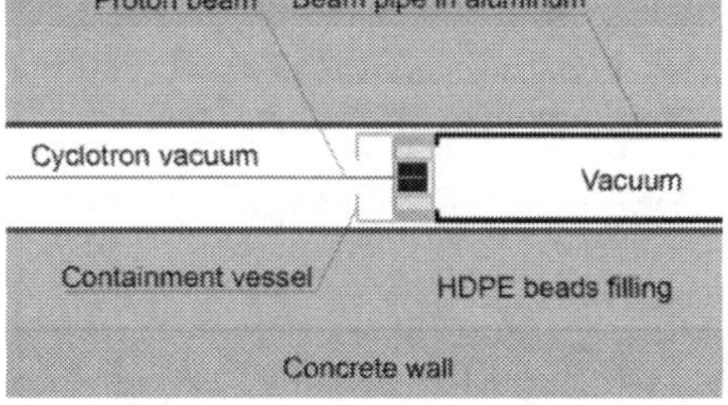

Fig. 2 Schematic representation of the target system inside the conduit through the cyclotron concrete shield wall (grey). In green, the polyethylene filling in the gap between the vacuum line and the wall. A composite material plug (in black) is used to insert the target system and then to hold it place.

The proton vacuum line passes through the conduit in the cyclotron shield wall and is accessible through a flange in the experimental hall. The gap between the vacuum pipe and the conduit (30 cm diameter) is filled with High Density PolyEthylene (HDPE) beads, both to limit neutron propagation to and from the cyclotron hall, and to act as a collimator to define the neutron beam into the experimental area.

The target system is inserted 1 m inside the conduit, measured from the cyclotron hall. At the end of the conduit, a 50 cm layer of concrete blocks is added to the cyclotron shield wall to ensure adequate containment of the neutrons produced by the target (visible in Fig. 1). A composite material plug (shown in black in Fig. 2) is used to insert the target system and its cooling system into the beamline through the flange in hall A9, and to hold it in place. This plug can be evacuated by a dedicated pumping system. At the standard test point, 2.6 m downstream of the target, the effective diameter of the fast neutron beam is 10 cm.

2.2 The neutron spectra of NEPIR-0

Figure 3 shows the expected neutron energy spectra (in the forward direction, with a 3° semi-angle) produced by the 30 mm thick Be neutron production target, for different values of the energy of the impinging protons. Simulations have been performed with the MCNPX code [9].

For a 70 MeV proton current of ~1 μA, the integral fast neutron flux at the standard test point, 2.6 m from the target, is $\Phi_n(1<E_n<70\text{ MeV}) \sim 3\times10^6\text{ cm}^{-2}\text{s}^{-1}$. At the most distant test point, 12.6 m from the target, the diameter of the neutron beam increases to 50 cm while the integral neutron flux drops to $\Phi_n(1<E_n<70\text{ MeV}) \sim 1.5\times10^5\text{ cm}^{-2}\text{ s}^{-1}$.

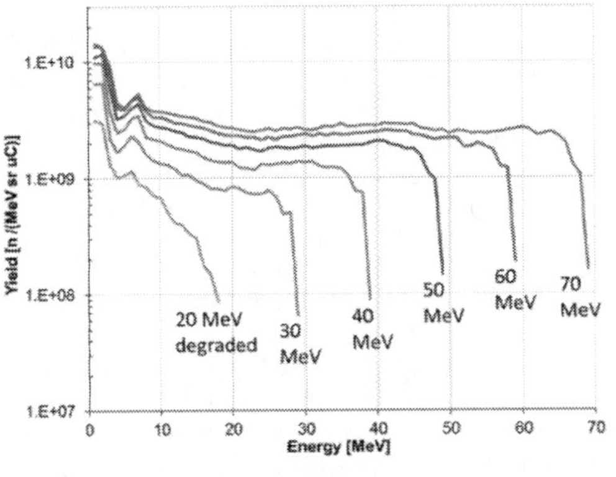

Fig. 3. Simulated (MCNPX) fast neutron spectra in the forward direction for different proton beam energies.

For proton energies greater than 30 MeV, the distributions are somewhat flat, up to a sharp cutoff value close to the energy of the proton beam. This feature will be exploited to give NEPIR-0 pseudo-QMN capabilities: by irradiating a sample with neutrons produced using proton beams with two different energies, one can calculate, by subtraction of the renormalized distributions, the effects due to neutrons in the energy interval defined by the two cut-off values. An example is shown in Figure 4, where a neutron spectrum obtained by a 60 MeV proton beam is normalized and subtracted from the spectrum by a 70 MeV proton beam. The effective neutron spectrum corresponds to a box-like distribution in the 58-68 MeV (FWHM) energy interval, with an integral flux of $\sim 2.8\times10^5\text{ cm}^{-2}\text{ s}^{-1}$.

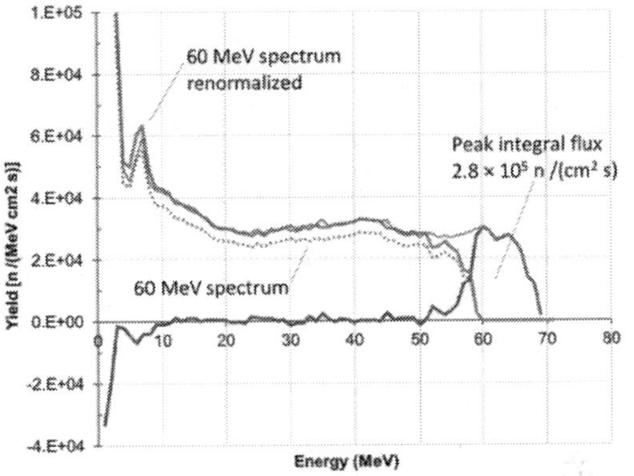

Fig. 4. The effective neutron spectrum obtained by subtraction of neutron spectra generated by 60 MeV (renormalized) and 70 MeV proton beams.

2.3 The CoolGAL target system of NEPIR-0

At NEPIR-0 the 70 MeV proton current is limited to 1 μA. For this reason we initially considered a simple low power (< 100 W) air-cooled Be cylinder (∅ = 30 mm, thickness = 30 mm) with an aluminum cladding and air cooling fins. However, any cladding must be designed carefully to account for thermal expansions all while maintaining good contact. In addition Be is well known to be problematic, as proton irradiation eventually causes swelling, the formation of bubbles and defoliation, with the risk of contaminating the cyclotron beam line. It is of paramount importance that the target be free to expand and that any debris be contained.

For these reasons we are proposing CoolGAL: a more sophisticated target system (Fig. 5 and 6). In CoolGAL, the Be cylinder is immersed in a static bath of a liquid metal eutectic alloy of Gallium, Indium and Tin (Sn), commercially known as Galinstan [10]. The liquid metal, contained by a copper cladding, ensures a good thermal contact with the Be and the internal surface of the Cu cladding, Au-coated to ensure tolerance to Galinstan. A 25 micron thick, robust havar [11] membrane separates the liquid metal from the beamline vacuum; this entrance window is cooled efficiently as it is in contact with the liquid metal. The outer surface of the cladding and a matching copper outer layer define a cooling circuit for a flowing coolant (water). To improve heat exchange, the outer surface of the cladding has cooling fins. The water flow in the enclosed circuit is ensured by input and output lines connected to a chiller in the experimental area.

The cladding is evacuated trough the reservoir tank prior to be filled with liquid metal and a pressure gauge is used to monitor the vacuum condition on top of the liquid. Any degassing reaction products that may leave the Be target during and after an irradiation would be confined in this tank, thereby keeping these contaminants from diffusing towards the cyclotron. If a threshold pressure value is exceeded, the proton beam is stopped, the target system extracted and inspected and the outgassed material disposed of in a proper way through a vent valve.

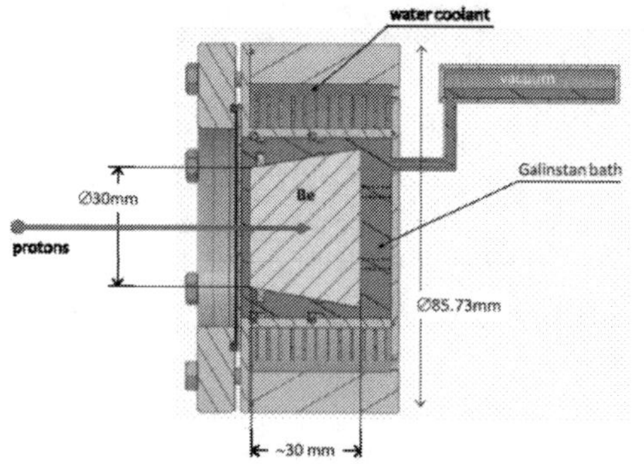

Fig. 5. Conceptual representation of the CoolGAL target system. The proton beam comes from the left, crosses a havar 25 micron thick window (visible in Fig.6) and a 1 mm thick liquid layer before it is stopped by the thick Be component (yellow). The Be is immersed in a Galinstan bath (purple) contained by a copper cladding.

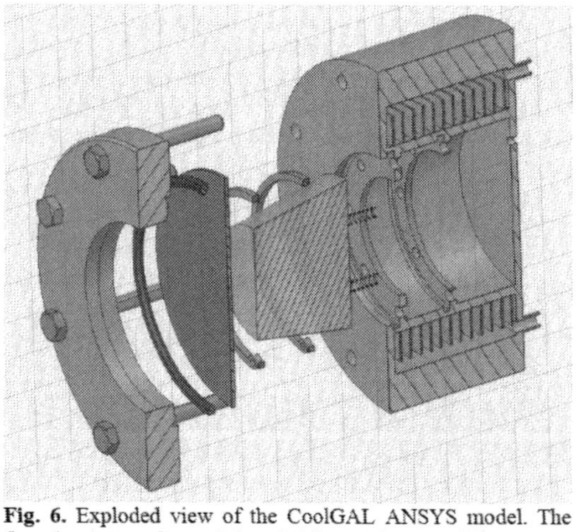

Fig. 6. Exploded view of the CoolGAL ANSYS model. The finned outer surface of the cladding and a matching copper outer layer define a cooling circuit. The surface of the cladding in contact with the coolant (flowing water) has cooling fins. Forced air cooling is not viable.

The working scheme of the Be target is shown in Fig. 7, together with the setup to fill it with liquid metal. The procedure to fill the target with the liquid metal is made up of four steps:

1. At the beginning valve V3 is closed and a vacuum pump is used to evacuate the target, down to a pressure of the order of 10^{-6} mbar. The pressure is monitored with the Pirani gauge located on of the expansion vessel. In this condition the pressure difference across the havar window is 1 atm, the thickness of the window is selected to enable it to withstand the provided stress. The slow pumping down time avoids any impulsive force.

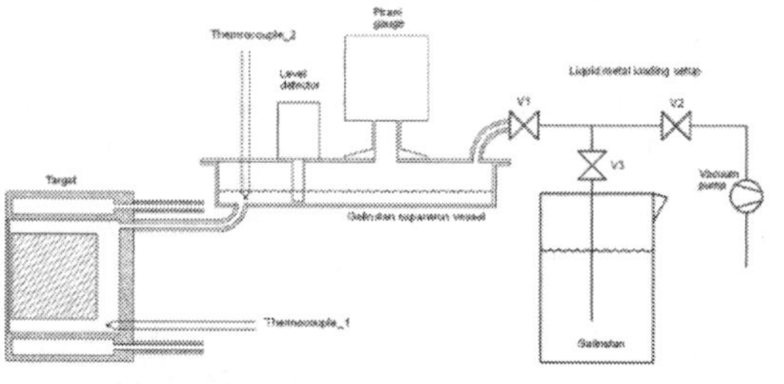

Fig. 7 Logic working scheme of the CoolGAL target system, including the setup to load it with Galinstan and provide the vacuum in the expansion vessel.

2. Valve V2 is closed to isolate the system from the outside environment.
3. Valve V1 is opened allowing the suction of the required amount of liquid metal in the evacuated expansion vessel and target cladding. By gravity, the liquid will completely fill the cladding, wetting all of the surface of the Be neutron converter and all the inner surface of the cladding. Also a small fraction of the inner volume of the expansion vessel will be filled with Galinstan, the rest will remain evacuated.
4. Valve V1 is closed, isolating the target system.

At this point the loading setup can be removed and the target installed. The target assembly is provided of four sensors to guarantee its safety both during working time and during rest time. Two temperature sensors (thermocouples) are used to measure the Galinstan temperature inside the cladding and inside the expansion vessel. A Pirani gauge will continuously monitor the vacuum in the expansion vessel to detect possible release of gasses (potentially radioactive) from the Be component. A level detector will be used to ensure the stability of the level of the cooling liquid.

In normal conditions, the pressure difference across the havar window will be only the small hydrostatic pressure provided by the Galinstan (much smaller than 1 atm). In case of rupture of the havar, the Galinstan will gently pour out and will be collected by a cylindrical containment vessel placed in front of the havar window (visible in Figure 2).

2.4 ANSYS calculations of CoolGAL

The final design of the target is currently under way and near completion.

Preliminary thermo-mechanical ANSYS [12] calculations of the minimal model show that, with a 1 μA proton current of 70 MeV protons (70 W), forced air cooling is not a viable solution; flowing water as coolant is, instead, very promising.

 https://doi.org/10.1051/epjconf/202023103002

ANSYS calculated regime temperature distributions in the Be and the Galinstan at 70 W for water cooling are shown in Figure 9 for a simple model: the power deposition of the beam is described as a cylinder 1 cm in diameter and the natural convection in the Galinstan is simulated by including buoyancy. The calculations show that there is a tendency of the temperature to be higher at the top of the Galinstan due to the natural movement caused by the density effect. This effect is evident in Figure 10 that shows the temperature distribution for 700 W; it is not apparent at 70 W, as the temperature differences are too small.

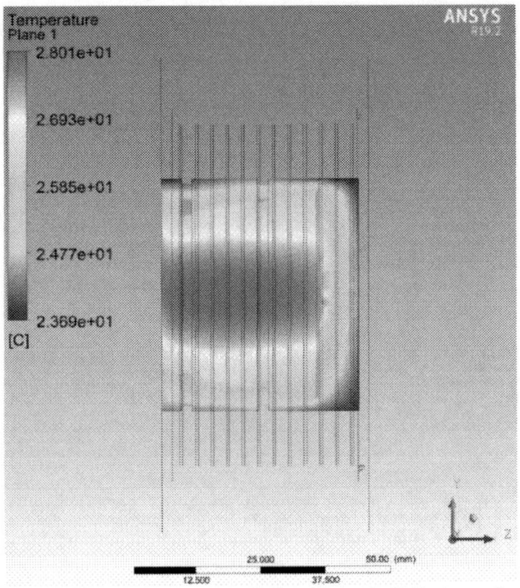

Fig. 9. Temperature map of the Be and Galinstan of a preliminary ANSYS model of CoolGal with water coolant for a 1 μA current of 70 MeV protons (70 W). The maximum temperature of the Be component is 28°C.

The very low maximum temperature (28 °C) of the Be component at 70 W guarantees the exceptional safety level required for a target located so close to the cyclotron (distance < 8 m), that no effective counter measures can be implemented to prevent contaminations in case of catastrophic failure.

The thermal reliability of CoolGAL with the present design stage is also very promising for future developments requiring higher proton currents: Figure 11 shows the maximum temperature of the Be component as a function of power. In the present design, there are only convective currents in the Galinstan liquid; to improve performance further, we plan to consider a flowing Galinstan circuit.

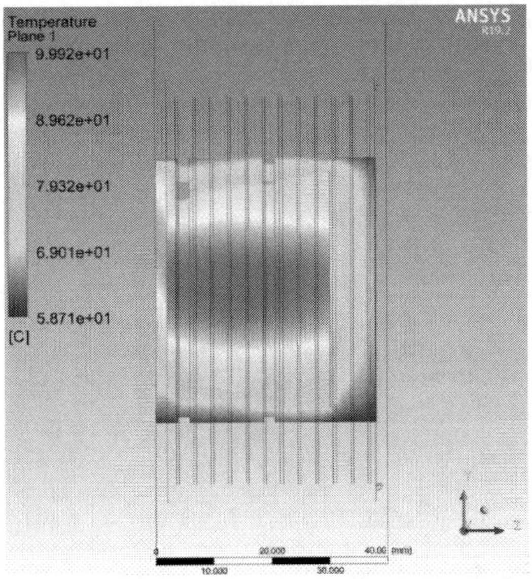

Fig. 10. Temperature map of the Be and Galinstan of a preliminary ANSYS model of CoolGal with water coolant for 700 W (10 μA current of 70 MeV protons). The maximum temperature of the Be component is 100°C.

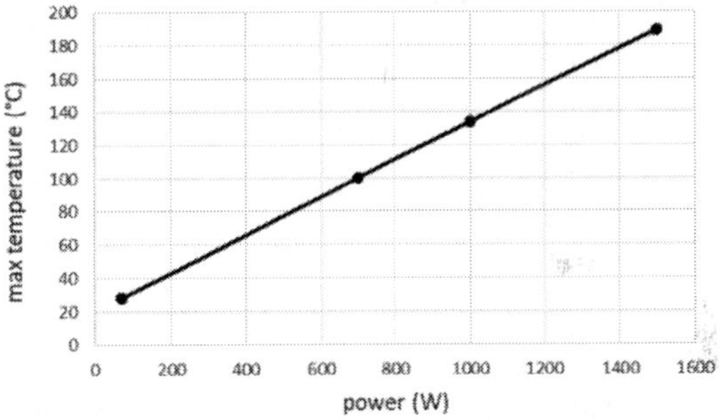

Fig. 11. Summary of the maximum temperature of the Be component versus beam power of the ANSYS model of CoolGAL using water coolant flowing at v = 2 m/s with an input temperature of 20°C.

3 Conclusions

In summary, the advantages of the liquid metal bath in CoolGAL are:

- it allows thermal dilatation of the beryllium;
- it ensures thermal contact and heat transfer to the copper cladding for conduction;
- thanks to natural convection, it distributes the heat on the whole surface of the cladding;
- it cools down the havar window;
- it contains any debris of beryllium if/when blistering occurs;
- it is viable for higher power applications.

Our goal is to commission the NEPIR-0 facility by the end of 2020-beginning 2021 and to open it to users from academia and especially industry in 2022.

The rapid implementation and success of NEPIR-0 with users, especially industrial ones, will catalyze the upgrade to the final NEPIR configuration, the first step of which will be the construction of a dedicated independent bunker and air containment system, followed by the installation of the beam optics allowing the facility to deliver fast QMN beams, while maintaining and improving the capability of the Be target to provide a high flux of continuous energy neutrons.

References

1. G. Prete et al., The SPES project at the INFN-Laboratori Nazionali di Legnaro, EPJ Web of Conferences 66, 11030 (2014)
2. M. S. Gordon et al., Measurement of the flux and energy spectrum of cosmic-ray induced neutrons on the ground, IEEE Trans. Nucl. Sci. V. 51 , I. 6 (2004)
3. T. Nakamura, M. Baba, E. Ibe and Y. Yahigi, Terrestrial Neutron-Induced Soft Errors in Advanced Memory Devices (World Scientific Publishing Co. Pte. Ltd, 2008)
4. JESD89A, Measurement and Reporting of Alpha Particle and Terrestrial Cosmic Ray-Induced Soft Errors in Semiconductor Devices, Oct. 2006.
5. M. Bagatin, S. Gerardin, Soft errors in floating gate memory cells: A review, Microelectron. Reliab. V. 55 I. 1, 2015, P. 24-30
6. W. Yang et al., Atmospheric neutron single event effect test on Xilinx 28 nm system on chip at CSNS-BL09, Microelectron. Reliab. V. 99, 2019, P. 119-124
7. J.W. Wilson et al., Deep space environments for human exploration, Advances in Space Research 34 (2004) 1281–1287
8. F. Tommasino et al., Proton beam characterization in the experimental room of the Trento Proton Therapy facility, NIM A, V 869, 2017, P. 15-20
9. D. B. Pelowitz, M. R. James, G. W. McKinney, J. W. Durkee et al., MCNPX 2.7.A Extensions (LANL, 2008).
10. https://en.wikipedia.org/wiki/Galinstan
11. https://en.wikipedia.org/wiki/Havar_(alloy)
12. ANSYS® Academic Research, Release 19.2

Development of a regenerated Beryllium target and a thermal test facility for Compact Accelerator-based Neutron Sources

Jean-François Muraz[1], Daniel Santos[1], Véronique Ghetta[1], Julien Giraud [1], Julien Marpaud[1], Marine Hervé[1], Pascal Sortais[2], Mauro Forlino[3]

[1] Laboratoire de Physique Subatomique et de Cosmologie 38026 Grenoble, France
[2] Polygon Physics 38000 Grenoble
[3] ARFITEC internship engineer

Abstract. Recently, the possibility to use compact accelerators coupled to high current ion sources for the production of intense low energy proton or deuteron beams has motivated many research laboratories to develop accelerator based neutrons sources for several purposes, including Neutron Capture Therapy (NCT). The NCT needs a high flux, about 10^{9} $n.cm^{-2}.s^{-1}$, of thermal neutrons (E<10 keV) at the tumour site. Up to now, the NCT required neutron flux was mainly delivered by nuclear reactors. However, the production of such neutron flux is now possible using proton or deuteron beams on specific targets able to stand a high power (~15-30 kW) on a small area (~10 cm²). This specific target design, materials and supports, has to cope with extreme physical constraints. The LPSC team has conceived an original solution formed by a thin (8 µm) rotating beryllium target deposited on a graphite wheel and coupled with a beryllium sputtering device for periodic ^{9}Be layer restoration. By means of $^{9}Be(d,n)\,^{10}B$ nuclear reaction, this target irradiated by a 10-20 mA deuteron beam (1.45 MeV) should produce the required neutron flux. In order to validate the target design of the neutron flux production and the beryllium target thermal capabilities, we built a 30 cm diameter rotating Beryllium target prototype and a compact electron beam line able to deliver a power density of 3kW/cm².

1 Context

Neutron sources are used in many domains such as isotopes production, non-destructive imaging, sub-surface exploration, silicon transmutation doping, and, in the medical, domain Neutron Capture Therapies profiting of the huge ^{10}B neutron capture cross section [1-2] producing an energetic alpha particle and Li nuclear recoil with a 478 keV gamma ray in 94% of cases [fig 1]

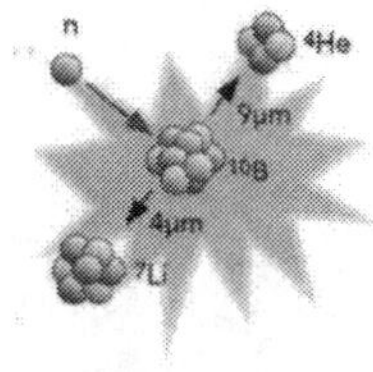

Fig. 1. Boron Neutron Capture nuclear reaction

Unlike nuclear reactors, accelerator-based neutron sources appear like safe, compact and cheap solutions to produce a high neutron flux [3]. This is especially true for Neutron Capture Therapies, which require the highest thermal neutron flux produced by the smallest compact neutron source suitable to sit in a hospital.

In this context, the LPSC team realized that one of the critical points for compact accelerator-based neutron source developments is the robustness of the target under a continuous high current beam.

2 The regenerated Beryllium rotating target

The LPSC team has developed an original beryllium rotating target where the beryllium layer can be regenerated. This target has been designed to cope with a power density of 3 kW/cm².

In a square aluminium vacuum chamber, a 15 mm thick circular graphite disc is turning at 1 to 3 Hz [fig 2]. This rotation is produced and controlled by a stepping motor and the rotation torque is transferred to the disc shaft using a magnetic shaft coupling to avoid rotary seals and so to reduce the target maintenance.

Fig. 2. Internal view of the beryllium target

* Corresponding author: muraz@lpsc.in2p3.fr

https://doi.org/10.1051/epjconf/202023103003

Fig. 2. Internal view of the beryllium target

The beryllium layer is deposited and regenerated on the graphite disc by Ion Beam Sputtering [fig 3]. This process consists of using a compact ECR ion source to produce a specific beam which hits a pure beryllium target oriented in such a way that the beryllium atoms are sputtered into the graphite surface.

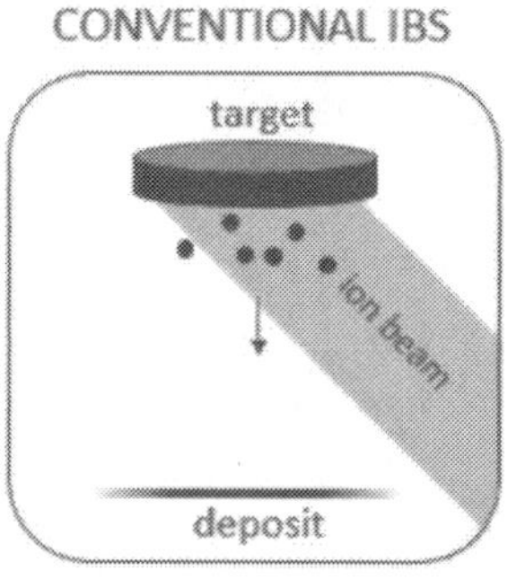

Fig. 3. Ion Beam Sputtering working principle

The heat, produced by the beam on the beryllium layer and the graphite disc, is transferred by radiation to the aluminium vacuum chamber, which is water-cooled on both sides [fig 4].

Fig. 4. Water cooling channels

The graphite disc temperature is measured by 2 optical pyrometers placed before and after the beam spot.

3 The electron beam thermal test facility

In order to test in situ our target developments, the LPSC team has developed a thermal test facility [fig 5].

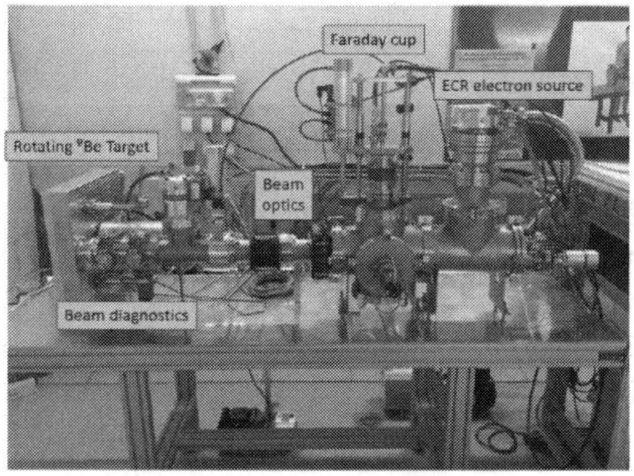

Fig. 5. The 3 kW thermal electron beam facility

This facility is based on an ECR electron source produced by the Polygon Physics Company. The ECR electron source [fig 6] is set to produce a 18 keV - 168 mA electron beam.

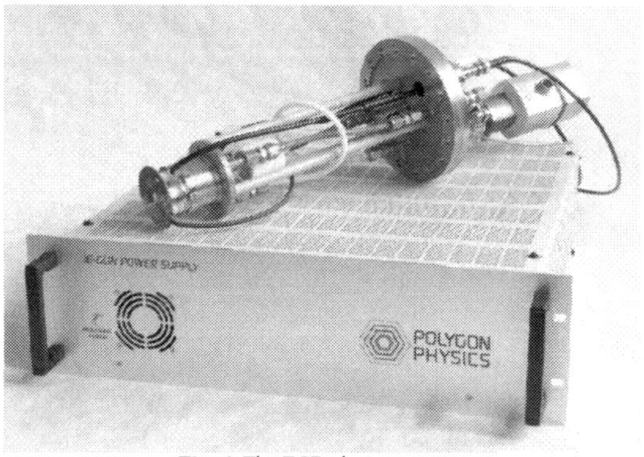

Fig. 6. The ECR electron source

The beam current is measured by an extractible Faraday cup, which is water-cooled [fig 7].

Fig. 7. The Faraday cup

The beam can be positioned using two magnetic steerers and focalized by acting on a magnetic solenoid [fig 8].

Fig. 8. The electron beam optics

Due to its high intensity, the electron beam produces fluorescence light in the visible range in the residual gas contained in the beam pipes. The beam size and position are then controlled online by a non-intersecting Beam Induced Florescence diagnostic device. Two perpendicular fast cameras will do the transverse profile measurement.

4 First measurements

We performed very preliminary measurements to check the target behaviour under power variation [fig 9]. These measurements were done up to 2.5 kW.cm^{-2}.

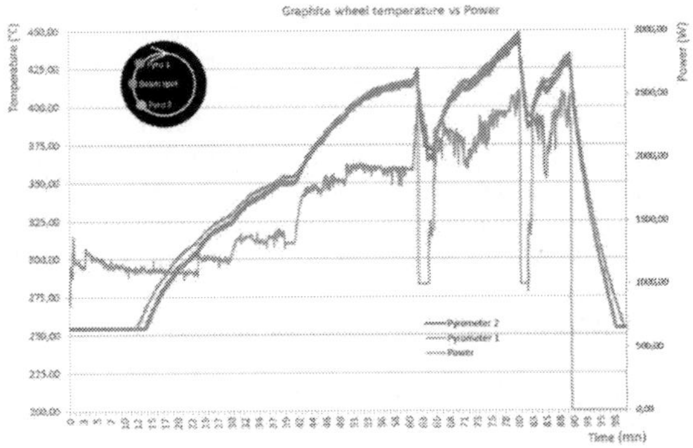

Fig. 9. Target thermal behaviour according to beam power

According to these preliminary measurements, we can clearly see the fast thermal response of the target under the beam power solicitations.

Conclusion

We have designed and built a regenerated rotating beryllium target and an electron beam facility to perform thermal target tests up to 3 kW.cm^{-2}. We performed some preliminary measurements up to 2.5 kW.cm^{-2} confirming the good thermal abilities of the target design. Our next goals are to complete the thermal tests and to study the ^{9}Be layer behaviour under the 3 kW.cm^{-2} power density as a function of working time. An NIM article on all these developments is being written and will be soon published.

References

[1] M.E.Capoulat, N. Sauzet, A.A. Valda, L. Gagetti, O. Guillaudin, L. Lebreton, D. Maire, P. Mastinu, J. Praena, Q. Riffard, B. Tampon, D. Santos, A.J Kreiner ***Neutron spectrometry of the ^{9}Be(d (1.45 MeV), n)^{10}B reaction for accelerator-based BNCT*** https://doi.org/10.1016/j.nimb.2019.03.005

[2] M.E. Capoulat, D.M. Minsky, A.J. Kreiner, ***Computational assessment of deep-seated tumor treatment capability of the 9Be(d, n)10B reaction for accelerator-based Boron Neutron Capture Therapy (AB-BNCT)***, Phys. Medica. 30 (2014), https://doi.org/10.1016/j.ejmp.2013.07.001.

[3] T.E. Blue, J.C. Yanch, ***Accelerator-based epithermal neutron sources for boron neutron capture therapy of brain tumors***, J. Neurooncol. 62 (2003) https://doi.org/10.1023/A:1023247222043

A 50 kW Liquid-Lithium Target for BNCT and Material-Science Applications

Michael Paul[1,*], *Ido* Silverman[2,*], *Shlomi* Halfon[2], *Semion* Sukoriansky[3], *Boris* Mikhailovich[3], *Tala* Palchan[1], *Arkady* Kapusta[3], *Arthur* Shoihet[4], *Daniel* Kijel[2], *Alexander* Arenshtam[2,†] and *Eli* Barami[2,3]

[1]Racah Institute of Physics, Hebrew University, Jerusalem, Israel 91904
[2]Soreq Nuclear Research Center, Yavne, Israel 81800
[3]Department of Mechanical Engineering, Ben-Gurion University of the Negev, Beer-Sheva, Israel
[4]NRCN, Beer-Sheva, Israel

Abstract. A compact Liquid Lithium Target (LiLiT) has been operating at SARAF for several years with beam power of several kW (1.9-2.5 MeV, up to 2 mA). When bombarding the lithium with low energy protons neutrons are generated. The neutron source, mainly used for nuclear astrophysics research, was decommissioned in 2016 towards an upgraded model - with possible applications to Boron Neutron Capture Therapy (BNCT) and material-science studies. The improved version has been designed to sustain 50 kW proton beam power (2.5 MeV, ~20 mA) to provide sufficient neutron flux required for clinical BNCT application. The new model has a 50 mm wide lithium jet to enable dissipation of the higher beam power and an improved heat exchanger to remove the power to a secondary cooling loop. A new Annular Linear INduction electro-magnetic pump (ALIN) has been designed and built to provide the required lithium flow rate. Other mechanical improvements facilitate the maintenance of the system and the robustness of operation. Radiological risks due to the 7Be produced in the reaction are reduced by using an integrated lead shielding of the lithium reservoir. An integrated neutron moderator is being designed to adjust the neutron energy to the spectrum best suited to BNCT. A low power (6 kW) model of the new design with a narrower nozzle (18 mm wide) and a rotating-magnet electro-magnetic pump is operating at SARAF to support the ongoing astrophysics and nuclear research program [1], [2]. To fulfill clinical BNCT, the upgraded LiLiT model will require an accelerator of appropriate energy and intensity. The design features of the new system are presented in this paper.

1 Background

Neutrons, penetrating non-ionizing particles, can be utilized in a form of radiotherapy generally termed today Neutron Capture Therapy (NCT). NCT was first suggested by G. L. Locher [3] as early as 1936, a few years only after the discovery of the neutron by J. Chadwick [4]. It is based on the radiological effects of radiations emitted following the in-situ neutron capture on a suitable nucleus. The target nucleus is best selected so that the capture results in energetic charged particles, which have a short range in tissues and a high specific energy loss (dE/dx) for maximal biological local damage. A high thermal neutron cross section and minimal residual activity are also important criteria in the selection of the target nucleus. The availability of ^{10}B, a stable isotope of boron, following the rich medical literature developed on the toxicology of boron and a large cross section of 3840 barns of the ^{10}B(n,α)^{7}Li reaction for thermal neutrons have made this nuclide appropriate in the modality called Boron Neutron Capture Therapy (BNCT). On absorption of a thermal neutron by ^{10}B, the excited ^{11}B nucleus promptly (~ 10-12 s) decays to two charged particles (α and ^{7}Li) accompanied with high probability by a 478 keV γ-ray. The emitted charged particles α and ^{7}Li sharing a total energy of 2.31 MeV (1.47 MeV and 0.84 MeV, respectively) are highly-ionizing particles with short range in organic matter (~ 9 μm and ~5 μm, respectively) that are similar to the size of a single cell. The implementation of Boron Neutron Capture Therapy (BNCT) as a cancer radiotherapy requires an intense and practicable source of low energy (epithermal) neutrons. For many years, suitable neutron sources for BNCT were constrained to nuclear reactors. A reactor can produce a sufficient neutron flux (estimated to ~10^9 n/cm^2/s [5]) at an irradiation facility beam port for therapy duration of 30-90 min (see [6]) but suffers significant drawbacks in suitability and availability.

Worldwide efforts to design an accelerator-based neutron converter have focused on the use of lithium through the reaction ^{7}Li(p,n)^{7}Be [7–13] at proton energies of 1.9-2.8 MeV, which produces average neutrons energies in the range 25 - 500 keV. However, a reliable conventional lithium target working under beam power levels (~ tens kW), as considered for therapy purpose in the energy range above, proves very difficult to build because of the mechanical, chemical and thermal properties of lithium (principally its low melting point at

* Corresponding authors: Ido Sliverman (ido@soreq.gov.il), Michael Paul (paul@vms.huji.ac.il)
† Deceased

181°C). The development of the liquid-lithium target LiLiT as a high-intensity epithermal neutron source, operated in conjunction with the SARAF linear accelerator has led us to investigate its application for neutron radiotherapy [1, 14, 15].

The source strength of epithermal neutron spectrum (0.1 eV - 10 keV) optimal for BNCT clinical use is considered to be of the order of 10^9 n/cm^2/s, together with minimizing fluxes and dose rate of thermal neutrons (causing collateral dose to surface tissue without penetrating to a deep-seated tumor) and of fast neutrons (causing collateral dose to healthy tissue through knock-out protons). A beam-shaping assembly is therefore necessary for the tailoring of the ^{7}Li(p,n) neutrons to an optimal spectrum, together with a shield for reducing the accompanying gamma dose. In order to respond to these demands, it is estimated that the proton beam power on the lithium target must be of the order of 30-50 kW for proton energy in the range 1.91-2.5 MeV.

The present project was aimed at the design and realization of a high-power liquid-lithium target compatible with the neutron fluxes required for BNCT, as far as both spectrum and intensity are concerned.

2 Thermal and mechanical design of the system

2.1. Heat transfer calculations

Heat transfer calculations of the Boron Neutron Capture Therapy (BNCT) target were performed using the commercial ANSYS-Fluent 17.2 software. The current simulations concentrated on the mitigation of the excessive evaporation risk. In order to examine the required velocity, enabling safe operation of the system, the inlet velocity and the values of the beam power were varied in the range of 1-20 m/s and 5-50 kW, and the temperatures and evaporation rates were examined for these conditions. A laminar solver was employed to solve the Navier-Stokes equations together with the energy equation for heat computations. Solving the equation with the assumption of a laminar flow is the worst case in terms of temperature analysis as turbulence enhances heat transfer in the fluid. Figure 1 presents the estimated mass evaporation rate as a function of lithium velocity and beam power. These results are for a 50 mm wide nozzle and gaussian beam parameters of $\sigma_x = 8$ mm and $\sigma_y = 12$ mm, which have been chosen from mechanical and neutronic considerations. Acceptable conditions regarding evaporation rate, up to 10 mg/h (established according to the experience with the original LiLiT system), are below the double black line. Based on these simulations, it was found that the system will be able to dissipate ~30 kW proton beam with lithium jet velocity of ~6 m/s which was achieved with the original LiLiT system. For 50 kW beam power a 13 m/s lithium jet velocity is required to maintain evaporation below ~10 mg/h. It will be available with the system upgraded design describe hereby. The lithium jet thickness was set to 1.5 mm, well above the stopping range of 2.5 MeV protons.

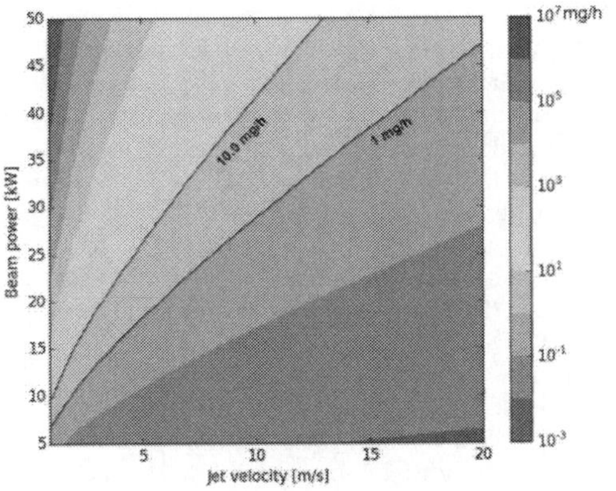

Fig. 1. Mass evaporation rate as a function of lithium velocity and beam power, for a gaussian beam parameters of $\sigma_x = 8$ mm and $\sigma_y = 12$ mm. Acceptable conditions regarding evaporation rate are below the double black line (evaporation rate below 10 mg/h). A more conservative operation limit of 1 mg/h is marked by a single black line

2.2 System final design and fabrication

The liquid-lithium nozzle, a critical component in the system, has been designed based on water experimental simulations, since water flow at 20°C and lithium flow at 225°C have almost identical Reynolds numbers. A water loop for the water simulation experiments was designed and built (Figure 2a) to create water flow velocities up to 20 m/s through the nozzle.

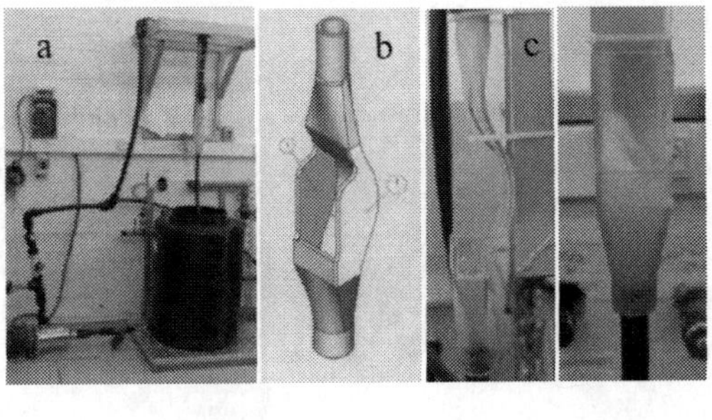

Fig. 2. a) A water loop for the nozzle design water simulation experiments; b) a preliminary wide nozzle design; c) wide nozzle during water experiments (water piles up below the nozzle at a velocity of ~10 m/s)

A preliminary wide nozzle was designed (Figure 2b) and a plastic model of the nozzle was built and tested in water flow simulation system (Figure 2c). A stable water film with a smooth surface was created when the jet velocity was below 10 m/s. At velocities above ~10 m/s (above ~0.75 l/s) water was seen piled up downstream the nozzle (Figure 2c), probably due to a bottleneck in the flow exit back to the 1" pipe. The final design has a 2" outlet pipe to provide enough through put for the low velocity return flow.

The 50 mm wide lithium nozzle optimized for high power but small source size has been designed based on the results of these water tests. The new design (see Figure 3) features an improvement in system maintenance by providing a method to replace the nozzle without changing the whole irradiation chamber and providing a backup method of heating the nozzle with internal heaters. The new design also features a method to deal with accidental lithium spillage into the irradiation chamber which had been an operating issue with the original design.

In addition, a revised version of the lithium heat-exchanger (HX) has been completed and produced. The new HX is designed to provide higher cooling rate, better response to thermal cycles and to reduce the risks of oil leak from the cooling loop, with long helical oil tubes and minimum welded joints. It also contains two connecting ports for electromagnetic pumps (EMP), one for the original rotating magnets EMP type at the middle of the HX height and a second at the bottom to connect to the new Annular Linear INduction (ALIN) type pump being commissioned. The new design also provides a method for lithium replacement. It is presented in Figure 4.

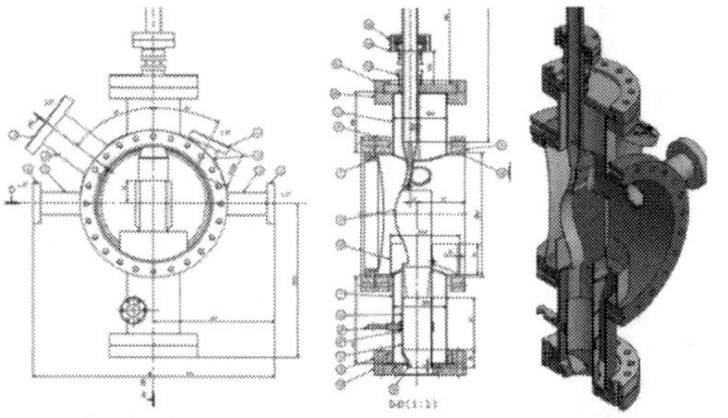

Fig. 3. The final design of the full power wide nozzle and the nozzle chamber

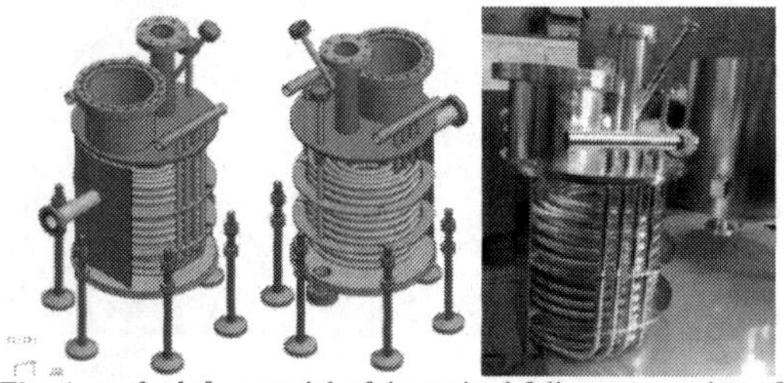

Fig. 4. on the left, a model of the revised full power version of the lithium heat exchanger, and at the right side a picture of the internal part before final welding

2.3 The ALIN pump design

For this project we reconsidered our choice of EMP lithium pump. The BNCT system requires a highly reliable, low-maintenance and efficient device (pump) for providing continuous circulation of liquid Li in the LiLiT system. The pump must be hermetically sealed and capable to work uninterruptedly for a long time without leaks, at high temperature and in contact with radioactive and chemically aggressive material. Mechanical pumps - centrifugal or positive displacement - were rejected as unable to satisfy the above requirements. An induction magneto-hydrodynamics (MHD) pump was eventually selected.

Based on the thermal design, the following pumping parameters were specified for the nominal working point (at maximal pump efficiency): volumetric flow rate $Q = 2\times10^{-3}$ m^3/s, pumping pressure $P = 2$ bar.

Three possible MHD pump types were considered:

1. Helical induction pump driven by fast rotating permanent magnets
2. Annular Linear Induction pump driven by 3-phase AC magnetic system (ALIN type)
3. Helical Induction pump driven by 3-phase AC magnetic system (HIP type)

Type 1 pump, having a rotating part, has been evaluated as requiring more maintenance than the induction types and hence as a possible cause of higher radiation dose exposure for the operating staff.

Reduced scale laboratory helical pump driven by AC inductor (HIP) was built and tested in the BGU MHD laboratory. The pump was designed and built for use with an existing laboratory 3-phase inductor of rotating magnetic field. The laboratory inductor could generate relatively weak magnetic field (up to 50 mT), such that the estimated pumping pressure was limited to 0.5 bar. The pump successfully produced the design pressure. However, the experiment revealed an inherent weakness of this design – high friction losses in the multi-turn helical channel.

A numerical code was developed for computation of the pump geometric, hydraulic and electric parameters. Some of the selected design parameters for the two ALIN pumps built (the prototype laboratory version and the final BNCT system version) are given in Table 1.

Table 1. Design and tested parameters of the two ALIN pumps

Parmeter	Symbol	Units	Value	
			Lab. pump	BNCT pump
Total pump diameter	D_{max}	m	0.35	0.378
Total pump length	L_{max}	m	0.72	0.72
Length of active zone of the inductor	L_a	m	0.684	0.684
External diameter of the annular channel	D_c	mm	82	112
Annular channel gap	Δ	mm	3	4
Diameter of ferromagnetic core	d_{fc}	mm	67	100
Number of electric coils	N_c		18	24
Magnetic field in the channel gap	B	T	0.167	0.247
Diameter of the coil wire	D_W	mm	3	2.76
Number of wire windings in a bundle	N_W	-	70	36
Mean LM flow velocity	V	m/s	2.73	3
Synchronic velocity of magnetic field	V_S	m/s	11.4	8.4
Mean slip	$S = 1\text{-}V/V_S$		0.76	0.64
Electric current in a phase	I	A	18	40
Magnetomotive force	MMF	A·turns	7560	11520
Total active electric power	W_E	W	3495	8350
Hydraulic efficiency of the pump	η	%	14	17

Conceptual representation of the ALIN type pump is shown on Figure 5. Linear induction pumps use a traveling magnetic field wave created by 3-phase currents. The induced azimuthal currents and the radial component of the traveling magnetic field generate a Lorentz force pushing the liquid metal along the channel annulus. The three-phase winding arrangement for the solenoids usually follows the sequence AA ZZ BB XX CC YY where A, B, C denote the balanced three-phase winding and X, Y, Z the opposite phase.

In order to test the new pump design, a low temperature version of the pump operating with Galinstan (Ga-In-Sn, liquid metal at >16°C) alloy was built and tested before the final high temperature version operating with liquid Lithium was built. Experimental circuit equipped with control valves, Venturi flow-meter, electronic pressure transmitters, three-phase transformer with variable voltage supply, programmable variable frequency power supply, frequency converter and gauss-meter for measurements of magnetic field induction was built to test the performances of this pump. The liquid metal flow-rate and pump pressure were measured at different electrical currents and frequencies. At any given set of electrical parameters, the flow rate is controlled by closing/opening control valve. The prototype pump was tested with two possible 3-phase connections – "star" and "delta". The characteristic Pressure-Flow rate curves at 50 Hz with "star" phase connections are shown on Figure 6.

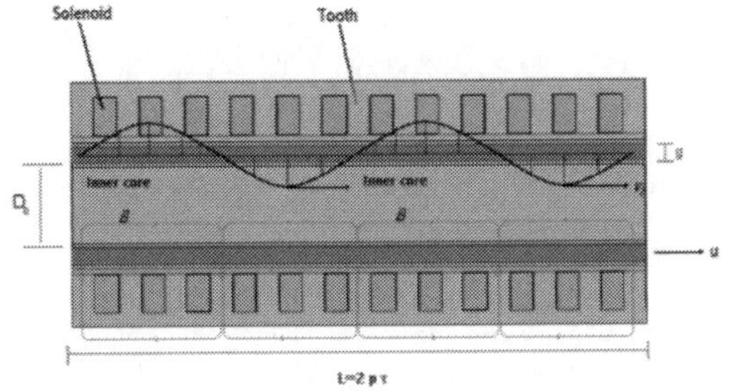

Fig. 5. Conceptual representation of the ALIN pump. The black curve schematically represents the magnetic induction (B) as a traveling wave of velocity vs. Liquid metal is shown in blue (reproduced from [16])

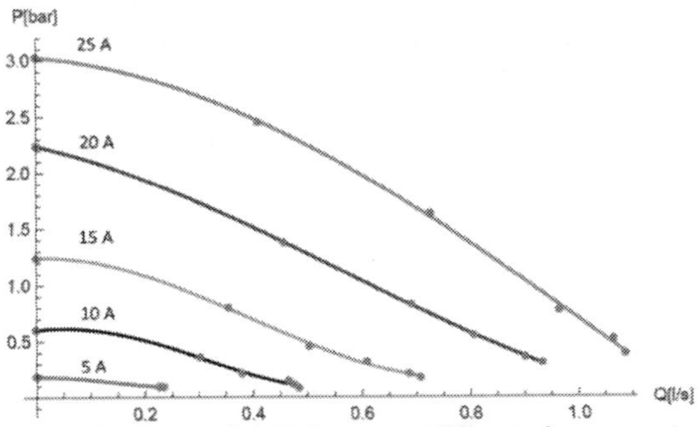

Fig. 6. The characteristic P-Q curves at different phase currents

The tests show that the pump did not achieve the required working head-flow-rate point with star connection. Delta connection was tested next. The line voltage with this connection is higher (380 V), thus the pump can operate at higher current. The maximal "shut-off" pressure of 10 bar was measured. However, the flow-rate remained below the design value. Based on these results new design parameters for the BNCT high temperature (300°C) lithium pump were specified (see right column in Table 1) and a new version has been designed and manufactured (shown in Figure 7).

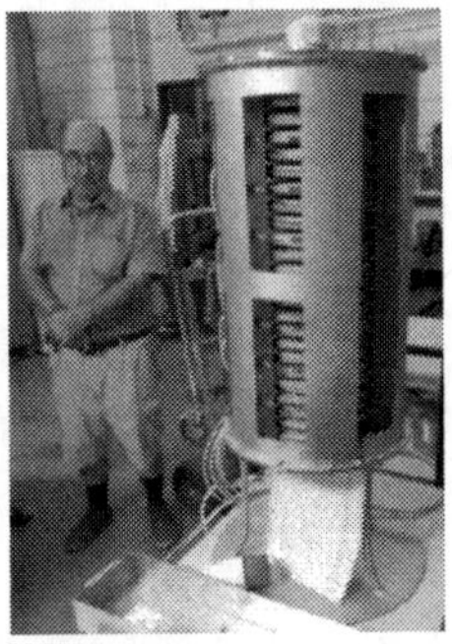

Fig. 7. A picture of the high temperature (300°C) liquid Lithium ALIN pump

3 LiLiT-BNCT loop system and operation safety

The LiLiT prototype target was decided to be decommissioned in 2016 following a leak developed in the lithium reservoir, in favor of the new design. In order to support the ongoing astrophysics and nuclear research program ([1, 2]) the construction of the new design has been divided into two stages. First, a low power (6 kW) model of the new design with a narrower nozzle (18 mm wide) and a rotating-magnets EMP was built and commissioned at SARAF in 2018. The system was designed to allow relatively easy replacement of loop components (nozzle, lithium chamber and pump). The new target as installed at SARAF phase-I target room is presented in Figure 8. First irradiation test took place in December 2018 and it is used since to produce neutrons. Then, as the design of the 50 kW nozzle, heat exchanger and ALIN pump matured they were manufactured and will be integrated into the operation system at a later date.

A major concern of operating a liquid lithium target is the safety issue of protecting a few kg inventory of radio-active liquid lithium. LiLiT-BNCT contains about 7 kg (15 liter) of metallic lithium. In case of leak in the vacuum vessel of the system, air might enter the vessel or liquid lithium might flow out. In both cases a reaction of the lithium with humidity in the air or with water might release hydrogen and heat. A combination which might cause fire or explosion which will disperse radio-active ^{7}Be contained in the lithium. Although the risks of this scenario are rather low, a defense-in-depth concept (borrowed from the nuclear industry) has been adopted in the design. The concept asks for evaluation of the failure scenarios of a specific design and providing mitigation plans and mechanisms to protect the environment assuming any one of these failures occur. Then, the evaluation is done again for the system with the additional protection level for scenarios when even the protection layer has failed. This sequence is repeated for several cycles until the estimated probability of total failure is as low as requested.

https://doi.org/10.1051/epjconf/202023103004

For the LiLiT target the primary failure scenarios which are of safety concern are that the particle beam heats an un-cooled part of the target or the nozzle when there is no lithium flow. Additional scenario is a break in a liquid lithium pipe which spills liquid lithium out (a low probability event). To mitigate the first scenario, the following protection measures are used:

1. A cooled collimator with current detection is placed in front of the nozzle to identify events when the beam steers away from the nozzle.
2. A flow meter is placed on the lithium pipe and the power and running of the EMP are monitored. All signals are required to be positive in order to enable the beam to bombard the target.
3. A secondary, low-vacuum chamber is located behind the nozzle. In case the beam penetrates through the nozzle into this secondary chamber, the low vacuum causes a signal in the target vacuum sensors to generate a shutdown signal for the beam.

In case this protection layer fails and liquid lithium does spill out of the target vessel it might react with the moisture in the air or some water in the vicinity of the target. To mitigate this scenario a second protection layer was designed and contains the following sub-systems:

1. The target is placed inside a stainless steel containment vessel to limit the immediate affected zone.
2. No water or water cooled systems are placed in this containment and a dry air is supplied to have a very low humidity environment (< 5% relative humidity) which reduces the risks of lithium reaction with water or fire if a spill occurs.
3. Cameras and smoke detectors are place in the containment to identify lithium leak if other sensor fail to identify it.

Then, if this second protection layer fails and the spilled lithium starts to react with water or moisture, a duplicate lithium fire extinguisher system using graphite power is available for manual operation from the control room. For any case, a ventilation system was installed to transfer lithium gaseous reaction products into a scrubber which will safely handle these products.

Fig. 8. Picture of the LiLiT-2 system installed at SARAF new target room with the frame of the containment vessel

This safety protection scenario was approved and a license to operate LiLiT as an intense neutron source was issued by the Israeli nuclear safety authority.

4 Neutron yield and energy spectrum

The energy of the generated neutrons in the liquid lithium target, although very close to the optimal BNCT neutron energies, should be slightly moderated. In order to provide a therapeutic neutron beam, a special moderator/reflector assembly named Beam Shaping Assembly (BSA) was designed to maximize the epithermal neutron flux Φ_{epi} while minimizing fast and thermal neutrons fluxes and shield from gamma radiation. The BSA should be installed between the neutron source and the patient.

Table 2. IAEA recommendations for a BNCT neutron source

Parameter	Symbol	Units	Value
Epithermal beam (0.5eV – 10keV)	Φ_{epi}	N cm^{-2} s^{-1}	$>10^9$
Fast neutron dose	D_{fn}/Φ_{epi}	Gy cm^2 /N$_{epi}$	$< 2.0\text{x}10^{-13}$
Gamma dose	D_γ/Φ_{epi}	Gy cm^2 /N$_{epi}$	$< 2.0\text{x}10^{-13}$
Thermal neutron flux ratio	Φ_{th}/Φ_{epi}		< 0.05
Directionality	J/Φ		> 0.7

As a preliminary assessment for the therapeutic effectiveness of the neutron beam, IAEA developed a set of recommended parameters for the neutron beam (ref. [5] and Table 2). Thermal neutron flux Φ_{th} (which cause radiation dose to surface tissues) must be reduced. Fast (>10 keV) neutron dose D_{fn} which accompany the incident beam, have a number of undesirable characteristics such as the production of high-LET protons, with a resulting energy dependence of their effect and must be reduced. Parasitic gamma dose D_γ accompanying the ^{7}Li(p,n) reaction must be reduced.

The beam-shaping assembly (BSA) includes mainly four parts, the moderator, the reflector, the filter for thermal neutrons and shielding for gamma radiation produced in the neutron source. The moderator has to slow down the fast neutrons yielded by the lithium target, without increasing significantly the fraction of thermal neutrons in order to get a net accumulation in the epithermal energy range (0.5 eV–10 keV). A reflector has to be included to either limit the neutron losses or scatter neutrons toward the beam port, while further improving the quality of the beam.

4.1 Simulation model

A simulation code to design a BSA that would fit the requirements for BNCT has been developed. The simulation is divided in two parts. The first of which, named SimLiT [15], calculates the neutron field produced in given conditions by LiLiT. The neutrons are then transported using the code GEANT4 [17] taking into account the geometry of the secondary target and surrounding components and materials. The SimLiT code, described in detail in Friedman et. al. [18], is a Monte Carlo program starting with a proton whose incident energy is sampled from a Gaussian distribution with given energy mean and standard deviation (energy spread) and whose position is sampled from a given radial Gaussian distribution. It then calculates the probability of a ^{7}Li(p,n)^{7}Be (and ^{7}Li(p,γn)^{7}Be* to the first excited 0.411 MeV state in ^{7}Be when the incident energy allows it) reaction within the Li thickness until neutron threshold energy is reached, using an energy-dependent stopping power dE_p/dx taken from the code SRIM [19]. The energy-dependent differential cross sections $(d\sigma/d\Omega)(E_{p,lab},\theta_{n,cm})$ are taken from Liskien and Paulsen [20] and from Gibbons and Macklin [21] above $E_{p,lab}$ = 1.890 MeV. Special care was taken for the treatment of the cross section between threshold and 1.890 MeV, where precise determination of the excitation function is crucial for correct reconstruction of the resulting neutron spectrum as discussed by Lee and Zhou [13]. The SimLiT code generates the outgoing neutron information (position vector, momentum vector and energy) which can be used as input for the following transport calculation. The SimLiT output neutron event file is used as input to a transport calculation with the code GEANT4 [22]. GEANT4 is an open source software toolkit of Monte-Carlo simulation for the passage of particles through matter. A realistic geometry and physics representation of the LiLiT and the different components of the BSA are built into the simulation. The neutron spectrum as seen at the beam port by a simulated detector is calculated. To evaluate the impact of the epithermal beam produced by the BSA in the human body a phantom head had been used, and in-phantom parameters were calculated through a set of detectors located in the phantom [22]. The phantom was designed as a simplified cylindrical phantom made from water surrounded by 0.5 mm PMMA, as reported in [22]. A full simulation of an (n,γ) reaction is impractical due to the small reaction probability; it is however highly important for a reliable simulation because of the different lengths of trajectories due to the neutron angle of emission and possible scattering inside the phantom. In order to improve the statistics for shorter calculation time we used the following method. For each neutron that enters any of the detectors, the neutron energy, $E_{n,i}$, the angle of trajectory $\theta_{n,j}$ and the length the neutron traveled inside the target, are recorded. If the energy of the simulated neutron changes inside a detector (e.g. from scattering), the new neutron energy and length are also recorded. The simulated neutron flux Φ in each detector is then calculated by:

$$\Phi_n\left(\frac{n}{cm^2 * mC}\right) = \frac{1}{AQ}\sum_j \frac{1}{\cos\theta_j}, \qquad (1)$$

where A is the surface of the detector, Q is the total simulated proton charge.

Four principal physical dose components should be considered (IAEA, [5]):

1. Fast neutron dose (D_{fn}) due to the proton recoil generated from ^{1}H(n,n)^{1}H interaction.
2. Thermal neutron dose (D_n) due to the energetic proton and the recoiling ^{14}C nucleus from the thermal neutron capture by ^{14}N via ^{14}N(n,p)^{14}C reaction.
3. Boron dose (D_B) from thermal neutron capture ^{10}B(n,α)^{7}Li reaction
4. Gamma dose (D_γ) which is a combination of photon dose derived from the BSA and dose from photons induced by neutron capture reactions in tissues.

The simulated neutron fluxes were converted to the BNCT absorbed dose components along the phantom based on kerma factors for the ICRU 46 adult brain composition [22].

$$D_{En}\left(\frac{Gy}{sec}\right) = \frac{I}{VQ}\sum_j K_j * l_j, \qquad (2)$$

where V is the volume of the detector, Q is the total simulated proton charge, I is the proton beam current, K_j is the kerma factor for the specific neutron energy and l_j is the length it traveled inside the detector. Boron dose rate distributions in healthy tissue and tumor tissue are calculated with different boron concentration of ~3.5 ratio between tumor and healthy tissue (40 μg ^{10}B/g and 11.5 μg ^{10}B/g typical for BPA compound, respectively).

4.2 Benchmarking of simulations

The SimLiT-GEANT4 simulations have been carefully benchmarked with two different papers [23], [24]. In [23] the neutron source for the BSA is a proton beam of 2.3 MeV and 10 mA combined with a solid Lithium target. It includes a Pb reflector, a moderator built out of MgF_2 and MgO and a filter made of Bi and enriched lithiated-polyethylene with ^{6}Li to avoid undesirable thermal neutrons and gamma rays contamination in the beam. Their calculations were carried out using the Monte Carlo MCNP code. We build our SimLiT-GEANT4 simulations according to the configuration presented in [23] (see Figure 9). Table 3 presents the neutron beam parameters calculated in [23] and through our simulation.

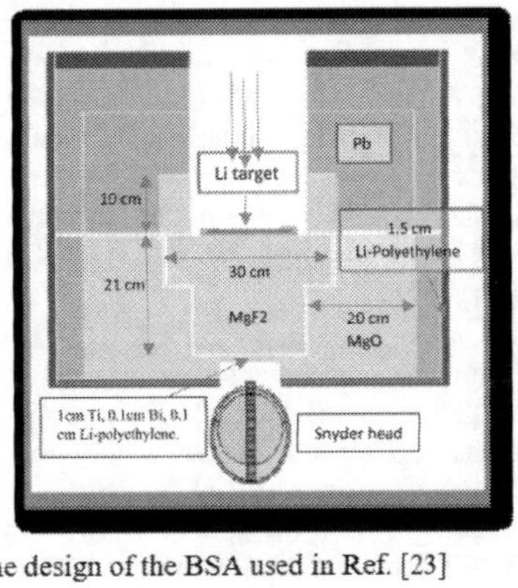

Fig. 9. The design of the BSA used in Ref. [23]

Table 3. Comparison of neutron beam quality parameters between MCNP and SimLiT-GEANT4 neutron beams designed as in [23], calculated for 2.3 MeV protons on lithium. The discrepancy in the D_{fn}/ϕ_{epi} values between the MCNP and SimLiT-Geant4 simulations is under study

Model	Φ_{epi} [10^9 n/cm²/sec]	D_{fn}/Φ_{epi} [10^{-13} Gy cm²]	$\Phi_{epi}/\Phi_{thermal}$	J/Φ
MCNP	1.04	1.25	29.4	0.657
SimLiT-Geant4	1.24	4.51	21	0.607

In ref. [24], two types of neutron generation reactions have been considered. One is p-Li reaction and the other is p-Be reaction. The p-Li reaction gives larger number of neutrons than the p-Be reaction at low proton energy region around 2-3 MeV, and the neutron energy is much lower than that produced by the p-Be reaction at higher proton energy. In their work they calculated, using the Monte-Carlo MCNP code, several moderator materials including fluorine, F. From these results they decided to use the moderator materials; MgF_2 for the p-Li reaction and LiF for the p-Be reaction. We choose to benchmark the configuration of the MgF_2 36 cm x 21 cm as moderator material for the p-Li reaction. Table 4 shows the comparison of the values required for the BNCT system.

Table 4. Calculated values for neutron epithermal flux, ratio of fast to epithermal neutrons and thermal flux for a ^{7}Li(p,n) setup after moderation [24]. The Table compares values calculated with two main established transport codes (MCNP, GEANT4).

Model	Φ_{epi} [n/cm²/sec/mA]	D_{fn}/Φ_{epi} [Gy cm²]	$\Phi_{thermal}$
MCNP	$2.79*10^7$	$1.0*10^{-12}$	$3.29*10^3$
SimLiT-Geant4	$4.51*10^7$	$8.9*10^{-13}$	$2.31*10^4$

Our own design for the BSA is presented in Figure 10. In this calculation we chose to use a 2.5 MeV ± 15 keV, 20 mA proton beam that hits a 2 mm thick liquid lithium target with a beam size of 16 mm diameter. The BSA consists of 25 cm diameter Pb reflector that surrounds the target and the moderator in order to reflect back the scattered neutrons. The moderator is build out of three different materials. The first is a 20 cm diameter 20 cm long MgF_2 disk, followed by a 20 cm diameter 10 cm long Al and finally 20 cm diameter 5 cm long AlF_3. In addition a 20 cm diameter 2 cm long Pb disk is added to shield the accompanied gamma ray from the source. At the end of the BSA there is a 1 mm enriched lithiated polyethylene with ^{6}Li filter to minimize the thermal neutron flux.

The phantom dimensions are 20 cm diameter and 20 cm long, with 0.5 cm polymethyl methacrylate (PMMA) walls and filled with water, simulating a targeted patient. Inside the phantom a set of 19 detectors are placed 1 cm apart in the center along the direction of the beam line. The fluxes and doses on these detectors were calculated, to give the flux/dose distribution as a function of depth in the phantom. Figure 11 shows neutron energy spectrum corresponding to the BSA at the entrance to the phantom. The energies centered at 10 keV, which is considered to be the ideal spectrum for treating deep-seated tumors. Table 5 shows current, non-optimised, calculation of the neutron parameters, to be compared with the IAEA recommendations (Table 2).

Table 5. Calculated neutron beam quality parameters for 20 mA, 2.5 MeV proton beam in a BSA configuration as shown in Figure 11

Model	Φ_{epi} [10^8 n/cm²/sec]	D_{fn}/Φ_{epi} [10^{-13} Gy cm²]	$\Phi_{epi}/\Phi_{thermal}$	J/Φ
SimLiT-Geant4	1.02	14.3	666.4	0.6

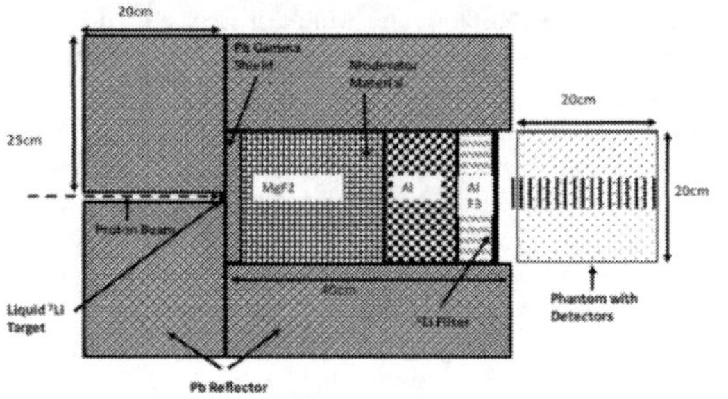

Fig. 10. BSA design for the SimLit-Geant4 calculation

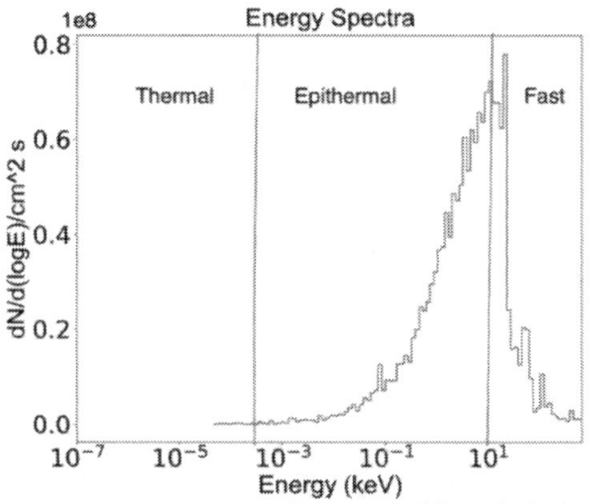

Fig. 11. Neutron spectrum at beam port of the optimized BSA

The boron dose and neutron dose distribution in the phantom were calculated according to equation (2). We rearranged them into two dose components, defined in the dose protocol prescribed for BNCT, based on clinical experience [25]. Namely, the heavy-particle dose (HCP) dose to tumor, (Boron dose to tumor + neutron dose to tumor) which should be above 15 Gy for therapeutic use and the HCP dose to healthy tissue (Boron dose to healthy tissue + neutron collateral dose to healthy tissue), which is recommended to be below 15 Gy.

The estimate of gamma-ray dose to tumor and healthy tissue is still in progress (gammas from the lithium target and gammas produced in $^1H(n,\gamma)^2H$) and is limited to 10 Gy.

Figure 12 shows the dose rate for the two components inside the phantom. The boron dose rate distributions in tumor were calculated with boron concentration of 40 μg ^{10}B/g and the healthy tissue with concentration of 11.5 μg ^{10}B/g.

The maximum HCP dose to healthy tissue is 15 Gy. According to Figure 12, the 15 Gy limit to healthy tissue can be reached after 21 minutes of irradiation time. The highest dose to tumor at this time is ~50 Gy. The advantage depth (AD), i.e. the depth in phantom at which the total therapeutic dose in tumor equals the maximum dose of the healthy tissue, indicating the depth of effective beam penetration, is 9.4 cm.

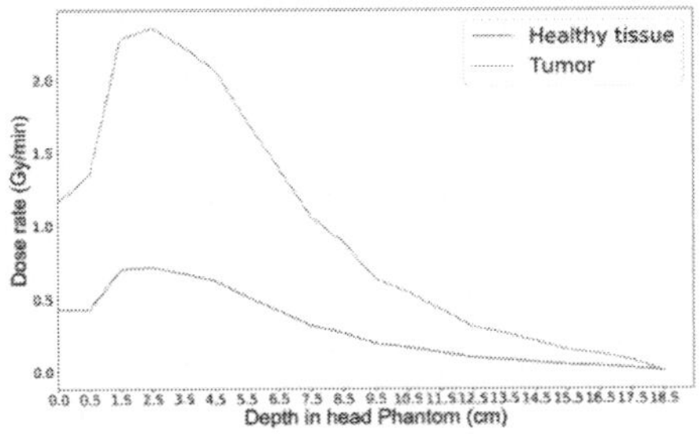

Fig. 12. Dose rate profiles in tumor and healthy tissue

5 Summary

A compact Liquid Lithium Target (LiLiT) has been operating at SARAF for several years with beam power of several kW (1.9-2.5 MeV, up to 2 mA). An upgraded model was considered to expand possible applications to Boron Neutron Capture Therapy (BNCT) and material-science studies. The improved version has been designed to sustain 50 kW beam power to provide sufficient neutron flux required for clinical BNCT application (up to 2.5 MeV, 20 mA). The new model has a 50 mm wide lithium jet and beam parameters of σ_x = 8 mm and σ_y = 12 mm to enable dissipation of the higher beam power. This design is based on CFD simulation to study the evaporation risk (achieving below 10 mg/h) and validated with water flow test with a mock-up. It also has an improved heat exchanger to remove the power to a secondary cooling loop and a new ALIN pump to provide the required lithium flow rate. Other mechanical improvements facilitate the maintenance of the system and the robustness of operation.

Radiological risks due to the ^{7}Be produced in the reaction are reduced by using an integrated lead shielding of the lithium reservoir. An integrated neutron moderator is being designed to adjust the neutron energy to the spectrum best suited to BNCT. A major concern of operating a liquid lithium target is the safety issue of protecting a few kg inventory of liquid lithium containing radioactive traces of ^{7}Be. In case of leak in the vacuum vessel of the system, air might enter the vessel or liquid lithium might flow out. In both cases a reaction of the lithium with humidity in the air or with water might release hydrogen and heat. A combination which might cause fire or explosion which will disperse the radio-active ^{7}Be contained in the lithium. Although the risks of this scenario are rather low, a defense-in-depth concept (borrowed for the nuclear industry) has been adopted in the design.

A simulation code to design a BSA that would fit the requirements for BNCT has been developed. The simulation is divided in two parts. The first of which, named SimLiT, calculates the neutron field produced in given conditions by LiLiT. The neutrons are then transported using the code GEANT4 taking into account the geometry of the secondary target and surrounding components and materials. This code system was used to design a beam-shaping assembly to fit around the target. The therapeutic HCP dose to the tumor is ~50 Gy for a limiting dose of 15 Gy to healthy tissue. The estimate of gamma-ray dose to tumor and healthy tissue is still in progress as well as full BSA optimization. The effective beam penetration depth, i.e. the depth in body at which the total therapeutic dose in tumor equals the maximum dose of the healthy tissue, is 9.4 cm.

A low power (6 kW) model of the new design with a narrower nozzle (18 mm wide) and a rotating-magnets electro-magnetic pump is operating at SARAF to support the ongoing astrophysics and nuclear research program. To take full advantage of the upgraded LiLiT model, an accelerator of appropriate energy and intensity (2.5 MeV, 20 mA, ~50 kW) is required.

This research was supported in part by the Ministry of Science, Technology & Space, Israel.
This paper is dedicated in memory of Dr. Alexander Arenshtam which was the chief designer of the original LiLiT and made important contributions to other projects in the SARAF facility.

References

1. M. Paul, M. Tessler, M. Friedman, S. Halfon, T. Palchan, L. Weissman, A. Arenshtam, D. Berkovits, Y. Eisen, I. Eliahu, G. Feinberg, D. Kijel, A. Kreisel, I. Mardor, G. Shimel, A. Shor, and I. Silverman, The Eur. Phys. J. A **55**, 44 (2019).
2. I. Mardor, O. Aviv, M. Avrigeanu, D. Berkovits, A. Dahan, T. Dickel, I. Eliyahu, M. Gai, I. Gavish-Segev, S. Halfon, M. Hass, T. Hirsh, B. Kaiser, D. Kijel, A. Kreisel, Y. Mishnayot, I. Mukul, B. Ohayon, M. Paul, A. Perry, H. Rahangdale, J. Rodnizki, G. Ron, R. Sasson-Zukran, A. Shor, I. Silverman, M. Tessler, S. Vaintraub, and L. Weissman, The Eur. Phys. J. A **54**, 91 (2018).
3. G. L. Locher, Am. J. Roentgenol. Radium Therapy, **36**, 1–13 (1936).
4. J. Chadwick, Proceedings of the Royal Society of London. Series A, Containing Papers of a Mathematical and Physical Character, **136**, 692–708 (1932).
5. IAEA-TECDOC-1223, International Atomic Energy Agency, 1223, (2001).
6. R. F. Barth, J. A. Coderre, M. G. H. Vicente, and T. E. Blue, Clinical Cancer Research, **11**, 3987–4002 (2005).
7. O. E. Kononov, V. N. Kononov, and N. A. Solov'ev, Atomic Energy, **94**, 417–420 (2003).
8. K. Tanaka, H. Yokobori, S. Endo, T. Kobayashi, G. Bengua, I. Saruyama, Y. Nakagawa, and M. Hoshi, App. Rad. and Iso., **67**, 259–265 (2009).

9. A. Burlon, A. Kreiner, A. Valda, and D. Minsky, App. Rad. and Iso., **61**, 811–815 (2004).
10. D. A. Allen and T. D. Beynon, Medical Physics, **27**, 1113–1118 (2000).
11. V. Aleynik, A. Burdakov, V. Davydenko, A. Ivanov, V. Kanygin, A. Kuznetsov, A. Makarov, I. Sorokin, and S. Taskaev, App. Rad. and Iso., **69**, 1635–1638 (2011).
12. C. Willis, J. Lenz, and D. Swenson, LINAC08, MOP063 (2008).
13. C. L. Lee and X.-L. Zhou, NIM B, **152**, 1–11 (1999).
14. S. Halfon, A. Arenshtam, D. Kijel, M. Paul, D. Berkovits, I. Eliyahu, G. Feinberg, M. Friedman, N. Hazenshprung, I. Mardor, A. Nagler, G. Shimel, M. Tessler, and I. Silverman, Rev. of Sci. Ins., **84**, 123507 (2013).
15. S. Halfon, A. Arenshtam, D. Kijel, M. Paul, L. Weissman, O. Aviv, D. Berkovits, O. Dudovitch, Y. Eisen, I. Eliyahu, G. Feinberg, G. Haquin, N. Hazenshprung, A. Kreisel, I. Mardor, G. Shimel, A. Shor, I. Silverman, M. Tessler, and Z. Yungrais, Rev. of Sci. Ins., **85**, 056105 (2014).
16. C.O. Maidana and J.E. Nieminen, Nuc. Eng. and Tech., **49**, 82–91 (2017).
17. S. Agostinelli, et. al., NIM A, **506**, 250–303 (2003).
18. M. Friedman, D. Cohen, M. Paul, D. Berkovits, Y. Eisen, G. Feinberg, G. Giorginis, S. Halfon, A. Krása, A. J. M. Plompen, and A. Shor, NIM A, **698**, 117–126 (2013).
19. J. F. Ziegler, M. D. Ziegler, and J. P. Biersack, NIM B, **268**, 1818–1823 (2010).
20. H. Liskien and A. Paulsen, Atomic Data and Nuclear Data Tables, **15**, 57–84 (1975).
21. J. H. Gibbons and R. L. Macklin, Phys. Rev., **114**, 571–580 (1959).
22. J. T. Goorley, W. S. Kiger III, and R. G. Zamenhof, Medical Physics, **29**, 145–156 (2002).
23. L. Zaidi, M. Belgaid, S. Taskaev, and R. Khelifi, App. Rad. and Iso., **139**, 316–324 (2018).
24. Y. Kiyanagi, K. Asano, A. Arakawa, S. Fukuchi, F. Hiraga, K. Kimura, H. Kobayashi, M. Kubota, H. Kumada, H. Matsumoto, A. Matsumoto, T. Sakae, K. Saitoh, T. Shibata, and M. Yoshioka, Physics Procedia, **26**, 223–230 (2012).
25. Y. Nakagawa, K. Pooh, T. Kobayashi, T. Kageji, S. Uyama, A. Matsumura, and H. Kumada, J. of Neuro-Oncology, **62**, 87–99 (2003).

High-power TR-24 cyclotron-based p-n convertor cooled by submerged orifice jet

Bém *Pavel*[1], Běhal *Radomir*[1], Götz *Miloslav*[1], Plíhal *Petr*[2], Poklop *Dušan*[1], Štefánik *Milan*[1], Štursa *Jan*[1]*, Zach *Václav*[1]

[1]Nuclear Physics Institute of the CAS, v. v. i., Husinec–Řež 130, 250 68 Řež, Czech Republic
[2]*HVM Plasma Ltd., Na Hutmance 2, 158 00 Prague 5, Czech Republic* Sciences,

Abstract. The TR-24 cyclotron (Advanced Cyclotron Systems Inc., Canada) of the Nuclear Physics Institute in Řež provides protons with variable energies from 18 MeV up to 24 MeV and beam current of 0.3 mA. For such parameters, the p +Be source reaction on thick Be target can produce a white-spectrum neutron field ($E_n \leq 22$ MeV) with the intensity of 5×10^{12} n/s/sr in forward direction. Present paper outlines the development of Be-target cooling system, devoted to remove the heat load of 7 kW (density up to 4 kW/cm^2) from the target. Due to novel "orifice-form" of jet cooling (resulting in the shortest source-to-sample distance of 20 mm) with extremely high cooling efficiency, the TR-24 p-n convertor can achieve neutron-flux up to 2×10^{12}n/cm^2/s nearby the target output.

1 Introduction

For accelerator-based neutron irradiation facilities the deuteron break-up on light-nuclei target (^{3}H, D_2O, Li, Be and C) presents most intensive source-reactions tool. However, high-power deuteron accelerators suffer from principal limitations (cyclotrons) and/or by high technical demands (linear accelerators), which result in expensive facilities (SARAF and future projects IFMIF, DONES [1]) – consequently multipurpose and far from dedicated compact solutions.
Commercially available and relatively low-cost high-power proton cyclotrons (mostly of medical purposes) incorporated with suitable proton-neutron convertor (usually with fixed and/or rotating Be discs) are widely proven in the BNCT.

Fig. 1. The TR-24 cyclotron of the Nuclear Physics Institute, Řež.

Present work points at relatively low-cost high-power medical proton cyclotrons, which could provide fast "fusion-like" neutrons at intensity which could not only supply in parallel the future irradiation facilities, but - actually - the material-research fission reactors in particular.
The NPI TR-24 cyclotron, (Advanced Cyclotron Systems Inc.)(Fig. 1) provides protons with variable energies up to 24 MeV and 0.3 mA of external beam current. For such parameters, the p+Be source reaction on thick Be target can produce a white-spectrum neutron field ($E_n \leq 22$ MeV) with an intensity of 5×10^{12} n/s/sr in forward direction (based on MCNPX calculations, experimental activation tests, and standalone investigation of the p+Be source reaction for 24 MeV proton beam using the multi-foil activation technique at the NPI [2]).
Most critical point in the effort to reach desirable high neutron flux density consists in removing a high local heat load from the target by cooling assembly with minimized dimensions in the direction of neutron emission.

2 Cooling assembly of the target

2.1 ANSYS simulations

Due to Gaussian-like profile of a cyclotron beam spot on the target ($\sigma \leq 20$ mm), the heat density up-to 4 kW/cm ^{2} resulting in overall heat load of 7.2 kW needs to be removed from the target of any proton-neutron convertor. Point-like form of accelerator based neutron sources leads to an inverse square dependence of the neutron flux density on source-to-sample distance. Therefore, the dimensions of cooling assembly are to be

* Corresponding author: stursa@ujf.cas.cz

minimized to use high flux density option in a vicinity of the target.
Taking advantages in well appropriate dimensions (small radial length) and commonly known high heat transfer coefficients, various types of orifice nozzles were considered and tested to form a submerged impingement jets in cooling arrangement of static Be target. Nozzles were investigated to form a cooling assembly with water flow of maximum available rate 2 l/s and a system pressure of 1.1 MPa on static Be target.
Special effort was devoted to minimize the distance between the target and output backing side of the cooling chamber.

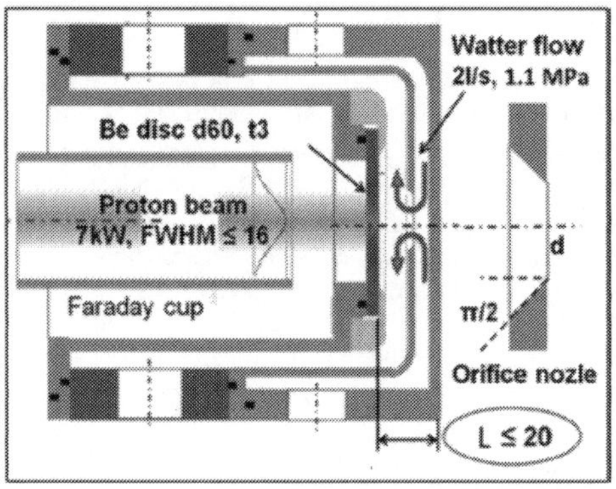

Fig. 2. Schematic arrangement of target chamber with Be disc, Faraday cup and orifice nozzle for ANSYS simulation. Dimensions are in mm.

The simulations were carried out in the forced convection-flow mode ANSYS (Fig. 2) to determine basic characteristics (heat transfer coefficient, fluid velocity at shear layer and pressure in the stagnation one) for different nozzle types (see Fig. 3).

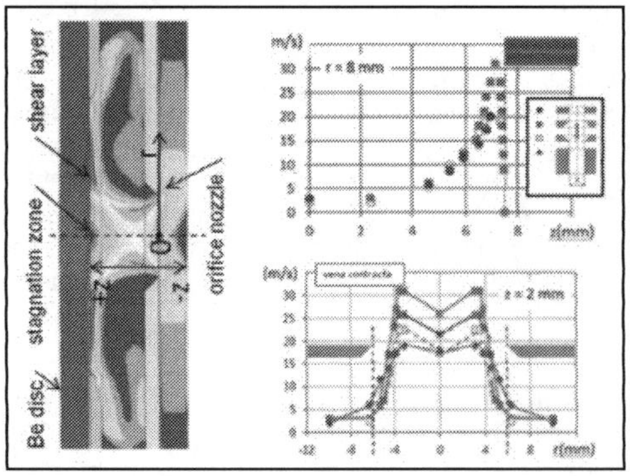

Fig. 3. Resulting velocity distribution of cooling fluid in the shear layer and vena-contracta region for various form of the orifice nozzle. The open sharp-edge orifice was selected as the optimal solution.

2.2 The prototype of Be target chamber

Considering similar empirical data on water wettability (contact angle) of the aluminum and beryllium surfaces, the Al disc (Fig. 4) – instead of Be disc was utilized in the pilot experiments to facilitate a set of three thermocouples inside the target (rather complicate in the case of Be material).

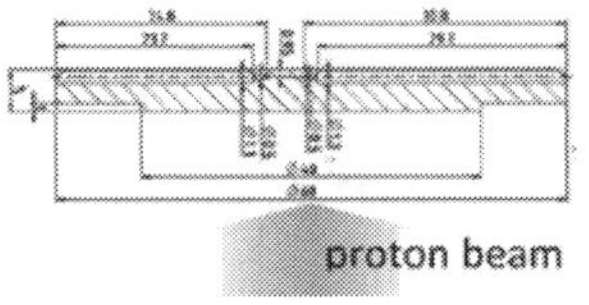

Fig. 4. The aluminum dummy-target for measurement of temperature distribution along beam spot using a set of thermocouples.

The thickness of target discs was determined by full stopping of 24 MeV protons. To take into account an expected hydrogen embrittlement (blistering effects) due to intensive proton beam, backing material with high hydrogen diffusion coefficient is considered instead of usually used not-stopping the protons within the target itself. The reason comes from possible disruption of thermal dynamics in the stagnation zone when part of thermal load is dissipated in the shear layer of cooling water.

2.3 Cooling tests

A mock-up of the target setup (Fig. 5) was manufactured to verify the ANSYS simulation and to determine empirical parameters of boiling mode of cooling.

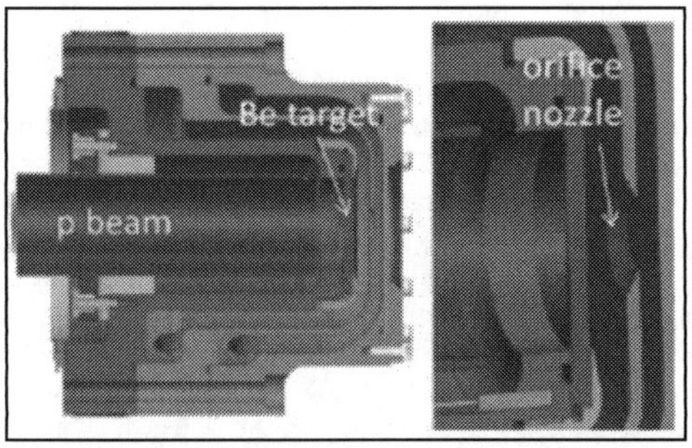

Fig. 5. The target chamber of p-n convertor assembly with orifice-nozzle for the water jet cooling - the prototype construction.

Temperature distribution in the target was measured by the set-up of three thermocouples at water flow rate of 2 l/s, the system pressure of 1.1 MPa, constant water-flow temperature of 20°C (ensured by a cooling unit) at different beam currents and spot dimensions during the irradiation with 24 MeV protons. In the Fig. 6, a typical behavior of measured temperature in various locations across the beam-spot area is given. Here, the linear dependence of temperature on beam current corresponds to heat transfer at convective single-phase mode. Clear evidence is seen for the onset of boiling, the area of nucleate and fully developed nucleate boiling and the presence of local critical heat flux as well.

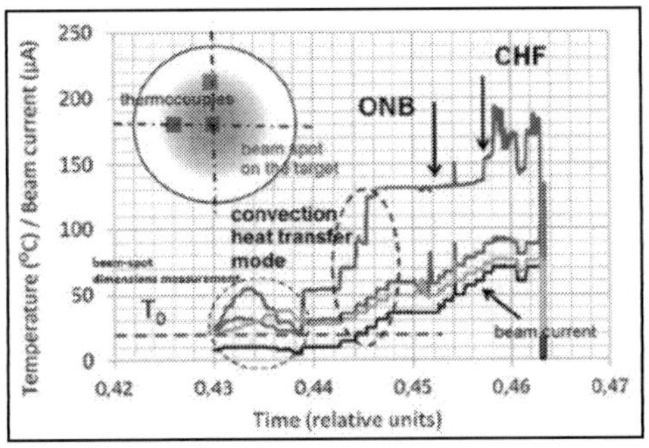

Fig. 6. Typical behavior of temperature in thermocouples located in various positions on the beam spot area during the irradiation by a proton beam.

3 Results and conclusions

Preliminary analysis of experimental data has a) confirmed the ANSYS simulation of transfer characteristics below the onset of boiling point (130°C) and b) provided original data regarding the boiling heat transfer mode and in particular the behavior of critical heat flux regime, see the Fig. 7. Measured steady-state heat fluxes ranged from 2.5 to 4 kW/cm ^{2} exceed several times the typical value of 0.5 kW/cm ^{2} for contemporary p+Be neutron generators usually operating with single-phase flow cooling of fixed Be targets [3,4]

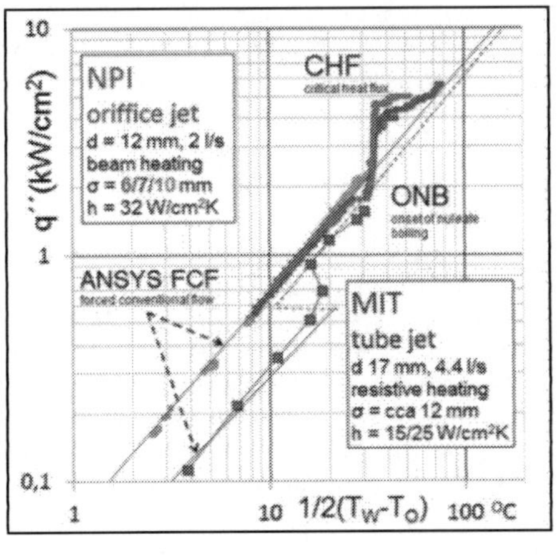

Fig. 7. Preliminary determined heat transfer characteristics of the submerged orifice jet cooling compared to long pipe-nozzle jet, developed at MIT BNCT neutron facility.

Due to novel "orifice-form" of cooling assembly (resulting in the shortest s-to-s distance of 20 mm), the TR-24 p-n convertor can achieve neutron-flux up to $2x10^{12} n/cm^2/s$ nearby the target output, the highest value of flux density for fast-neutron irradiation purposes until now. Target station with the open area at forward direction is developed to provide the irradiation under non-perturbed arrangement of different samples and associated hardware. Remotely controlled manipulators of irradiated components are being developed to ensure basic operation in the large induced activity ranging up to Sv/h. The methods to minimize the blistering efect during operation are under investigation.

Submerged water jet cooling assembly was tested at BNCT neutron facility of the Massachusetts Institute of Technology [5]. Removing of 5 kW heat loaded by resistive heating into fixed steel-dummy target has been reported. However, long pipe-nozzle derogates a possibility to reach high density of the neutron flux in irradiated samples.

Acknowledgements

This work was supported by OP RDE, MEYS, Czech Republic under the project CANAM, CZ.02.1.01/0.0/0.0/16_003/0001812.

References

1. A. Ibarra et al., *The IFMIF-DONES project: preliminary engineering design*, Nuclear Fusion 58 (2018) 105002.
2. M. Stefanik, P. Bem, E. Simeckova, J. Stursa, V. Zach, M. Majerle: *Neutron field study of p(24)+Be source reaction using the multi-foil activation technique*, Fusion Engineering and Design **146** (2019), pp. 2053-2056
3. T. Rinckel, D.V. Baxter, J. Doskow, P.E. Sokol, T. Todd, Physics Procedia **26** (2012), pp. 168-177
4. P. Bém, V. Burjan, M. Götz, M. Honusek, U. Fischer, V. Kroha, J. Novák, S.P. Simakov and E. Šimečková, Int. Conf. on Nucl. Data for Science & Technology., Nice, (Apr. 2007).
5. B.W. Blackburn, Thesis (M.S.), Massachusetts Institute of Technology, Dept. of Nuclear Engineering, (1997)

Energy and target material dependence of the neutron yield induced by proton and deuteron bombardment

Paul Zakalek[1,*], *Paul-Emmanuel Doege*[1], *Johannes Baggemann*[1], *Eric Mauerhofer*[1] and *Thomas Brückel*[1]

[1]Jülich Centre for Neutron Science, Forschungszentrum Jülich GmbH, 52425 Jülich, Germany

Abstract. The neutron yield for compact accelerator driven neutron sources depends on the target material, the ion type and its energy. When such sources are operated with low energy proton beams below 30 MeV, typical target materials are lithium and beryllium. New developments indicate that higher energies or a deuteron beam might be useful to increase the neutron yield at constant accelerator power. Here we present the total neutron yield analytically calculated for protons and deuterons at energies up to 100 MeV for various target materials. The total neutron yield depends on the involved cross sections and the stopping power of the target material. This study shows that for energies lower than 30 MeV light target materials with a deuteron beam are preferable whereas for energies above 30 MeV heavy target materials show a high neutron yield with little difference for a proton or deuteron beam.

1 Introduction

Today, neutron scattering and neutron analytical methods e.g. prompt gamma activation analysis or imaging, are mostly operated at fission or spallation neutron sources. Compact Accelerator-driven Neutron Sources (CANS) operating at low energies are an attractive and cost-effective alternative to spallation and fission sources. Several CANS are operated worldwide like the LENS facility in USA [1], the SARAF facility in Israel [2] or the established CANS network in Japan [3]. The recent shutdown of several research reactors in Europe leads to a corresponding reduction in neutron beam days [4]. Projects such as SONATE [5] in France and the High-Brilliance neutron Source project (HBS) [6, 7] in Germany work on the development of high-power CANS to counter act the reduction in beam days.

The neutron production at a CANS is based on (p, n) or (d, n) reactions in a suitable target material initiated by impinging ions with energies in the MeV range. The energy is below the threshold of spallation reactions (< 100 MeV). The neutron yield depends on the ion type (proton, deuteron, etc.), the ion energy and the target material. As all existing CANS are mostly operated with a proton beam at similar energies (3 – 13 MeV), they use either a beryllium or a lithium target, in order to maximize the neutron yield at these energies [8, 9, 10].

The development of a high-power CANS with an average deposited power of up to 100 kW in the target and a peak power up to 10 MW can either be realized by increasing the current or the ion energy. As the ion current is limited to 100 - 200 mA with accelerator technologies currently existing, an increase in energy is inevitable. This leads to other target materials than beryllium or lithium for an optimized neutron yield at ion energies above 30 MeV.
In the following we will present a systematic study of the neutron yield depending on the target material and the energy of a proton or deuteron beam up to 100 MeV.

2 Neutron yield calculations

Neutrons are produced at a CANS by ions in the MeV range impinging onto a suitable target i.e. production of many neutrons, high radiation resistance and good thermo-mechanical properties. When the energy of the ions is above the threshold for the (p, n) or (d, n) reaction, neutrons are produced. As the ions are decelerated when traveling through the target material, their energy decreases and the corresponding cross section for neutron production changes. The neutron yield for a specific target material therefore depends on the neutron production cross section and the stopping power of the target material.

2.1 Theory

The probability $\omega(E)$ that a single ion of energy E undergoes a nuclear reaction with N_T target atoms into a reaction channel with the cross section $\sigma(E)$ is given by

$$\omega(E) = \frac{\sigma(E) N_T}{A}, \tag{1}$$

with A as the illuminated target surface. With the Avogadro constant N_A, the molar mass M and the density ρ, the number of target atoms inside a volume V represented by the area A and the thickness x is

* Corresponding author: p.zakalek@fz-juelich.de

$$N_T = \frac{N_A}{M}\rho V = \frac{N_A}{M}\rho Ax. \quad (2)$$

As the ions lose energy while traveling through the target, the probability density for a reaction can be written as

$$\frac{d\omega}{dx}(E) = \frac{N_A}{M}\rho\sigma(E). \quad (3)$$

The formula above must be integrated over the stopping range R or the total loss of projectile energy to get the probability of the reaction

$$\omega(E \rightarrow 0) = \int_0^R \frac{N_A}{M}\rho\sigma(E)\,dx. \quad (4)$$

The cross section for neutron production for a specific ion type at an energy E and a given target element are given in the TENDL data library [11], the ENDF data library [12] or can be calculated using the TALYS nuclear code [13]. The corresponding penetration range for a given energy interval dE can be calculated using the SRIM 2013 toolkit [14]. A python program to calculate the neutron yield was written and uploaded to a GitHub server [15]. The following analysis was done using the TALYS nuclear code using the default parameters as cross sections for all elements can be calculated. The neutron yield was calculated for the natural abundance and a full stop of the ions inside the target material.

2.2 Neutron production cross section

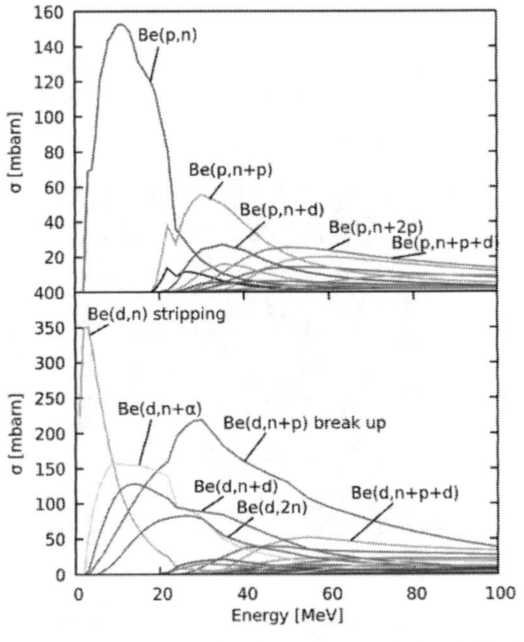

Figure 1: Cross sections for (p, n) reactions (top) and (d, n) reactions (bottom) in beryllium according to the TALYS nuclear code [13].

For a (p, n) or (d, n) reaction, the ions need to approach the target atoms close enough. Therefore, they need to overcome the Coulomb barrier. As the target is at rest, the ions need to be accelerated above a specific threshold energy which is depending on the atoms involved. The resulting (p, n) or (d, n) reactions depend on the nuclear structure of both atoms, is stochastic in nature and is therefore described by a cross section representing its probability.

In Figure 1 the cross sections for the (p, n) and (d, n) reactions in a beryllium target are presented. The cross sections for the ^{9}Be(d, n) reaction are generally larger than the cross sections for the ^{9}Be(p, n) reactions. This is due to the additional neutron of the deuteron and the possible stripping reaction. In addition, the stripping reaction for deuterons sets in and peaks at low particle energies. The total neutron production cross section for a deuteron beam is in general for all target elements larger than the total neutron production cross section for a proton beam. But it is also necessary to consider the stopping range and therefore the stopping power according to equation 4.

2.3 Stopping power and range

When particles travel through matter, they lose energy due to inelastic scattering. This is described by the linear stopping power

$$S(E) = -dE/dx, \quad (5)$$

where dE is the energy loss at initial energy E in the path length dx. This energy loss is due to two processes. First, the electronic stopping power due to inelastic collisions of the ion with the electrons of the target and an ionization of the target atoms. This process is described well with the Bethe-Bloch formula. Second, the nuclear stopping power due to a repulsive Coulomb interaction with the target nuclei. This process can be described with a Coulomb interaction potential

$$V(r) = \frac{Z_1 Z_2 e^2}{e\pi\varepsilon_0 r}\Phi(r/a), \quad (6)$$

with the atomic numbers Z_1 and Z_2, the electron charge e, the radius r, the screening length a and a screening function $\Phi(r/a)$ with the reduced radius r/a. A commonly used semi-empirical screening function also used in SRIM was proposed by Ziegler, Littmark and Biersack [16]. Figure 2 shows the electronic and nuclear stopping powers for typically used target materials for protons and deuterons.

As can be seen from Figure 3, the electronic stopping power is orders of magnitude above the nuclear stopping power for light ions like proton and deuteron. Additionally, it is evident that the stopping power for deuterons is higher than for protons resulting in a lower penetration range for the same energy. The stopping range can be determined from the stopping power with

$$R = \int_0^E \frac{1}{S(E)}dE. \quad (7)$$

https://doi.org/10.1051/epjconf/202023103006

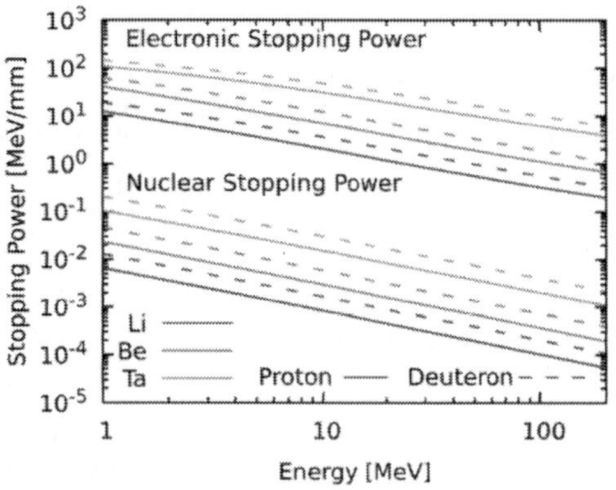

Figure 2: Stopping power, i.e. energy loss per path length traveled for protons and deuterons in Li, Be and Ta. The stopping power was calculated using SRIM 2013 [14].

Stopping ranges for typically used target materials and energies for protons and deuterons are presented in Table 1. As mentioned, the probability for a neutron-producing reaction scales linearly with the stopping range. Because of this, the larger cross sections for deuterons in comparison to protons are partly compensated with the larger stopping range for protons.

Table 1. Stopping range for protons and deuterons for various target materials and energies calculated with SRIM 2013 [14].

	Energy [MeV]	**Proton stopping range** [mm]	**Deuteron stopping range** [mm]
Li	5	0.91	0.53
Be	10	0.86	0.50
Be	30	6.00	3.42
Ta	50	3.02	1.87
Ta	70	5.34	3.28

2.4 Limitations

The analysis of the neutron yield relies on the quality of the used cross sections calculated with the TALYS nuclear code [13] or the used database [11, 12] and the stopping power calculated with the SRIM toolkit [14].

The electronic stopping power which is well described by the Bethe-Bloch equation is validated well and produces reliable results. The nuclear stopping power calculated with the Ziegler-Biersack-Littmark (ZBL) universal screening potential inside the SRIM toolkit shows deviations for various ions [17]. As the nuclear stopping power is about three orders of magnitude smaller than the electronic stopping power, the stopping powers calculated are still quite reliable.

The relevant cross sections were calculated using the TALYS nuclear code which uses nuclear models to calculate the relevant cross sections. Although the code is tested against experimental data, the quality of the calculated cross sections depends on the isotopes and produces inaccurate cross sections for light elements. Therefore, it is recommended to use specific cross sections from the ENDF/B-VIII database library [12]. The errors for cross sections can be estimated to a factor of three when comparing the calculated cross sections to experimental validations e.g. for beryllium [18]. However, as we are mainly interested in the general trends as function of beam energy and target material, the use of the TALYS nuclear code seems justified.

3 Results

Using the method presented, the neutron yield was calculated for protons and deuterons impinging on various elements in the naturally occurring isotopic composition. The neutron yield for the (p, n) and (d, n) reactions are presented in Figure 3 and Figure 4, respectively.

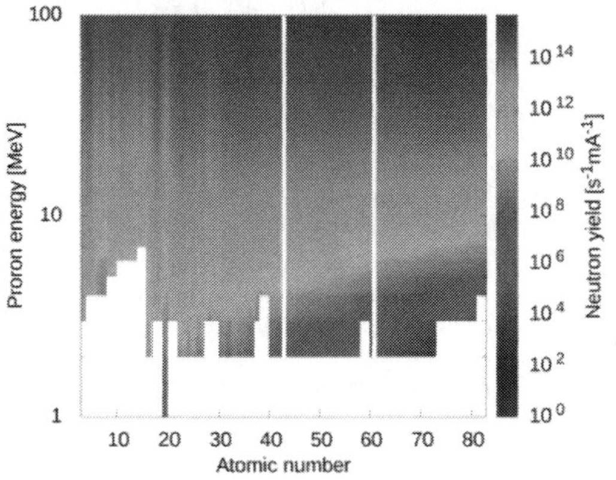

Figure 3: Proton induced neutron yield for various target materials determined from the cross sections calculated with the TALYS nuclear code [13] and the stopping powers calculated with the SRIM 2013 toolkit [14].

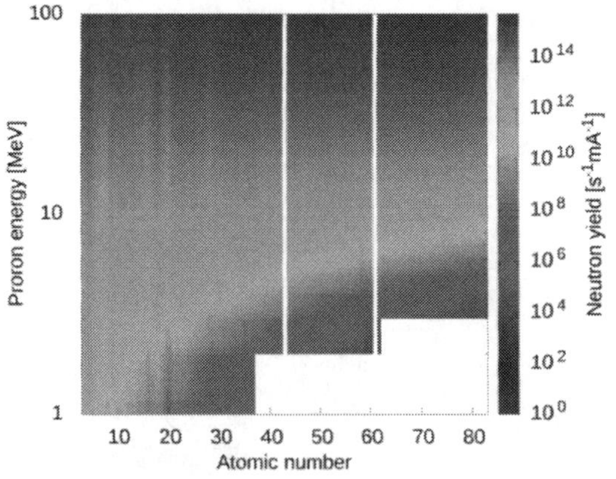

Figure 4: Deuteron induced neutron yield for various target materials determined from the cross sections calculated with the TALYS nuclear code [13] and the stopping powers calculated with the SRIM 2013 toolkit [14].

The colour code in the images shows the neutron yield as a function of the ion energy and the target element represented by the atomic number. The elements with the atomic numbers 43 (Tc) and 61 (Pm) have no stable isotope and have therefore not been investigated. The (p, n) reaction is endothermic for all elements as the protons need to pass the Coulomb barrier to engage in the reaction and to remove a neutron from the target nuclei, whereas for deuterons, this reaction is exothermic for lighter elements and becomes endothermic for heavier elements due to the stripping reaction channel where the deuteron can easily lose its neutron.

The neutron output increases with increasing energy for protons as well as for deuterons. At energies below 10 MeV, low Z-materials like lithium (Z=3), beryllium (Z=4) or argon (Z=18) produce a high number of neutrons for proton induced reactions. Argon is not used as a target material due to the gases state but lithium and beryllium are commonly used target materials. All other elements have a lower neutron output. At intermediate energies around 10 MeV, the elements with atomic numbers between 20 and 40 form a transition region in terms of total neutron yield. In this range vanadium (Z=23) shows a high neutron yield and might be an interesting target material. Above ~ 30 MeV energy, the heavy elements exhibit a high neutron yield without much variation from element to element.

The neutron yield produced by impinging deuterons shows a similar behaviour as for the protons. Below an energy of 30 MeV, beryllium has the highest neutron yield. Above this energy, the heavy elements show a similar neutron yield which then exceeds the neutron yield of the light elements.

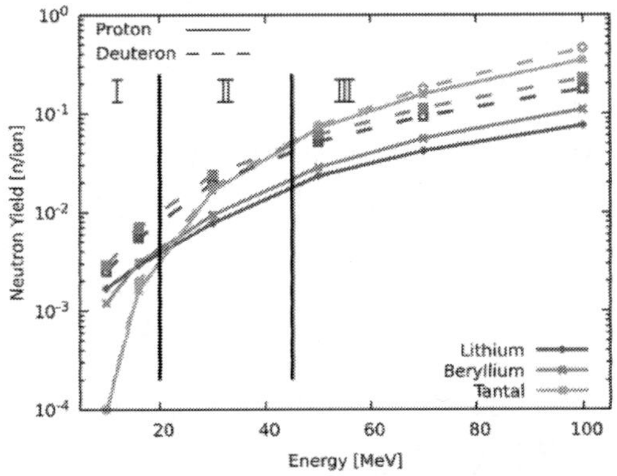

Figure 5: Neutron yield for Li, Be, and Ta as a function of the primary ion energy. The energy can be seperated into three regions, I, II, and III showing a different behaviour for the neutron yield.

Figure 5 shows the neutron yield induced by proton or deuteron for selected target materials as a function of their energy. For the low energy region I, below 20 MeV, the neutron yield is dominated by light elements especially lithium and beryllium. In the transition region II, heavy and light elements show similar neutron yields. The high energy region III is dominated by heavy elements with respect to the neutron yield. As the difference in neutron yield for the heavy elements is not significant, the choice of the material in the high energy region should be taken considering other factors like lifetime, blistering, machinability, etc. A stack of several materials e.g. heavy and light with adopted layer thickness will show a higher neutron yield than a single elemental target, especially in region II.

Figure 6 shows the ratio of the neutron yield for deuterons and protons for various elements. This ratio shows a dependency on the target material as well as on the primary ion energy. For light elements the gain factor when using deuterons is very large but decreases with increasing energy whereas for heavy elements it is lower and rises with the ion energy. The large changes between different target materials for low atomic numbers is most likely due to the inaccurate cross sections calculated with the TALYS nuclear code for light elements.

It can still be deduced that deuterons are preferred at low ion energies or light target elements whereas protons are preferred for high ion energies or heavy target elements.

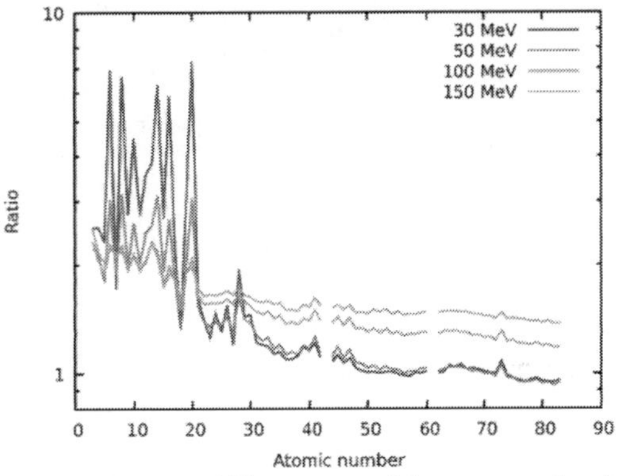

Figure 6: Neutron yield for protons and deuterons as a function of the target element expressed as the deuteron/proton neutron yield ratio.

4 Discussion

The analysis of the neutron yield shows a complex behaviour depending on the primary ion type, the energy and the target material.

In general, the choice of the ion type, its energy and the target material depends on the accelerator available. For very low energies (< 5 MeV) a lithium target is preferable and is generally used for the neutron production for e.g. boron neutron capture therapy (BCNT) [9, 10]. The neutron yield can be increased significantly by changing the ion type from protons to deuterons.

For higher energies up to 30 MeV, a beryllium target [8, 19] is the material of choice and is used at different neutron sources [1, 20] or future facilities [5]. The gain factor when changing the ion type from protons to deuterons would still be relevant for such sources but needs careful considerations as the complexity of the accelerator increases. The target shows additionally a different thermo-mechanical behaviour when irradiated with a proton or deuteron beam [21].

At high energies, a heavy metal target is the best choice. It offers a large variability regarding the target material and has a high neutron yield. This is the reason why within the HBS project [6, 7] a tantalum target was chosen. It has a high neutron yield, can be manufactured well, has good thermo-mechanical properties and has a high blistering threshold [22]. The small gain when using deuterons does not justify the increased accelerator complexity and the larger investment costs. Estimated total neutron yields are presented in Table 2.

Table 2. Calculated total neutron yield for different targets, ion types and energies normalized to the ion current (columns 3, 4) as well as the source strength normalized to the average power at the target (columns 5, 6).

	Energy [MeV]	p [$s^{-1}mA^{-1}$]	d [$s^{-1}mA^{-1}$]	p [$s^{-1}kW^{-1}$]	d [s^{-1} kW^{-1}]
Li	5	3.7e+12	5.4e+12	7.4e+11	1.1e+12
Be	10	7.7e+12	1.8e+13	7.7e+11	1.8e+12
Be	30	5.9e+13	1.5e+14	2.0e+12	5.0e+12
Ta	50	4.4e+14	4.7e+14	8.8e+12	9.3e+12
Ta	70	9.8e+14	1.1e+15	1.4e+13	1.6e+13

5 Summary and Outlook

It was shown that the neutron yield for CANS using (p, n) and (d, n) reactions depends on the cross sections for the specific reaction channels and the stopping power / stopping range for the ion type and the target material. A light target material and a deuteron beam are preferable for low energy ion beams, whereas heavy target materials and a proton beam are preferable for high energy ion beams.

Independently of the quality of the cross section calculations, this analysis shows interesting target materials for specific ion types and energies. Therefore, we are planning to perform experiments to determine the total neutron yield for e.g. beryllium, vanadium and tantalum and to compare it with the calculations.

The approach presented in this paper and the algorithm uploaded to GitHub can easily be adapted to calculate the yield of secondary particles e.g. tritium. This will help to screen interesting target materials avoiding time consuming MCNP simulations.

References

1. C. M. Lavelle, D. V. Baxter, A. Bogdanov, V. P. Derenchuk, H. Kaiser, M. B. Leuschner, M. A. Lone, W. Lozowski, H. Nann, B. v. Przewoski, N. Remmes, T. Rinckel, Y. Shin, W. M. Snow, P. E. Sokol, Nucl. Instrum. Meth. A, **587**, 324–341 (2008)
2. I. Mardor, O. Aviv, M. Avrigeanu, D. Berkovits, A. Dahan, T. Dickel, I. Eliyahu, M. Gai, I. Gavish-Segev, S. Halfon, M. Hass, T. Hirsh, B. Kaiser, D. Kijel, A. Kreisel, Eur. Phys. J. A **54**, 91 (2018)
3. Y. Kiyanagi, Eur. Phys. J. Plus **131**, 132 (2016)
4. C. Carlile, C. Petrillo, *Neutron scattering facilities in Europe: present status and future perspectives*, ESFRI Physical Sciences and Engineering Strategy Working Group, June 2016
5. A. Menelle, F. Ott, F. Prunes, B. Homatter, B. Annighöfer, F. Porcher, C. Alba-Simionesco, N. Chauvin, J. Schwindling, A. Letourneau, A. Marchix, N.-H. Tran, arXiv:1612.00237 (2016)
6. U. Rücker, T. Cronert, J. Voigt, J. P. Dabruck, P. -E. Doege, J. Ulrich, R. Nabbi, Y. Beßler, M. Butzek, M. Büscher, C. Lange, M. Klaus, T. Gutberlet, T. Brückel, Phys. J. Plus **131**, 19 (2016)
7. P. Zakalek, J. Baggemann, T. Brückel, S. Böhm, T. Cronert, P.-E. Doege, T. Gutberlet, J. Li, E. Mauerhofer, O. Meusel, H. Podlech, M. Rimmler, U. Rücker, M. Schwarz, J. Voigt, International Conference on Heavy Ion Accelerator Technology **14**, 117-121 (2019)
8. C.M. Lavelle, D.V. Baxter, A. Bogdanov, V.P. Derenchuk, H. Kaiser, M.B. Leuschner, M.A. Lone, W. Lozowski, H. Nann, B.v. Przewoski, N. Remmes, T. Rinckel, Y. Shin, W.M. Snow, P.E. Sokol, Nucl. Instrum. Meth. A **87**, 324–341 (2008)
9. B. Bayanov, V. Belov and S. Taskaev, J. Phys.: Conf. Ser., **41**, 460–465 (2006)
10. R. Fujii, Y. Imahori, M. Nakakmura, M. Takada, S. Kamada, T. Hamano, M. Hoshi, H. Sato, J. Itami, Y. Abe, M. Fuse, AIP Conf. Proc., **1509**, 162 (2012)
11. A. J. Koning, D. Rochman, J.-Ch. Sublet, N. Dzysiuk, M. Fleming. S. van der Marck, Nucl. Data Sheets, **155**, 1-55 (2019)
12. D.A. Brown, et al., Nucl. Data Sheets, **148**, 1-142 (2018)
13. A. J. Koning, D. Rochman, Nucl. Data Sheets, **113**, 2841-2934 (2012)
14. J. F. Ziegler, A. D. Ziegler, J. P. Biersack, Nucl. Instrum. Meth. B, **268**, 1818-1823 (2010)
15. https://iffgit.fz-juelich.de/zakalek/neutron-yield.git
16. J. F. Ziegler, J. P. Biersack, M. D. Ziegler, *SRIM: The Stopping and Range of Ions in Matter*, SRIM Co., Chester, MD (USA), (2008)
17. H. Paul, AIP Conf. Proc. **1525**, 309 (2013)
18. I. Tilquin, P. Froment, M. Cogneau, Th. Delbar, J. Vervier, G. Ryckewaert, Nucl. Instrum. Meth. A, **545**, 339 – 343 (2005)
19. P.-E. Doege, P. Zakalek, J. Baggemann, E. Mauerhofer, U. Rücker, T. Cronert, T. Gutberlet, Y. Beßler, J. Wolters, M. Butzek, S. Böhm, R. Nabbi, G. Natour, Th. Brückel, Journal of Neutron Research, **20**, 47-54 (2018)

20. M. Takamura, Y. Ikeda, H. Sunaga, A. Taketani, Y. Otake, H. Suzuki, M. Kumagai, T. Hama, Y. Oba, J. Phys.: Conf. Ser., **734,** 032047 (2016)
21. P. Zakalek, P.-E. Doege, J. Baggemann, T. Cronert, Y. Beßler, M. Butzek, J. Wolters, E. Mauerhofer, U. Rücker, T. Gutberlet, G. Natour, Th. Brückel, Physica B, **551**, 484 – 487 (2018)
22. V. T. Astrelin, A. V. Burdakov, P. V. Bykov, I. A. Ivanov, A. A. Ivanov, Y. Jongen, S. G. Konstantinov, A. M. Kudryavtsev, K. N. Kuklin, K. I. Mekler, S. V. Polosatkin, V. V. Postupaev, A. F. Rovenskikh, S. L. Sinitskiy, E.R. Zubairov., J. Nucl. Mater., **396**, 43 – 48 (2010)

Cryostat for the provision of liquid hydrogen with a variable ortho-para ratio for a low-dimensional cold neutron moderator

Sebastian Eisenhut[1,*], *Marcel* Klaus[1], *Johannes* Baggemann[2], *Ulrich* Rücker[2], *Yannick* Beßler[3], *Alexander* Schwab[1], *Christoph* Haberstroh[1], *Tobias* Cronert[2], *Thomas* Gutberlet[2], *Thomas* Brückel[2], and *Carsten* Lange[4]

[1] TU Dresden (Bitzer-Chair of Refrigeration, Cryogenics and Compressor Technology), Dresden, Germany
[2] Forschungszentrum Jülich (JCNS), Jülich, Germany
[3] Forschungszentrum Jülich (ZEA-1), Jülich, Germany
[4] TU Dresden (Chair of Hydrogen and Nuclear Energy), Dresden, Germany

Abstract. A significant contribution to the enhancement of the neutron brilliance achievable with Compact Accelerator-driven Neutron Sources (CANS) can be made by an optimized cold moderator design. When using liquid para-H_2 as the moderating medium, the concept of low-dimensional cold moderators can be employed to increase the neutron brightness (as currently foreseen at the European Spallation Source ESS). Para-H_2 shows a drop in the scattering cross section by about one order of magnitude around 15 meV, resulting in a large deviation between the mean free paths of thermal and cold neutrons. Taking advantage of this effect, the cold moderator geometry can be optimized to allow the intake of thermal neutrons through a relatively large envelope surface and then extracting them in an efficient way towards the neutron guides. One drawback of this solution is the lack of thermalization of the cold neutrons. In the context of the HBS (High Brilliance Neutron Source) project, efforts are made to overcome this problem by increasing the scattering cross section of the H_2 in a defined way. The idea is to admix small amounts of ortho-H_2, which maintains its large scattering cross section in the region below 15 meV. Like this, the neutron spectrum can be shifted towards lower energies and adjusted for the needs of the respective instruments. In a cooperation between TU Dresden and FZ Jülich, an experimental setup has been created to prove the feasibility of this concept. The main component of the experimental setup is a LHe-cooled flow cryostat that enables the separate condensation of a para-H_2 and a normal-H_2 flow and a subsequent mixing of the two in precise proportions. The resulting LH_2 mixture at 17 - 20 K is fed into a small cold moderator vessel (approx. 200 ml). In this work, the current status of the setup is presented. The construction and commissioning of the mixing cryostat have been completed and first test runs show that different ortho-para-H_2 mixtures can be produced. In the near future, the system will be ready for measurements at a neutron source.

1 Introduction

The progressive shutdown of existing research reactors in Europe causes an increasing shortage of neutron availability for scattering experiments. To overcome this problem, currently there are efforts being made to establish a network of Compact Accelerator-driven Neutron Sources (CANS). CANS represent a cost-effective and versatile way to ensure the neutron provision and establish local facilities with easy access. The Jülich Centre for Neutron Science (JCNS) is currently developing a future CANS, the High Brilliance neutron Source (HBS), in which the whole chain from neutron production to detection is to be optimized according to the experimental needs. The design of the cryogenic moderator is of great importance for an optimized cold neutron provision [1].

Several materials, among them mesitylene, methane, and hydrogen, are currently being investigated as possible candidates for the cryogenic moderators used at HBS. Liquid para-H_2 has been used in cold sources since many years, but recently there has been a new trend in para-H_2 moderator design: Thanks to the drop in the scattering cross section of para-H_2 at a neutron energy around 15 meV (see Figure 1), it is possible to extract neutrons from the cryogenic moderator in a very efficient way due to low self-absorption, if the moderator is implemented as a so-called low-dimensional cold moderator [2].

On the other hand, a major drawback of the extremely low scattering cross section in the cold neutron range is that a good neutron thermalization cannot be reached inside a low-dimensional moderator [1]. As slower neutron spectra are preferred for certain scattering experiments, there is a great interest in further reducing the neutron temperature.

* Corresponding author: sebastian.eisenhut@tu-dresden.de

As shown in Figure 1, ortho-H_2 does, in contrast to para-H_2, not exhibit a drop in the scattering cross section. The admixture of (usually small) well-defined amounts of ortho-H_2 is, therefore, an effective way of increasing the scattering cross section of the moderator medium and expected to enable a more complete thermalization. This, however, bears a trade-off, as with growing ortho-H_2 content also the neutron extraction properties deteriorate to some extent.

When using optimized ortho-para mixtures instead of pure para-H_2, the result could be a cryogenic moderator of an unprecedented performance, featuring the advantages of the low-dimensional design approach but mitigating the problem of an incomplete thermalization.

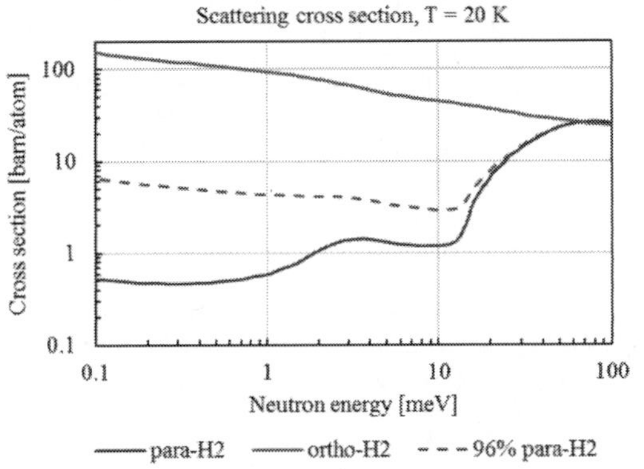

Figure 1. Scattering cross section of para-H_2, ortho-H_2 [3] and a mixture containing 96% para-H_2 as a function of neutron energy. The data for the creation of this diagram has been extracted from plots presented in [4].

Currently, the design of cold moderators using hydrogen is subject to uncertainties due to the ambiguity in the required nuclear data [4]. Furthermore, Monte Carlo simulations of moderators using ortho-para mixtures are especially hard to manage and prone to errors [5, 6]. To overcome this problem and enable a reliable moderator design, there are endeavors by JCNS to measure the neutron spectra resulting from different ortho-para mixtures (and pure para-H_2) experimentally. In this work, the cryogenic system required to provide liquid hydrogen with a variable ortho-para ratio is presented.

2 System overview

2.1 Cold moderator

The design of the cold moderator vessel follows the concept of low-dimensional cold moderators as proposed by Mezei et al. [2] and was designed for pure para-H_2. The MCNP-based parameter studies required for dimensioning the LH_2 (liquid hydrogen) volume in the cold moderator vessel were done at the Chair of Nuclear Engineering and Technology Transfer (NET) at RWTH Aachen in cooperation with JCNS-2 [5, 6], whereas the final engineering design was a contribution of ZEA-1 at Forschungszentrum Jülich [7]. The result is a quasi-one-dimensional cylindrical moderator vessel that is fed with thermal neutrons from all sides. Due to the high scattering cross section at thermal neutron energies (see Figure 1), most incoming neutrons are scattered down to cold neutron energies within the mean free path $l_{thermal} = 1$ cm. When using pure para-H_2, the scattering cross section for neutrons of energies below 15 meV is more than one order of magnitude lower, resulting in an increased mean free path of about $l_{cold} = 11$ cm up to approx. 1.5 m. As a consequence, the extraction of cold neutrons along the longitudinal axis of the cylindrical moderator vessel is particularly efficient, as a big moderator volume can be used to provide cold neutrons within the limited divergence required. The small diameter of the cylindrical moderator vessel allows feeding the all of its volume efficiently with thermal neutrons without the risk of depletion in the center of the moderator. The optimal moderator dimensions are a trade-off between neutron flux and brilliance at a considered sample position. The smaller the illuminated area of the cold moderator, the higher the brilliance but the smaller the flux. However, a first approximation is a diameter d corresponding to $l_{thermal}$ and a length l corresponding to l_{cold}, though for reasons of the achievable total neutron flux and manufacturing, the dimensions $d = 2.5$ cm and $l = 10$ cm have been chosen [6], resulting in a total liquid hydrogen volume of 200 ml. The operating temperature and absolute pressure are $T = 17$ to 20 K and $p = 1.3$ to 1.5 bar.

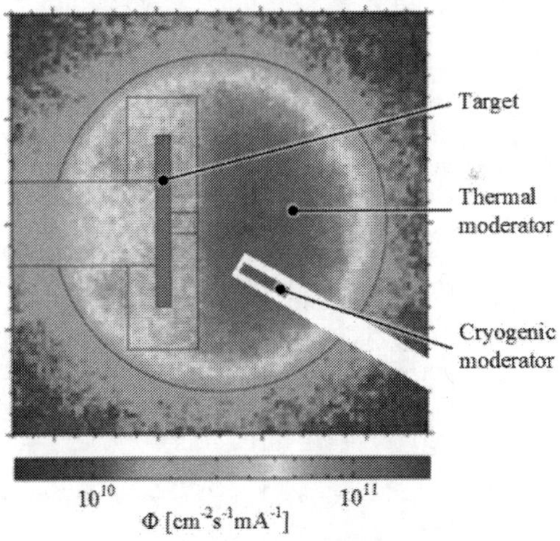

Figure 2. Schematic view of target, thermal moderator and cryogenic moderator positioned inside the thermal neutron density maximum.

Figure 2 contains a schematic view of the potential position of the cryogenic moderator inside the extraction channel of the thermal moderator and Figure 3 shows a CAD model of the engineering design of the moderator vessel with the He cooling channels milled into its shell. The latter allow maintaining the LH_2 within a narrow temperature band inside the subcooled regime. The vessel is positioned inside a long, cylindrical vacuum jacket which is connected to the main cryostat and can be inserted into the thermal moderator.

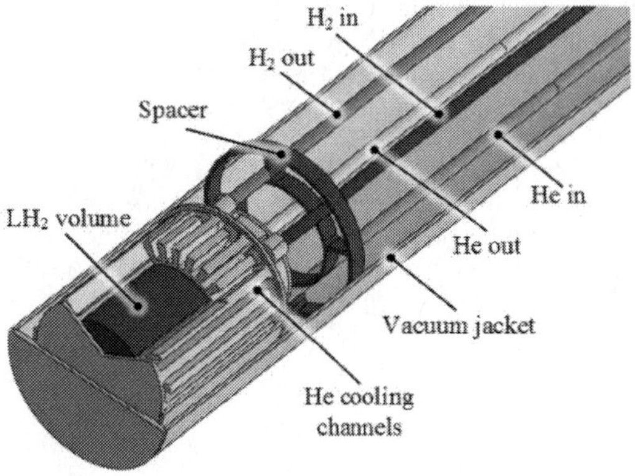

Figure 3. CAD model of the cryogenic moderator assembly.

2.2 Process and cryostat design

The cryogenic system is required to condense hydrogen and maintain it within the desired temperature and pressure range so that it is in a subcooled state. Filling the moderating volume shall take about 30 min at the design mass flow. A further central, and very challenging, requirement is the adjustability of the concentration in the range from 25 to 100% para-H_2 with an accuracy of ±0.1% para-H_2. This represents a novelty from a cryogenics point of view - the presented cryostat is the first hydrogen liquefier with an adjustable ortho-para ratio. The concentration shall furthermore be monitored externally by a suitable measurement device.

to different ortho-para ratios, but not as a final solution used for scattering experiments on a larger scale.

Figure 4 shows a schematic view of the cryogenic system. It is based on the following concept: two separate hydrogen flows, regulated by mass flow controllers (MFCs), are condensed in a counter flow heat exchanger (HX) cooled by helium cold gas. One of the hydrogen flows is converted to nearly 100% para-H_2 using a ferric oxide catalyst bed contained inside the heat exchanger, whereas the other flow maintains its initial composition of 75% ortho-H_2 and 25% para-H_2 (also referred to as normal-H_2). After passing the heat exchanger, the two flows are mixed together at the desired ratio, as imposed by the mass flow controllers. The mixture is fed into the moderator vessel, where it is cooled with helium cold gas passing through the cooling channels inside the outer shell.

In order to be able to control the para concentration the intended way, the ortho-para conversion in the catalyst needs to be complete and the system as a whole needs to be kept in a stable thermal equilibrium. These are significant challenges that needed to be considered during the design and commissioning process.

The moderator vessel is constantly overfilled with a defined hydrogen mass flow in order to compensate for any spontaneous para-ortho or ortho-para conversion occurring in the cryostat, to achieve reasonable mixing ratio errors associated to the mass flow controllers, and to enable the continuous monitoring of the hydrogen composition at the exhaust.

To prevent the risk of freezing hydrogen (at temperatures below 13.8 K) and to stabilize the system thermally, an electrical heater is located right after the LHe inlet inside the cryostat. It evaporates and heats up the helium to about 15 K. A second heater is located just before the triple heat exchanger (He, para-H_2 and normal-H_2), which can be used to control and stabilize the helium entry temperature at the heat exchanger independently from the helium temperature at the outlet of the moderator vessel. A helium-cooled radiation shield at about 50 K makes use of the remaining cooling capacity of the helium cold gas just before it leaves the cryostat.

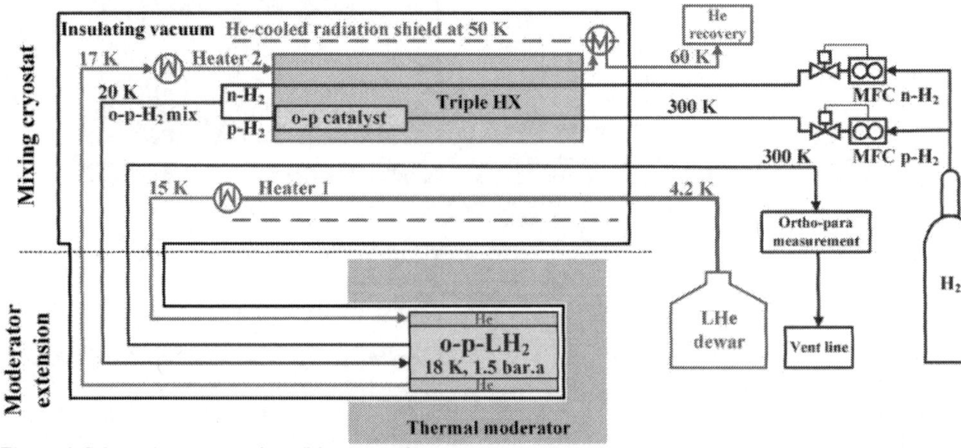

Figure 4. Schematic representation of the cryogenic system.

The process and the resulting cryostat have been designed and built over the course of the last years, mainly at TU Dresden [8]. The present system is understood as a proof-of-principle prototype for MCNP validation purposes and measurements of neutron spectra associated

 https://doi.org/10.1051/epjconf/202023104001

2.3 H_2 composition monitoring

It is of crucial importance to monitor the para concentration in real time by means of an external device independent from the mixing cryostat. Even though the mix composition, in theory, can be predicted quite easily knowing the separate para-H_2 and normal-H_2 mass flows and their corresponding para concentrations, transient thermal states in the cryostat, as well as undesired ortho-para (or para-ortho) conversion, can potentially cause a mix concentration that differs from the prediction.

Currently, a speed of sound-based approach is being pursued at TU Dresden. Good experiences have been made with the binary gas analyzer BGA244HP by Stanford Research Systems, which allows an uncomplicated online monitoring of the speed of sound in gaseous phase. The para concentration is obtained from linear interpolation between two reference points: normal-H_2 (25% para-H_2, directly from the bottle) and hydrogen with the equilibrium composition corresponding to 77.3 K (51% para-H_2, produced in an oversized reference catalyst immersed in a liquid nitrogen bath).

Because ortho- and para-H_2 differ only very slightly in most physical properties, measuring the para concentration is a rather challenging task. Due to the tough accuracy requirement, the concentration measurement is repeatedly calibrated by feeding it alternately with either one of the two reference gases over the course of an experiment. This accounts for small drifts in the measured speed of sound that can occur over the course of several hours. Furthermore, great efforts are made to keep any contaminants in the analysed hydrogen gas to a minimum, as they can significantly affect the measurement accuracy. The device and all pipes are extensively pumped and purged with pure hydrogen and the residual water content of the gas leaving the speed of sound system is continuously monitored using a dew point transmitter.

Determining a definite value for the absolute error in terms of para concentration is an intricate and ongoing task. However, a significant amount of experience using the device for measuring concentrations between 25% and 51% para-H_2 has been gathered, indicating a repeatability of $<\pm 0.3\%$ para-H_2 when used in the above described way. At concentrations higher than 51%, accuracy and repeatability of the measurements are expected to deteriorate with an increasing para concentration if no third reference gas with $\approx 100\%$ para-H_2 can be used for calibration purposes. At the moment, there is still no independent device for the production of nearly pure para-H_2 existing at the lab.

A promising way to increase the sensitivity of such speed of sound-based concentration measurement is reducing the gas temperature. The difference between the sound speeds of ortho-H_2 and para-H_2 spreads up at lower temperatures until it reaches a maximum at around 130 K [8] (see Figure 5). For this reason, a measurement rack that allows an active cooling of the device down to temperatures around 230 K is being developed, promising three- to four-fold sensitivity in terms of para concentration compared to the current setup at room temperature.

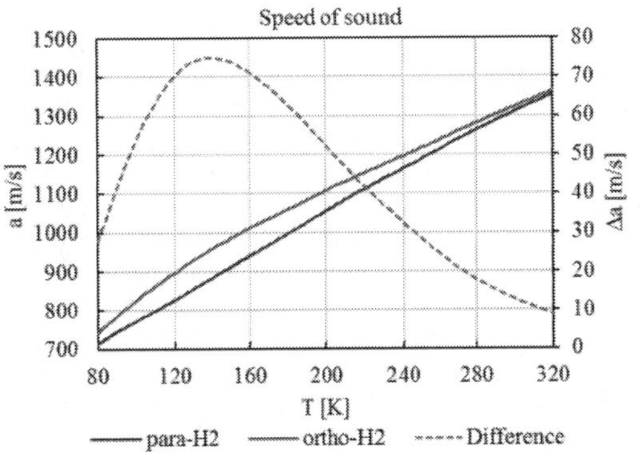

Figure 5. Speed of sound a of ortho-H_2 and para-H_2 and their difference Δa vs. temperature T.

3 Commissioning

3.1 General status

First test runs of the system were done already in August 2017 at AKR-2 [8] (a zero-power training reactor located at the campus of TU Dresden). However, several problems caused an unreliable operation of the system, most importantly insufficient vacuum pressure and cold leaks in the process lines. Over the course of the last year, the system has been reworked extensively to ensure leak tightness and a good insulating vacuum. Furthermore, the cryostats periphery has been improved and controls and data acquisition methods have been revised entirely.

Now, the system can in principle be operated in the intended way. For the first time, test runs have been made where a steady state could be maintained for several hours. A cool-down from room temperature takes about 3 hours, followed by 1 to 2 hours required to establish a stable operating point producing LH_2 at the desired pressure, temperature and para concentration. When maintained cool overnight by using a small LHe flow, re-establishing operating temperature is a matter of about 20 min the next day.

Commissioning is still ongoing, but great progress has been made during the last months. It could be shown that the cryostat works as designed.

3.2 Demonstration of mixing operation

Several tests have been run in order to demonstrate that the mixing cryostat is capable of producing different mix compositions. For reasons of safety, reduced complexity and experimentation time, these tests were done using short circuits instead of connecting the moderator vessel to the mixing cryostat. The cryostat was cooled down, brought into a stable thermal state, and the He temperature was reduced such that hydrogen condensation could take place inside the heat exchanger. Afterwards the setpoints of the two hydrogen MFCs were adjusted in such way that several different para concentrations could be produced subsequently, in the here presented test from 60% to 100% in steps of 10%.

To allow a sensible prediction of the mix concentration by using the mass flows controlled via the MFCs, it is a prerequisite to know the para concentrations of the two hydrogen flows. In particular, this means that the ortho-para conversion towards the temperature-dependent equilibrium state in the ferric oxide catalyst needs to be complete. Prior tests on the catalyst performance have shown that full conversion takes place in all relevant operating points (i.e. combinations of He and H_2 mass flows and He heater temperature setpoints).

Figure 6 contains measurement data taken during one of the mixing tests, showing the predicted and measured mix concentration over the course of almost three hours.

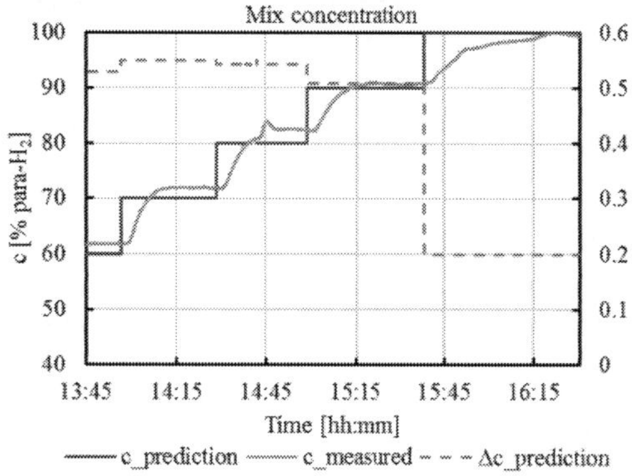

Figure 6. Para concentration during mixing test. The prediction is based on the measured partial mass flows and assumed concentrations of 99.9% para-H_2 and 25.0% para-H_2 respectively. The absolute prediction error is based on errors in the partial mass flows and an assumed absolute error of 0.2% para-H_2 after catalyst. The measured values are based on the speed of sound measurement calibrated using normal-H_2 and para-H_2 as reference gases.

The prediction is based on measurements of the hydrogen mass flows and assumed para concentrations (99.9% and 25.0%, respectively) of the two partial flows. The measured concentrations are based on the speed of sound readings.

It could be shown that the mix concentration can be controlled by the user as intended. The hydrogen MFCs respond very quickly and operate in an extremely stable way, which explains the perfectly step-like shape concentration prediction in Figure 6. When the mixing ratio of the partial streams is changed to adjust the ortho-para ratio, the measured concentration lags behind the predicted one, exhibiting a settling time of about 30 min using 30% of the maximum hydrogen flow. On the one hand, this is because the hydrogen inventory in the cryostat needs to be replaced entirely. Especially volumes of condensed hydrogen, e.g. contained in the process pipes after the mixing location, cause a long response time. On the other hand, changing the partial hydrogen mass flows imposes a disturbance on the system, causing a thermal transient in the heat exchanger and catalyst bed, which in turn influences the mix composition. To promote a high mixing accuracy during nominal operation as neutron moderator, the system should be run in only one operating point after H_2 condensation has started, and emptied entirely from condensate before a different operating point is targeted.

After the system had settled in each operating point, prediction and measurement coincided within about +2.7% para-H_2.

This is still far from the desired accuracy of ±0.1%. However, the goal of the present test was not to achieve the highest possible mix composition accuracy, but rather to demonstrate mixing operation for the first time and examine the systems response to changes in the concentration setpoint.

One of the reasons for the notable deviation between measurement and prediction is probably a poor measurement accuracy during the test. In order to allow a continuous tracking of the concentration without interruptions, no recalibration of the concentration measurement was done over the course of the experiment. There was only one reference measurement with normal-H_2 before the test, and one with nearly pure para-H_2 afterwards, with a time span of several hours in between. Usually, reference measurements should be done every 30 min to account for drifts in the speed of sound device. Currently, the only way of obtaining pure para-H_2 as a reference gas for calibration is producing it in the cryostat itself. Because this would mean an interruption in the operation at the desired mix composition and impose a large disturbance on the system, reference measurements like this can, at the most, be done in the very beginning or end of an experiment.

However, it could be shown that the mixing cryostat works as intended and it is ready for use with the moderator vessel.

Conclusion

In the frame of the HBS project, a holistic optimization of the chain from neutron production to neutron detection is a key objective, and the performance of the cryogenic moderator plays a vital role. In a joint project, FZ Jülich and TU Dresden are developing an experimental setup to test a low-dimensional liquid hydrogen moderator with a variable ortho-para ratio for the first time. Usually, pure para-H_2 is used for cryogenic moderators. However, by adding small, well-defined amounts of ortho-H_2, the neutron spectrum can potentially be shifted towards lower energies while maintaining the good neutron extraction properties that the low-dimensional (para-H_2) moderator approach allows. In this work, the cryogenic system for the provision of LH_2 with adjustable ortho-para ratio and its metrological monitoring via speed of sound measurements are presented. The commissioning of the system has advanced to a point where it is ready to be used with the cryogenic moderator vessel. It could be shown that the main cryostat can produce liquid hydrogen with an arbitrary para concentration between 25% and nearly 100% para-H_2.

However, several steps have to follow to complete the commissioning. As a next step, the system needs to be tested with the moderator vessel, and investigations

https://doi.org/10.1051/epjconf/202023104001

regarding the achievable mixing accuracy have to be conducted. Furthermore, the speed of sound-based concentration measurement will be refined. Afterwards, first tests under neutron radiation are to follow soon.

References

1. U. Rücker, T. Cronert, J. Voigt, J.P. Dabruck, P.-E. Doege, J. Ulrich, R. Nabbi, Y. Beßler, M. Butzek, M. Büscher, C. Lange, M. Klaus, T. Gutberlet, T. Brückel, Eur. Phys. J. Plus **131**, 19 (2016)
2. F. Mezei, L. Zanini, A. Takibyev, K. Batkov, E. Klinkby, E. Pitcher, T. Schönfeldt, J. Neutron Res. **17**, pp.101-105 (2014)
3. M. B. Chadwick et al., Nucl. Data Sheets **112**, pp. 2887–2996 (2011)
4. K. B. Grammer et al., Phys. Rev. B **91**, 180301(R) (2015)
5. J. P. Dabruck: *Simulation Studies on the Target Station for the High-Brilliance Neutron Source HBS*, PhD Thesis, NET – RWTH Aachen (2018)
6. T. Cronert, J. P. Dabruck, P.-E. Doege, Y. Beßler, M. Klaus, P. Zakalek, U. Rücker, C. Lange, M. Butzek, W. Hansen, R. Nabbi, T. Brückel, J. Phys.: Conf. Ser. **746**, 012036 (2016)
7. M. Hofmann: *Auslegung eines kryogenen Flussfingers im Rahmen einer Parameterstudie am AKR-2 als Voruntersuchung zur Entwicklung eines hoch-brillianten Neutronenquelle*, BSc Thesis, FH Aachen (2016)
8. M. Klaus, A. Schwab, Ch. Haberstroh, E. Eckhardt, Y. Beßler, J. Baggemann, T. Cronert, IOP Conf. Ser.: Mater. Sci. Eng. **502**, 012161 (2019)

Studies on Reflector Materials for Cold Neutrons

J. Rolando Granada [1,*], *J. Ignacio* Márquez Damián [2], and *Christian* Helman [1]

[1]Centro Atómico Bariloche and Instituto Balseiro, 8400 S.C. de Bariloche (RN), CNEA, Argentina
[2]Centro Atómico Bariloche and Instituto Balseiro, 8400 S.C. de Bariloche (RN), CONICET, Argentina

Abstract. Neutron scattering techniques and neutron applications in general are powerful and well-established tools for research in science and technology. However, those are highly intensity-limited techniques and the need to reduce the neutron loses between their source to detection is then an imperative endeavor. In particular, great effort has been devoted to the development of efficient ways to transport and guide neutron beams, as well as to reduce the flux leakage in the non-emitting sides of moderator systems. In this work we investigate and confirm by calculation in a real material the great quality of diamond nanoparticles as reflector for ultra-cold and very-cold neutrons, and propose the use of Magnesium Hydride at a low temperature as a very good reflector for cold neutrons

1 Introduction

The tremendous potential of neutrons as a probe of matter or as a research object by itself is limited by the relatively low-flux intensity of the neutron sources, as compared with photon sources. In addition, the different processes (production, slowing-down) and devices (moderators, transport systems, collimators, energy-selectors, detectors, etc.) reduce by several orders of magnitude the actual neutron intensity that eventually conveys the experimental information of interest. One of the research lines devoted to minimize those loses has been oriented to the search for efficient reflector materials, that may improve the efficiency of guiding surfaces or the actual reflection of neutrons on a containment walls to reduce their leakage. A large body of work has been done in the past, particularly concerning the interaction of slow neutrons with diamond nanoparticles. It has been demonstrated the high reflectivity of this material for UCN and VCN on the basis of calculations for ideal systems as well as by scattering experiments, and proposed that such capacity may extend to higher neutron energies, thus bridging the "reflectivity gap" in the neutron spectrum [1, 2].

In this work we present calculations aimed at evaluating the thermal neutron cross sections of bulk diamond and the small angle scattering contribution due to the nanostructure, using recent experimental information. With those ingredients we generated a cross section library over an extended neutron energy range for a real nanodiamond (ND) powder. Reflectivity calculations using our library confirm the excellent performance of that material for ultra-cold and very-cold neutrons, and also its rapid decrease for neutron velocities larger than about 150 m/s.

A different approach was employed to explore the characteristics of potential cold neutron reflectors, based on the requirement of a large scattering cross section combined with a small value of the average cosine of the scattering angle over the energy range of interest. Those conditions led us to consider metal hydrides, and eventually select magnesium hydride as a good candidate to fulfill those requirements.

2 Diamond nanopowders

During the last two decades, Nesvizhevsky and coworkers have been developing a large body of calculated and experimental results concerning the use of diamond nanopowder as a very good reflector for ultra- and very-cold neutrons [3,4] and its capacity as a good quasi-specular reflector for cold neutrons [5,6]. The calculations of cross sections were also done [7] for ideal ND systems always in the first Born approximation, and experimental SANS results were obtained for typical detonation nanodiamonds [8]. More recently, a research project [9] involving several groups was established aiming at a complete characterization of those materials and determination of their neutronic properties. One of those groups was able to perform SANS and transmission measurements, thus producing the first experimental total cross section results for a diamond nanopowder over the cold neutron energy region [10].

* Corresponding author: granada@cnea.gov.ar

2.1 Bulk diamond

The description of the neutron cross sections of nanodiamond (ND) over the thermal and cold energy regions demands a proper account of the bulk diamond properties, as their contribution will be significant for thermal neutrons.

In order to evaluate the elastic and inelastic components of the cross sections, we first calculated the phonon density of states (PDOS) of diamond, using *ab initio* algorithms [11, 12]. The result is shown in Fig.1, which is in good agreement with previous data [13, 14]; also, calculations of the specific heat based on our PDOS are in excellent agreement with the experimental results of A.C. Victor and W. DeSorbo [15, 16].

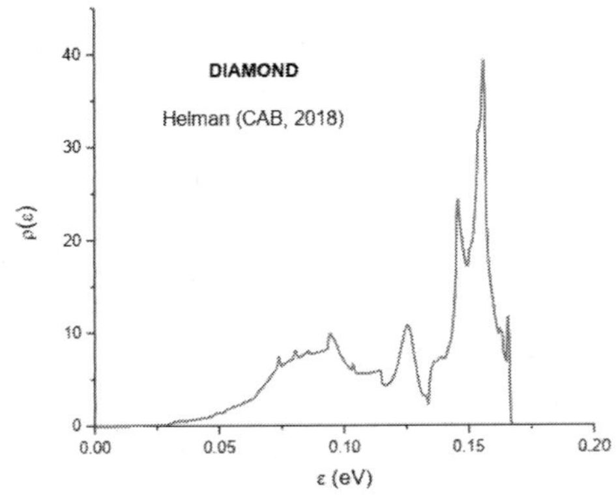

Fig. 1: Our calculated phonon density of states of diamond obtained by DFT calculation.

We developed a scattering kernel and generated cross section libraries over the thermal and cold neutron energy ranges, using the codes NJOY [17] for the inelastic contributions and NCrystal [18] for the elastic ones. The bound-atom scattering and absorption cross section values were taken from Ref.[19]. The calculated total cross section of polycrystalline bulk diamond at room temperature (heavy line) and their components (thin dashed line: elastic; thin continuous line: inelastic) are displayed in Fig.2, but surprisingly we found no experimental results for this magnitude to compare with.

2.2 Nano diamond

The characteristics of the diamond nanopowder and the interactions of slow neutrons with such system are given in [7]. The diffraction of the neutron wave on structural and density inhomogeneities gives rise to coherent elastic effects that superimpose upon the bulk crystal cross sections.

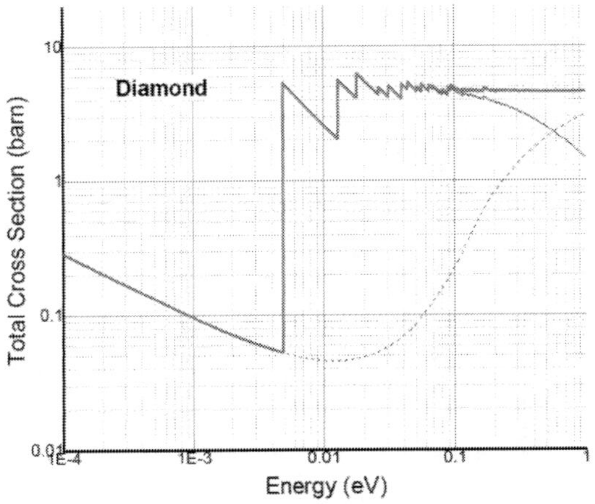

Fig. 2: The total cross section of bulk diamond over the thermal neutron energy range (red line). The black dashed line denotes the inelastic contribution, whereas the black solid line represents the elastic coherent component.

In order to produce a library corresponding to a real material, we used the experimental SANS results obtained by Teshigawara and coworkers [10], and fit them using the unified exponential/power-law approximation [20] as expressed in Eq.3 of Ref. [8]. In Fig.3 we show the results obtained for the scattered intensity I(Q) comparing the data points [10] with the fitted curve. This structure factor can be integrated to give the scattering total cross section corresponding to the coherent elastic effects caused by the nanopowder which, when added to the nanospheres' bulk diamond contributions produces the total scattering cross section for this system with apparent radius of gyration of 2.5 nm. Those calculated contributions together with the experimental points [10] are shown in Fig.4.

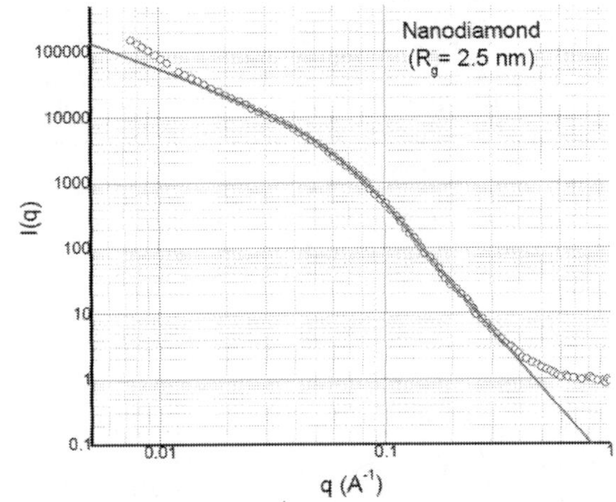

Fig. 3: Small angle scattering intensity as experimentally determined by Teshigawara *et al.* [10] (open circles) and our fitted curve using the unified exponential/power-law approximation [20].

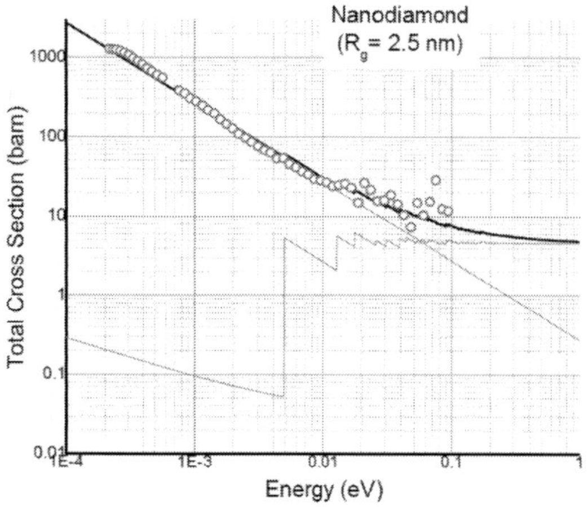

Fig. 4: Total cross section of nanodiamond, comparing our calculated curve (black line) with experimental data from Ref.[10]. Red and green lines denote the crystal and SANS components, respectively.

2.3 ND reflectivity

The scattering kernel, including small angle scattering effects, was implemented in the Monte Carlo program OpenMC [21], and this tool was used to study the reflective properties of nanodiamond layers. OpenMC was modified to include the total coherent scattering cross section component caused by SANS, and to sample the scattering angle from an I(Q) distribution parameterized as a two part power law.

Using this tool we modeled the experiments performed by Nesvizhevsky et al. [4], with a monoenergetic beam of neutrons impinging on a plate of variable thickness (Fig. 5). The reflectivity was computed as the ratio between the backscattered neutron current and the incident neutron current.

The results (Fig. 6) show a good agreement with experimental results, showing a good reflectivity for very cold neutrons that falls sharply for energies above 0.1 meV. Calculation of thicker ND slabs and removal of impurities allowed a modest increase of the energy at which the reflectivity decreases, but none of the calculations showed high reflectivities above 0.5 meV, in contrast of the calculations by Nesvizhevsky [2, 4].

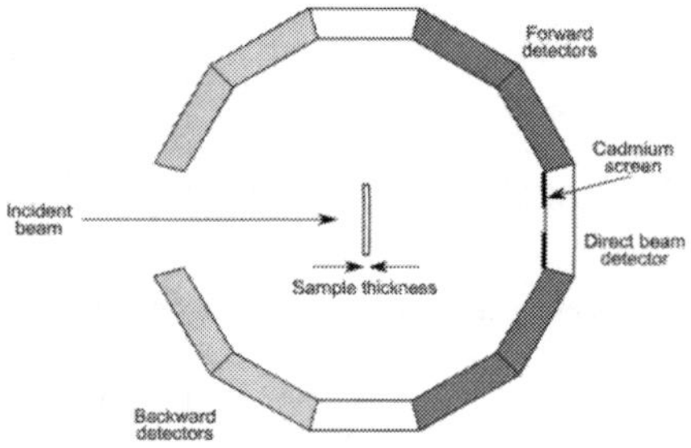

Fig. 5: Scheme of the modeled experiment. The sample thickness is the main parameter to analyse.

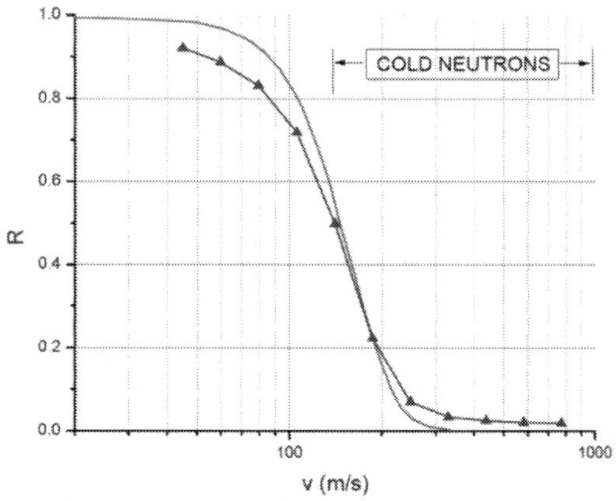

Fig. 6: Reflectivity vs. neutron velocity, for a 10 mm thick ND sample. The continuous line (red) is the result from Ref. [4], while the curve with symbols (blue) corresponds to our calculation (density = 0.6 g/cm³).

3 Magnesium hydride

As shown in the previous section, the reflectivity of ND reduces rapidly towards the cold and thermal neutron regions, as compared to that one presented for ultra- and very-cold neutrons. This is due to the fact that when the characteristic size of the scattering object is much larger than the radiation wavelength, in this case the neutron wavelength, the coherent scattering from individual objects occurs at small angles, and the probability for backward reflection becomes very small, although specular reflection processes can still be very intense.

At neutron energies where coherent elastic interaction with the scattering system cannot produce an intensity much larger than the bound-atom coherent cross section value (see Fig.4 for diamond), other kind of interactions should be explored as mechanisms to produce a significant reflection capacity. To drive our search for a material with such property, we must look for: a) large scattering cross section, b) large elastic cross section, c) highly isotropic angular cross section, d) small inelastic cross section, and e) small absorption cross section. These requirements can be realized by the following options: a) hydrogen, b) within a solid, c) with large effective mass, d) at low temperatures, e) with an absorption cross section due only to hydrogen. Several years ago we proposed that a material with these properties is Magnesium Hydride (MgH_2) at a low temperature [22], because Mg has the smallest absorption cross section amongst the common metals forming hydrides [19].

3.1 Cross sections of MgH_2

The structure and dynamics of the material are the necessary pieces of information to generate a scattering kernel and a cross section library. In Fig.7 we show our

https://doi.org/10.1051/epjconf/202023104002

adopted frequency spectrum for MgH_2 (line), together with experimental results obtained by Santisteban et al. [23] and Kolesnikov et al. [24]. We then calculated the components due to the metallic lattice and the hydrogen, employing again NJOY [17] and CRIPO [25].

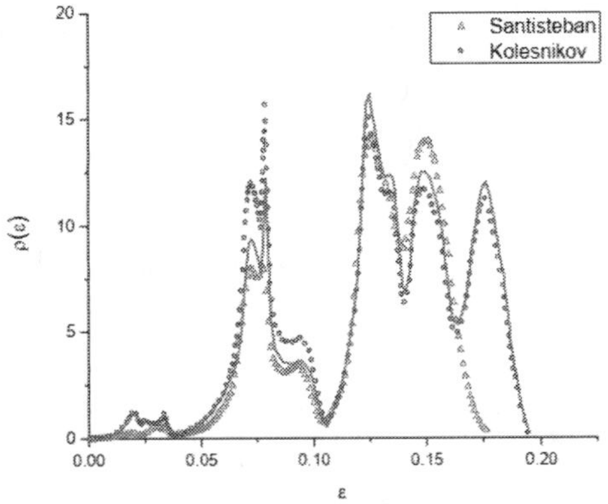

Fig. 7: Adopted density of states for MgH_2 (solid line), compared with the experimental results (symbols) of Refs.[23] and [24].

As a first validation test we compared the calculated total cross section with experimental data at room temperature [26], obtaining an excellent agreement as shown in Fig. 8. The elastic and inelastic (up and down) scattering components of the total cross section of H in MgH_2 at 20K is displayed in Fig.9, to emphasize the large area where elastic processes dominate the interaction with cold neutrons. This characteristic behaviour is due to the absence of low-energy excitation modes as displayed in Fig. 7, that causes the inelastic down-scattering processes to be reduced rapidly below thermal neutron energies. In addition, at low temperatures the small-weight acoustic modes are unpopulated, and so the up-scattering processes are very much reduced.

A simple figure of merit (FoM) to gauge the potentiality of a given material as a good neutron reflector is the product of its scattering capacity by the probability of a large scattering angle in the process. Specifically, we defined

$$\text{FoM} = \sigma_{el} * (1 - \mu) ,$$

where σ_{el} is the total elastic cross section and μ is the average cosine of the scattering angle. This magnitude, which is essentially the transport cross section, is displayed in Fig.10 for both ND and MgH_2 at 20K. It is apparent that this FoM is larger for the hydride at neutron energies above 0.0001 eV, in a large extent originated by $\mu \sim 1$ for ND corresponding to the onset of quasi-specular reflection over that energy region.

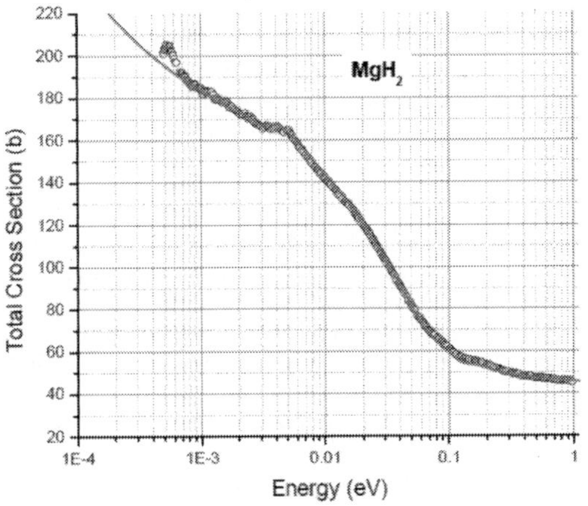

Fig. 8: Calculated total cross section of room temperature MgH_2 (solid line) and the experimental points of Ref.[26].

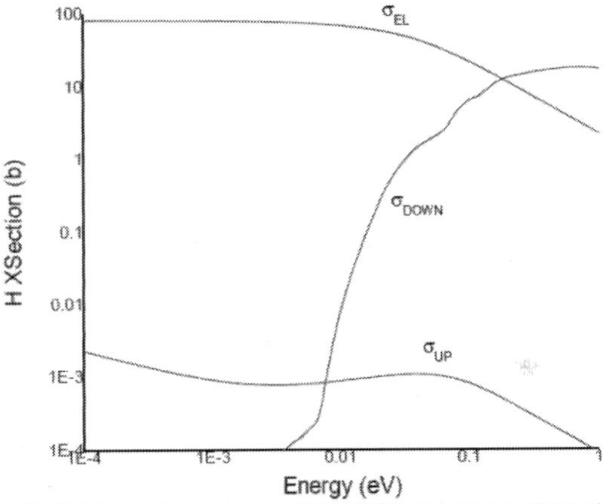

Fig. 9: The scattering components of the total cross section of H in MgH_2 at 20K.

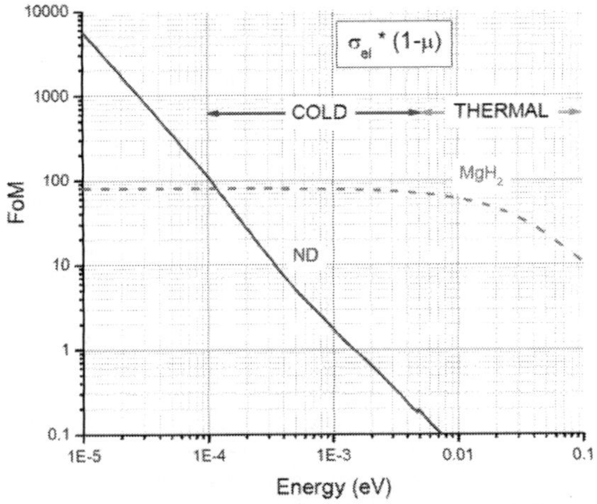

Fig. 10: Comparison of the indicated figure of merit for nanodiamond and MgH
neutron energy range.

3.2 MgH_2 reflectivity

OpenMC was used to simulate the configuration used by Nesvizhevsky [4] (Fig. 5) to compare the reflectivity of nanodiamonds and magnesium hydride. The scattering kernel calculated with NJOY and formatted into a standard ACE library was used to model the material.

The results (Fig. 11) show a reflectivity profile which remains almost constant at all energies, with a value of ~50% for 10 mm thickness with an apparent density of 1.0 g/cm^3. For very cold neutron energies the absorption of hydrogen reduces the reflectivity, but the large and almost constant elastic scattering cross section of hydrogen (Fig. 9) maintains a high reflectivity from very cold to thermal energies.

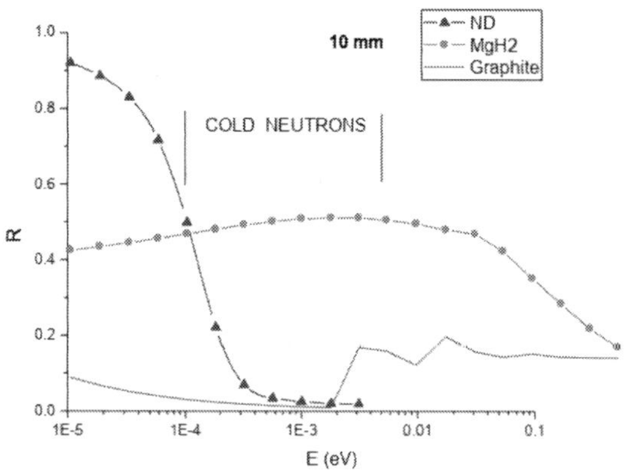

Fig.11: Reflectivity profile of MgH_2. Comparative with known moderators/reflectors shows constant performance for a wide energy range.

4 Reflectors in simple cold moderators

Using OpenMC we developed a simple but realistic model of a para-hydrogen cold neutron source to compare the efficiency of ND and cold MgH_2 as reflectors. The system consists of a 10x10x10 volume of pure liquid para-hydrogen at 19 K, with a 2-cm thick slab of room temperature water acting as premoderator. The source is embedded in a beryllium block acting as a thermal neutron reflector on four of the remaining sides, and one of the lateral sides is reserved for the output of the cold neutron source. In the four sides of the cold neutron source a 5 mm gap allows the introduction of a cold neutron reflector. The geometry of the model is shown in Fig. 12.

Using this system we computed the outgoing current for a monoenergetic 1 MeV, 20x20 cm distributed source impinging on the top for four configurations:

- No premoderator, no cold neutron reflector (5 mm void between the cold neutron source and the beryllium block).
- 2 cm premoderator, no cold neutron reflector.
- 2 cm premoderator, 5 mm nanodiamond cold neutron reflector.
- 2 cm premoderator, 5 mm 20 K magnesium hydride cold neutron reflector.

As expected, the use of a premoderator produces a significant improvement in the output of the cold neutron source (Fig. 13a), but little additional gain is observed when the nanodiamond reflector is introduced except for the very cold neutron energy range. When the nanodiamond reflector is replaced by a layer of cold magnesium hydride, a significant increase in the cold neutron output is observed. Computing the ratio of the spectra to the unreflected case (Fig. 13b) this gain is quantified to 20-25% over the cold and thermal energy ranges.

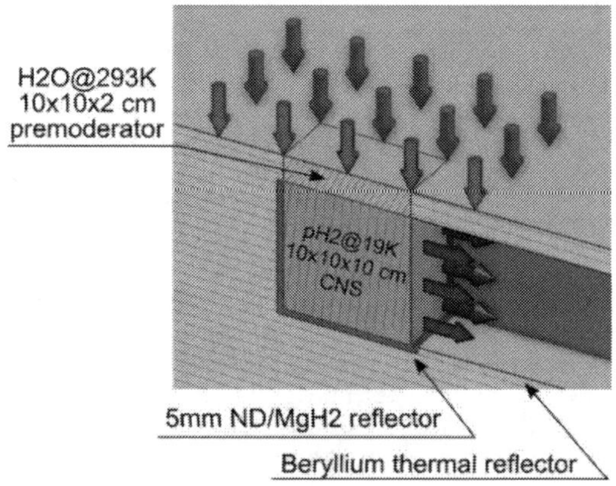

Fig.12: Geometry used to test the efficiency of nanodiamods and cold magnesium hydride as reflectors.

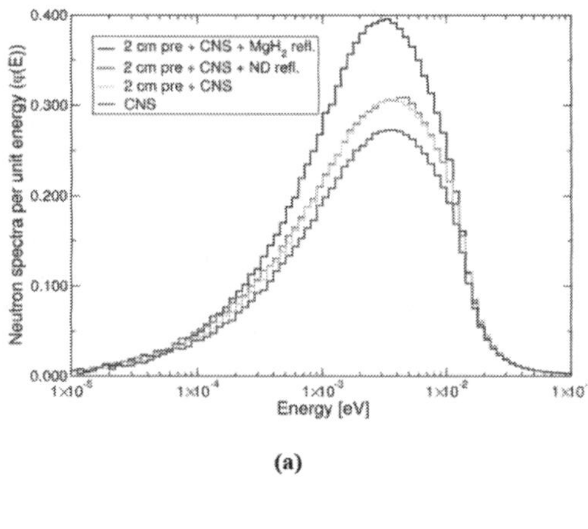

(a)

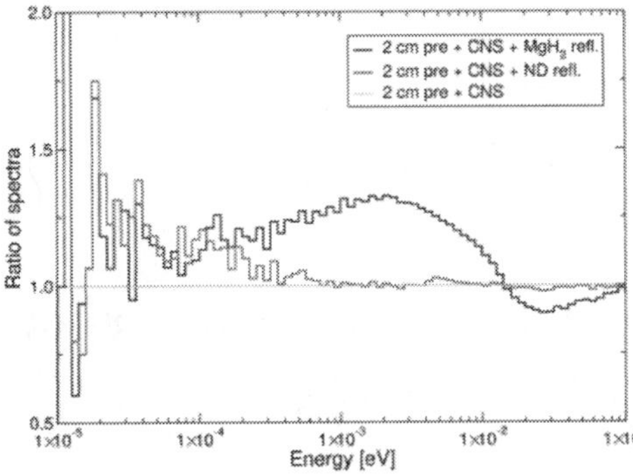

(b)

Fig.13: Spectra of the outgoing neutron current from the cold neutron source model (a). Ratio of the spectra, using as reference the case with premoderator and without cold neutron reflector (b).

5 Conclusions

We have developed a new scattering kernel to describe the interaction of slow neutrons with diamond, with special reference to nanopowders, and generated a cross section library that includes both the bulk contribution as well as the elastic coherent one due to the nanostructure. Reflectivity calculations using our library confirmed the excellent performance of that material for ultra-cold and very-cold neutrons.

On the other hand, to deeply explore a proposal we made few years ago concerning a potential good cold neutron reflector, we produced a new scattering kernel for MgH_2 which was validated against available transmission experiments. A cross section library for this material at low temperatures was generated, and used to perform reflectivity calculations that confirm its expected good performance as reflector of cold and thermal neutrons. An experiment to determine the intensity at backscattering angles for neutrons over that energy range is being prepared at this time.

We expect that this new reflector system could help to improve the efficiency of advanced cold neutron sources, at pulsed or stationary facilities.

References

[1] E.V. Lychagin *et al.*, Nuclear Instruments and Methods in Physics Research **A 611** (2009) 302
[2] V. V. Nesvizhevsky *et al.*, Carbon **130** (2018) 799e805
[3] V. V. Nesvizhevsky, Phys. Atomic Nuclei **65** (2002) 400
[4] V. V. Nesvizhevsky *et al.*, Nuclear Instruments and Methods in Physics Research **A 595** (2008) 631
[5] R. Cubitt *et al.*, Nuclear Instruments and Methods in Physics Research **A 622** (2010) 182
[6] V. V. Nesvizhevsky *et al.*, Phys.Rev. **A97,** 023629 (2018)
[7] V. A. Artemèv *et al.*, Crystallography Reports **61** (2016) 84
[8] M. V. Avdeev *et al.*, J.Phys.Cond.Matter **25** (2013) 445001
[9] IAEA CRP
[10] M. Teshigawara *et al.*, Nuclear Instruments and Methods in Physics Research **A 929** (2019) 113
[11] P. Giannozzi *et al.*, J.Phys.:Condens.Matter **21**, 395502 (2009)
[12] P. Giannozzi *et al.*, J.Phys.:Condens.Matter **29**, 465901 (2017)
[13] P. Pavone *et al.*, Phys.Rev. **B48**, 3156 (1993)
[14] A. Bosak and M. Krisch, Phys.Rev. **B72**, 224305 (2005)
[15] A. C. Victor, J.Chem.Phys. **36**, 1903 (1962)
[16] W. DeSorbo, J.Chem.Phys. **21**, 876 (1953)
[17] Macfarlane, R. et al LANL Report LA-UR-17-20093 (2017)
[18] Cai, X.X. and Kittelmann, T. NCrystal: A library for thermal neutron transport. Comp. Phys. Comm. (in press).
[19] F. Fernández Alonso and D. Price (Eds.) *Neutron Scattering, Vol 44,* Appendix (2013)
[20] G. Beaucage, J.Appl.Crystallogr. **29**, 134 (1996)
[21] Paul K. Romano, Nicholas E. Horelik, Bryan R. Herman, Adam G. Nelson, Benoit Forget, and Kord Smith, "OpenMC: A State-of-the-Art Monte Carlo Code for Research and Development," Ann. Nucl. Energy, **82**, 90–97 (2015).
[22] J.R. Granada *et al.*, ICANS XX, March 4-9, 2012 (Bariloche, Argentina)
[23] J.R. Santisteban *et al.* Phys.Rev. **B62**, 37 (2000)
[24] A.I. Kolesnikov *et al.*, J.Alloys Comp. **509S** (2011) S599
[25] F. Kropff and J.R. Granada, "CRIPO: a computer code to evaluate the cross section of Polycrystalline materials", Unpublished (1974).
[26] G. Muhrer *et al.*, Nuclear Instruments and Methods in Physics Research **A 629** (2011) 251

Directional reflection of cold neutrons using nanodiamond particles for compact neutron sources

Mostafa Jamalipour*,[1], *Luca* Zanini[2], and *Giuseppe* Gorini[1]

[1]Dep. of Physics, University of Milano-Bicocca, Milano, 20126, Italy
[2]European Spallation Source ESS ERIC, P.O Box 176, SE-221 00 Lund, Sweden

Abstract. Nanodiamond Particles (NDP) are new candidates for neutron reflection. They have a large scattering and low absorption cross-sections for low-energy neutrons. Very Cold Neutrons (VCN) are reflected from NDP with large scattering angles while Cold Neutrons (CN) have a quasi-specular reflection at small incident angles. A new scattering process has been added in Geant4 in order to examine the directional reflection of CN in an extraction beam made of NDP layer. Impurities in NDP are responsible for the up-scattered neutrons, especially hydrogen which has a large cross-section. Other impurities are also considered in Geant4 in order to produce a more accurate model for NDP scattering. The new scattering process is used to model a possible configuration of target-moderator-reflector in Compact Neutron Sources (CNS). A 13 MeV proton beam striking a beryllium target is chosen. Parahydrogen is placed as a cold moderator in order to produce CN. NDP are placed around the extraction beam for scattering the CN toward the exit of the beam. The results show that CN exiting the extraction beam can be increased thanks to the implemented NDP layer.

1 Introduction

Nanodiamonds were firstly produced by explosive shock more than 50 years ago [1]. These particles have the size of a few nanometers. Their diamond nucleus is surrounded by an onion-like shell consisting of impurities. Recently, it has been shown that nanodiamonds are attractive materials that can be used as reflectors against low-energy neutrons [2, 3]. They have a large total cross-section (two order of magnitude more than carbon for neutrons as the energy decreases to 0.2 meV) [4]. Very Cold Neutrons (VCN) can be trapped by nanodiamonds while Cold Neutrons (CN) may have quasi-specular reflection at small angles [2, 3].

Nanodiamonds Scattering Process (NSP) has been added as a discrete process in Geant4 Monte Carlo simulation toolkit. Nanodiamonds quasi-specular reflection was experimentally measured at the Institut Laue-Langevin (ILL) in 2009 [2]. The NSP implemented in Geant4 is validated with this experiment in order to see its feasibility. Moreover, nanodiamond total cross-section was measured by transmission at J-PARC [4]. This experiment is also validated in order to benchmark the code.

In this work we used Geant4 with the implemented nanodiamonds physics to investigate the applicability of nanodiamond reflectors for compact neutron sources. A beam of proton with the energy of 13 MeV strikes the beryllium target producing neutrons. Liquid parahydrogen is used as a moderator for CN production, while beryllium is used as reflector. Nanodiamonds are simulated as they are placed around the extraction beam in order to increase the chance of neutron reflection toward the CNS exit.

2 Compact Neutron Source (CNS)

A beam of proton having the energy of 13 MeV strikes the beryllium target with the dimensions of 1.5×0.75×0.2 cm^3. The neutrons produced from (p,n) reaction enter the parahydrogen cylinder with the temperature of 20 K having the radius of 3 cm and length of 14 cm. The beam extraction channel is about 1 m long starting from the center of cold moderator. Beryllium is also used as a

* Corresponding author: M.jamalipour@campus.unimib.it

https://doi.org/10.1051/epjconf/202023104003

reflector while steel is modeled as a shield. **Fig. 1** shows the proposed CNS for simulation study. The length of cold moderator has been optimized by testing different dimensions in order to have the highest possible cold flux after the moderation. However, the dimensions of the reflector, and overall geometry setup, was not optimized. Additionally, the target should be considered as ideal, as in a realistic configuration the target would probably have to have a larger area. The chosen geometry is apt to study the effect of nanodiamond reflector on the outgoing flux. **Fig. 2** shows the variation of CN production as a function of the moderator length for the wavelengths longer than 2.6 Å. **Table 1** illustrates the material and the dimensions of proposed CNS.

3 Nanodiamond Scattering Process (NSP) in Geant4

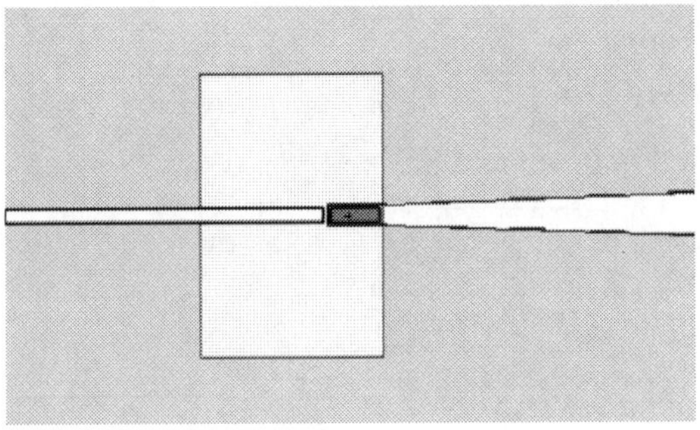

Fig. 1. Proposed CNS for simulation study.

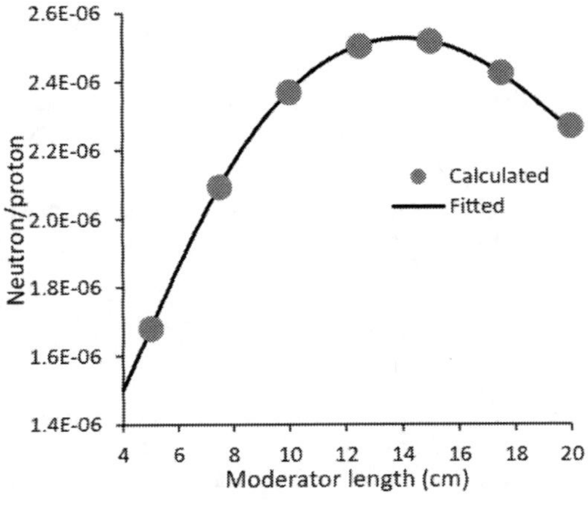

Fig. 2. CN production versus moderator length in CNS.

Table 1. CNS structure specifications.

Component	Material	Dimension
Target	beryllium	1.5×0.75×0.2 cm³
Moderator	liquid parahydrogen	r = 3cm h = 14 cm
Reflector	beryllium	66×66×66 cm³
Shield	steel	50×50×102 cm³
Clad	aluminum	-

A discrete process has been added in Geant4, a Monte Carlo simulation toolkit in order to implement the scattering of neutron in nano-dispersed media [5]. **Fig. 3** shows this implementation procedure. The particle size is selected, from which the cross-sections are calculated. Neutron position and momentum are updated based on the new scattering angle and cross-sections. This procedure is used as long as neutron is in the nano-dispersed media. Shell impurities are also added as a normal material in order to consider the up-scattered neutrons. Carbon, oxygen, nitrogen, and hydrogen are the most common impurities available in the nanodiamond shell. The scattering cross-section of neutron in nanodiamonds was calculated in [6, 7] using the first Born approximation. Considering the center-of-mass system (c.m.s), the scattering amplitude can be defined as:

$$f(\theta) = -\frac{2m}{\hbar^2} V_0 R^3 \left(\frac{\sin(qR)}{(qR)^3} - \frac{\cos(qR)}{(qR)^2}\right) \quad (1)$$

where V_0 is the real part of NDP Fermi potential, R is the particle size and q is the momentum transfer. The total elastic scattering cross-section is provided by the following formula:

$$\sigma_s = \int |f|^2 d\Omega = 2\pi \left|\frac{2m}{\hbar^2} V\right|^2 R^6 \frac{1}{(kR)^2} I(kR) \quad (2)$$

where,

$$I(kR) = \frac{1}{4}\left(1 - \frac{1}{(2kR)^2} + \frac{\sin(4kR)}{(2kR)^3} - \frac{\sin^2(2kR)}{(2kR)^4}\right) \quad (3)$$

where k is the incident neutron wave vector. Finally, with the help of optical theorem, the total absorption cross-section in the nanodiamond particle can be calculated as:

$$\sigma_a = \frac{4\pi}{3} \frac{2m}{\hbar^2} V_1 R^4 \frac{1}{kR} \quad (4)$$

Where V_1 is the imaginary part of the NDP Fermi potential.

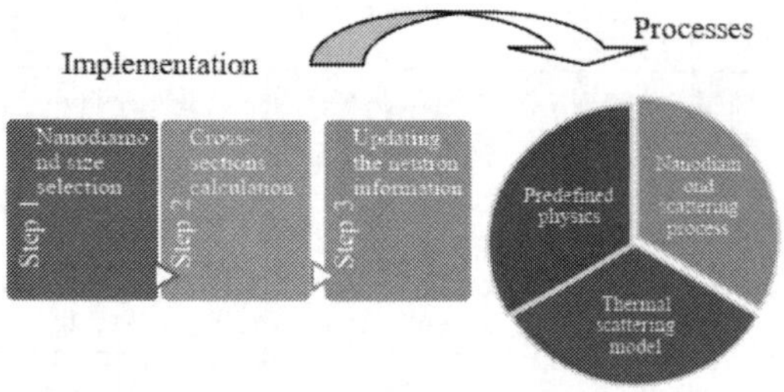

Fig. 3. Nanodiamond scattering process in Geant4.

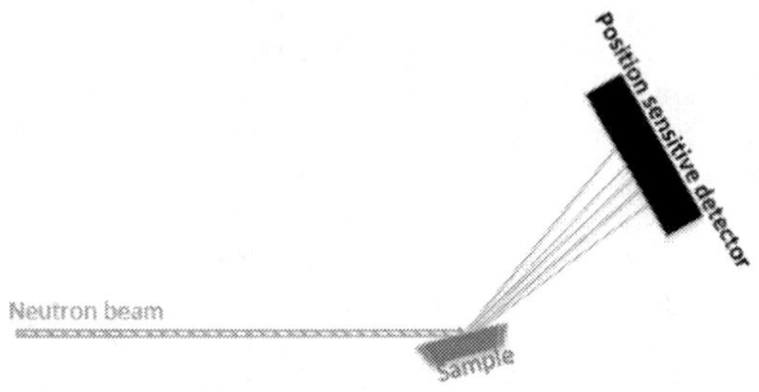

Fig. 4. Scheme of CN quasi-specular reflection experimental setup.

4 Quasi-specular reflection of Cold Neutrons (CN) in nanodiamonds

The quasi-specular albedo of CN in NDP at small incident angles was tested for the first time at Institut Laue-Langevin (ILL) in 2009. This quasi-specular reflection is due to the multiple-scattering of CN in nano-dispersed media. CN with short wavelengths compared with the particles' size have small scattering angles in NDP. Therefore, they are going to penetrate deep into NDP and do not come back to the surface until they are absorbed if they have large incident angles. In the ILL experiment, a beam of CN with the wavelength in the range of 2-14 Å stroke a sample of NDP having the size of 2 nm with small incident angles. The reflected neutrons were detected by position sensitive detectors placed 110 cm far from the center of the sample with the height of 5 cm. Different small incident angles were tested to observe the quasi-specular reflection of CN from NDP. **Fig. 4** shows the scheme of the first experiment for benchmark. For more information related to this experiment refer to Ref. [2].

The newly implemented process in Geant4 was used in order to benchmark the CN quasi-specular reflection from NDP. In this simulation, CN were irradiated on the surface of the sample containing NDP with the angles of 2° and 3°. **Fig. 5** and **6** show the simulation results of neutron Probability (Pr) detection in position sensitive detectors. Due to the incomplete modelling of interplanar distance in the nanodiamond lattice, the simulation results are different from the experiment for the neutrons with wavelength in the range of 2-4 Å. It has already been shown that this distance is equal to 3.57 Å [8]. However, the residuals drop dramatically for neutrons having wavelength larger than 4 Å indicating a good agreement between data and simulations.

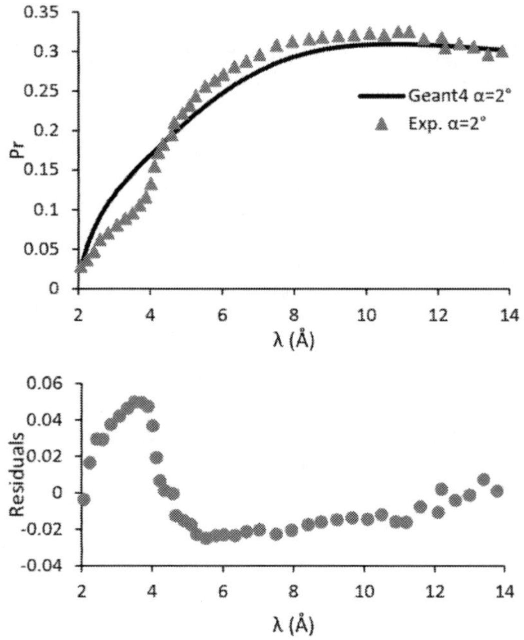

Fig. 5. Quasi-specular reflection of CN with incident angle of 2°.

5 Nanodiamonds total cross-section measurement

Total cross-section of NDP produced using detonation explosives was calculated using the transmission measurement in the energy range of 0.1 – 100 meV at J-PARC [4]. Moreover, it was shown that the inelastic scattering contribution compared to the elastic scattering cross-section is negligible. Having removed the NDP

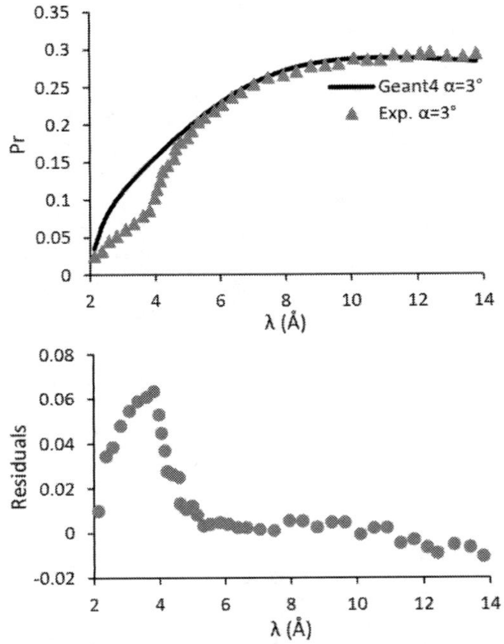

Fig. 6. Quasi-specular reflection of CN with incident angle of 3°.

moisture by means of vacuum heat treatment, a plate of NDP was prepared using a pressing jig in a glovebox filled with helium gas at atmospheric pressure. The average size of nanoparticles was 5.1 nm. The plate was placed in experimental setup, after its homogeneities were examined.

Neutron transmission *R(E)* was obtained as the ratio of neutron detection with or without the sample as:

$$R(E) = \frac{\varphi_{WND}\,(E)}{\varphi_{WOND}\,(E)} \qquad (5)$$

Where the $\varphi_{WND}(E)$ and $\varphi_{WOND}(E)$ are the neutrons measured by detector with NDP and without NDP plate.

The total cross-section was determined using the attenuation equation as follows.

$$\sigma(E)_{total} = -\frac{1}{N_{ND}\,d_{ND}} \ln\left(\frac{\varphi_{WND}\,(E)}{\varphi_{WOND}\,(E)}\right) \qquad (6)$$

Where N_{ND} and d_{ND} are the number density and thickness of the sample respectively.

Fig. 7 and **8** show the neutron transmission and total cross-section respectively. The simulation shows that there is a good agreement between the prepared model in Geant4 and the experimental data.

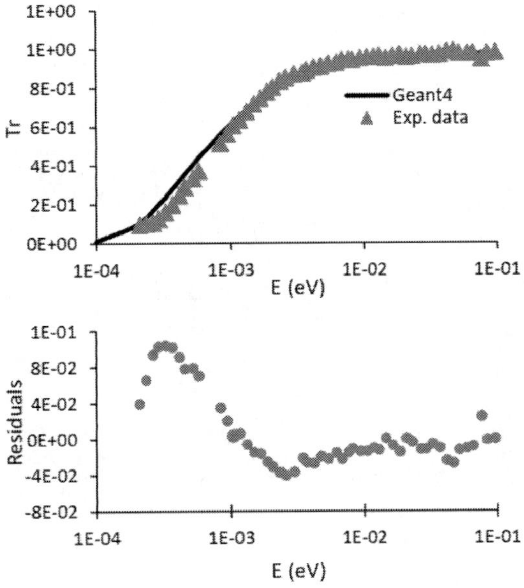

Fig. 7. Simulation of neutron transmission of nanodiamonds in Geant4.

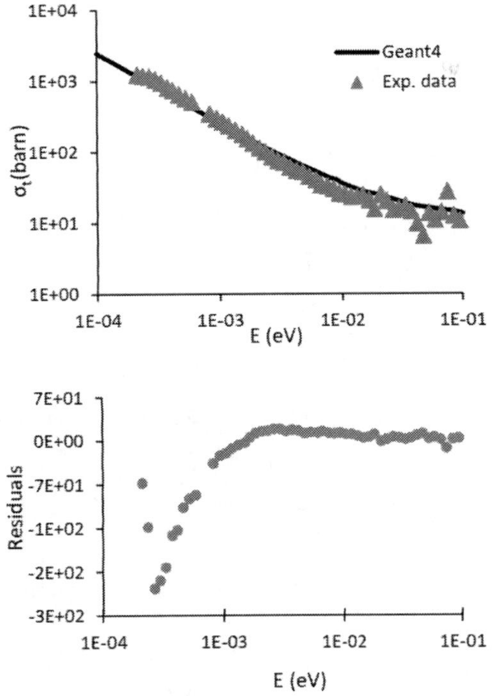

Fig. 8. Simulation of nanodiamonds total cross-section in Geant4.

6 Nanodiamonds application in CNS

The proposed CNS was modeled in Geant4 Monte Carlo simulation toolkit as shown in **Fig. 9**. The neutron spectrum exiting from the extraction beam is shown in **Fig. 10**. A MCNP6 model of the same geometry was built and the Geant4 simulations were compared with MCNP6.

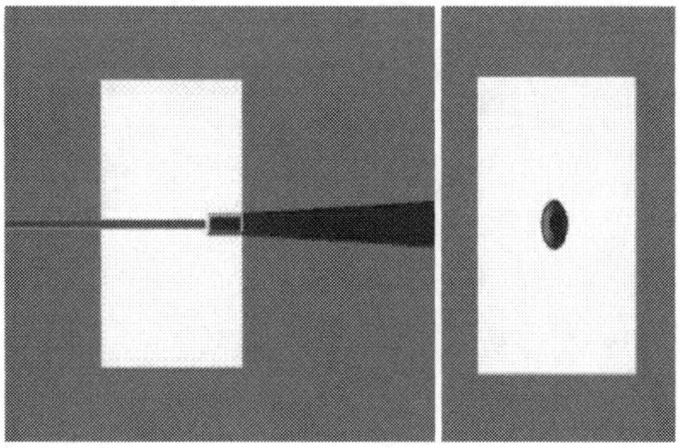

Fig. 9. CNS model prepared in Geant4.

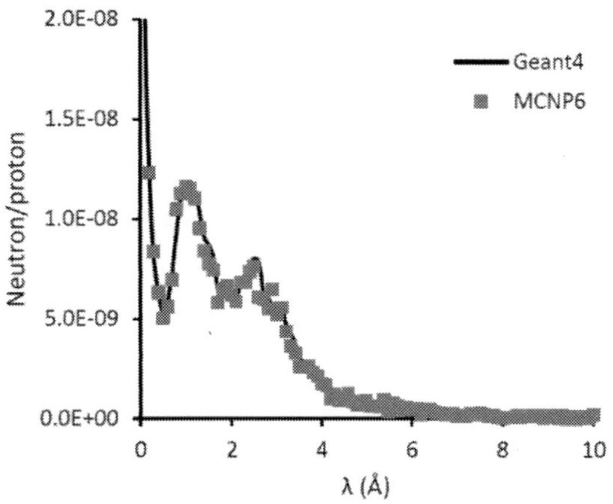

Fig. 10. CNS neutron spectrum at exit.

In order to see the effect of using NDP for directional neutron reflection, a NDP layer was placed instead of aluminum layer in the beam extraction cone. The thickness of this layer was 3 mm. Two sizes of NDP were tested in the simulation. The density of NDP in the simulation was considered 0.6 g/cm ^{3}. **Fig. 10** shows the neutron spectrum at the exit of CNS. Since the Bragg scattering has not been implemented yet, the simulation is not reliable for the neutrons in the range of 2-4 Å. Therefore, in **Fig. 11**, the difference of using diverse materials, namely, Aluminum (Al), Carbon (C), NDP with the size of 5 nm and 10 nm is shown only above 4 Å. The integral number of neutrons in different wavelength ranges is shown in **Fig. 12,** while the ratio of flux increase using NDP with respect to carbon is shown in **Fig. 13**. NDP help increase the flux more than 5 times for the neutrons in the range of 4-10 Å while this value reaches ten times more for neutrons having the wavelength longer than 10 Å. **Fig 14** shows the angular distribution of neutrons exiting the beam. We observe that the CN flux at low divergence, i.e., of interest in neutron scattering, is increased by about the factor of two. For larger divergence, the flux increase is much larger. Such an increase could be used in applications where the neutron divergence is not of interest (e.g. irradiation experiments), or, for instance, to transport CN to a second moderator placed at a distance from the target, such as a UCN convertor.

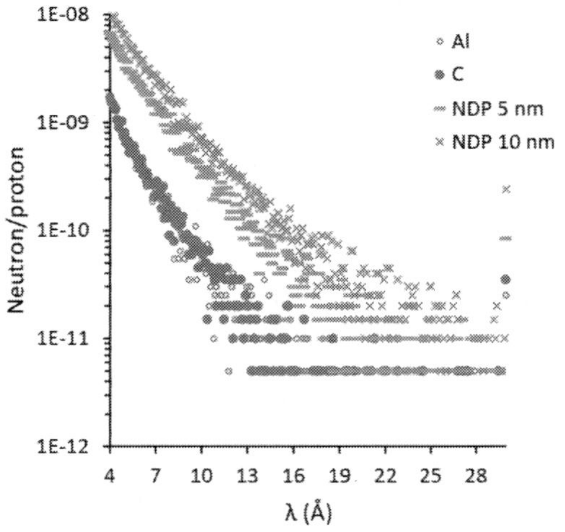

Fig. 11. CNS neutron spectrum with the wavelength more than 4 Å at exit.

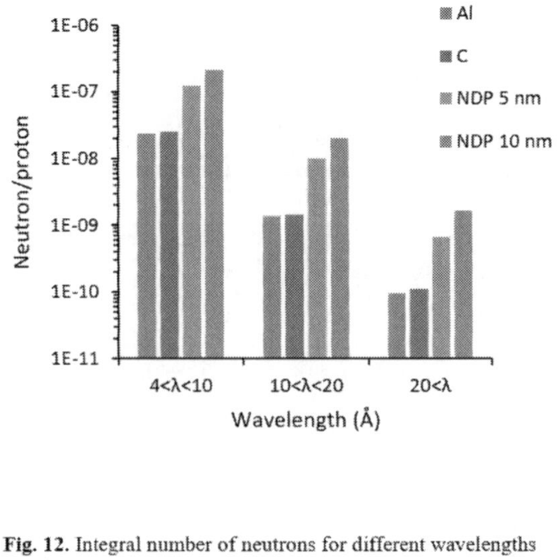

Fig. 12. Integral number of neutrons for different wavelengths at exit.

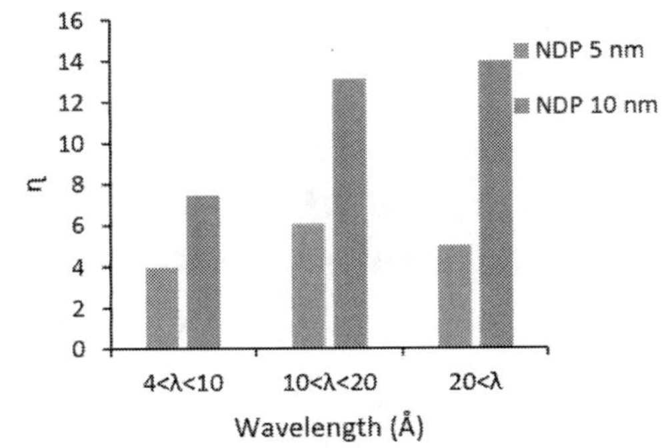

Fig. 13. Ratio of neutron increase using NDP in CNS.

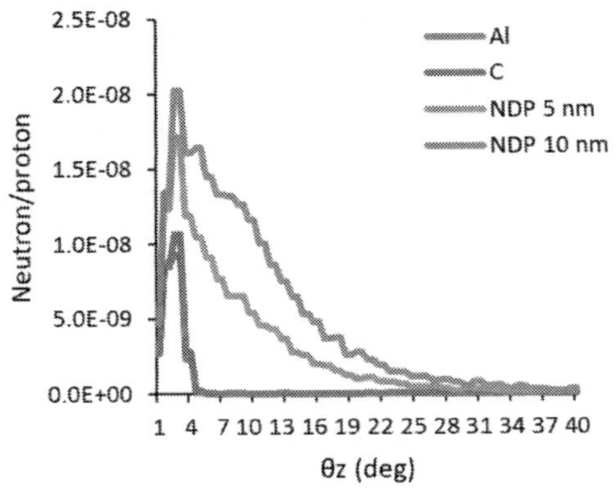

Fig. 14. CN angular distribution in CNS (4 Å < λ <10 Å).

7 Conclusions

Nanodiamond particles (NDP) were used in a case study of Compact Neutron Source (CNS) in order to increase the number of Cold Neutrons (CN) at exit. CN have a quasi-specular reflection from NDP if the incident angles are small. The scattering process of CN from NDP was implemented in Geant4 Monte Carlo simulation code as a discrete process. The new implemented model was benchmarked with two experiments previously performed in order to observe the quasi-specular reflection of CN and to measure the total cross-section of NDP by transmission. NDP were placed as a layer in the beam extraction cone of the CNS to increase the CN directional reflection toward the exit. The simulation results showed that the number of CN were increased significantly when the NDP layer was used. However, the model lacked the Bragg scattering process for the neutrons in the range of 2-4 Å. Moreover, it was observed that the NDP with a larger size are more efficient in terms of CN increase at exit. The implementation of NDP in extraction beam helps optimizing the CNS while providing the possibility of carrying out experiments where the neutron divergence is not of primary importance.

References

1. P. S. DeCarli and J. C. Jamieson, Science **133** (3467), 1821-1822 (1961).
2. R. Cubitt, E. Lychagin, A. Y. Muzychka, G. Nekhaev, V. Nesvizhevsky, G. Pignol, K. Protasov and A. Strelkov, Nuclear Instruments and Methods in Physics Research Section A: Accelerators, Spectrometers, Detectors and Associated Equipment **622** (1), 182-185 (2010).
3. V. Nesvizhevsky, E. Lychagin, A. Y. Muzychka, A. Strelkov, G. Pignol and K. Protasov, Nuclear Instruments and Methods in Physics Research Section A: Accelerators, Spectrometers, Detectors and Associated Equipment **595** (3), 631-636 (2008).
4. M. Teshigawara, Y. Tsuchikawa, G. Ichikawa, S. Takata, K. Mishima, M. Harada, M. Ooi, Y. Kawamura, T. Kai and S. Ohira-Kawamura, Nuclear Instruments and Methods in Physics Research Section A: Accelerators, Spectrometers, Detectors and Associated Equipment **929**, 113-120 (2019).
5. M. Jamalipour, L. Zanini and G. Gorini, Implementation of neutron reflection with nano-dispersed media in Geant4. (European Conference on Neutron Scattering, ECNS, Saint Petersburg, Russia, 2019).
6. V. Nesvizhevsky, G. Pignol and K. Protasov, International Journal of Nanoscience **6** (06), 485-499 (2007).
7. V. Nesvizhevsky, G. Pignol and K. Protasov, International Journal of Nanoscience **7** (02n03), 179-179 (2008).
8. V. Nesvizhevsky, M. Dubois, P. Gutfreund, E. Lychagin, A. Y. Nezvanov and K. Zhernenkov, Physical Review A **97** (2), 023629 (2018).

https://doi.org/10.1051/epjconf/202023104004

Slab geometry type cold neutron moderator development based on neutronic study for Riken Accelerator-driven compact Neutron Source (RANS)

Baolong Ma[1,2,*], *Yujiro* Ikeda[2,3], *Yoshie* Otake[2,1], *Makoto* Teshigawara[3], *Yasuo* Wakabayashi[2], *Masahide* Harada[3], *Motoki* Ooi[3], *Takao* Hashiguchi[2], *Yutaka* Yamagata[2], and *Shin* Takeda[2]

[1]School of Energy and Power Engineering, Xi'an Jiaotong University, Xi'an, China
[2]RIKEN center for advanced photonics, RIKEN, Wako, Saitama, Japan
[3]J-PARC Centre, Japan Atomic Energy Agency, Naka-gun, Ibaraki, Japan

Abstract. Cold neutrons with energy less than several meV are good probes for material research, and they have been available on large neutron facilities, whereas it is not commonly available on compact accelerator-driven neutron source. RIKEN Accelerator-driven Neutron Source (RANS) is a pulsed neutron facility which provides thermal neutrons and high energy neutrons at several MeV. We started a project to implement a cold neutron moderator for RANS to broaden cold neutrons applications. A cold neutron moderator system with a mesitylene moderator at 20K and a polyethylene pre-moderator at room temperature in the slab geometry was designed for RANS. So far, the thickness of the pre-moderator and mesitylene have been optimized to get the highest cold neutron flux by using a Monte Carlo simulation code, PHITS. Graphite reflector dimensions were also proven to have significant effect to increase the cold neutron intensity.

1 Introduction

Cold neutrons are important in neutron scattering investigation for material science because they have suitable energies for materials structure analysis. However, cold neutron sources are usually available at spallation neutron sources like J-PARC (Japan Proton Accelerator Research Complex) [1], ESS (European Spallation Source) [2], and SINQ (The Swiss Spallation Neutron Source at PSI) [3] and reactor neutron sources like ILL (Institut Laue–Langevin) [4], HFIR (High Flux Isotope Reactor) [5], and JRR-3 (Japan Research Reactor No.3) [6]. Recent advanced accelerator technology dramatically increases the availability of neutrons, especially with compact accelerator-driven neutron sources. RANS is an accelerator-based pulsed neutron facility using the $^{9}Be(p,n)^{9}B$ reaction [7] in a beryllium target with 7 MeV proton injection.

The slab geometry was taken for RANS cold neutron moderator to couple the neutron production target and the moderator. Polyethylene (PE) was adopted as a pre-moderator and mesitylene as a moderator. To maximize the cold neutron intensity as much as possible, pre-moderator thickness and mesitylene thickness were optimized by a Monte Carlo simulation, Particle and Heavy Ion Transport code System (PHITS) [8]. The graphite reflector effect on cold neutron intensity was investigated.

2. Overview of RANS

At RANS, pulsed neutrons are produced via the $^{9}Be(p,n)^{9}B$ reaction with 7 MeV proton injection. RANS has a beryllium target which is brazed to a vanadium backing material to let hydrogen diffuse and to mitigate hydrogen embrittlement. RANS target is cooled by water which flows through a titanium cavity. Currently, RANS uses 40 mm-thick polyethylene moderator for thermal neutrons. The core area of RANS target is surrounded by a graphite reflector to enhance the neutron beam intensity. RANS target station adopts a multi-layer shielding structure of BPE, Pb, BPE, and Pb with the thickness of 25 cm, 10 cm, 25 cm, and 5 cm, respectively. A schematic diagram of RANS target station is shown in Fig.1.

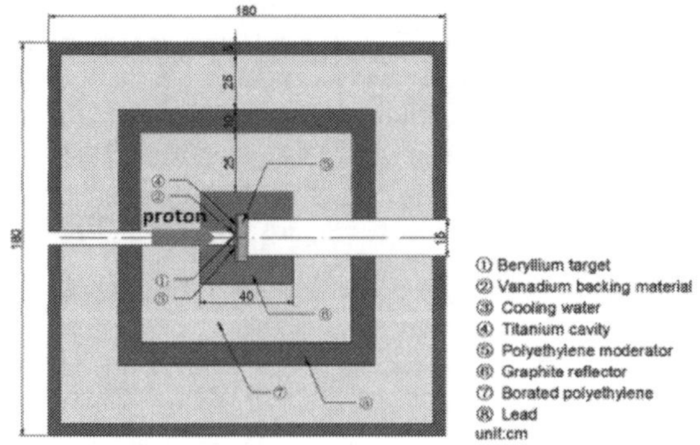

Fig. 1. RANS target station.

* Corresponding author: baolong.ma@riken.jp

With current RANS target assembly, the calculated neutron spectrum by PHITS at 3 m away from the target along the neutron extraction direction is shown in Fig.2. It shows that both thermal neutron flux and high energy neutron flux can be obtained.

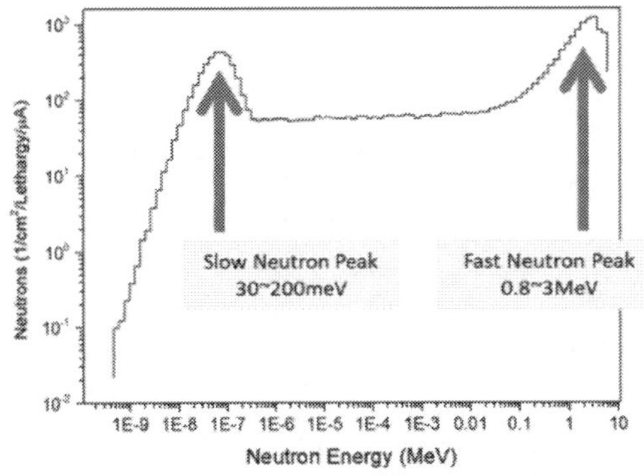

Fig. 2. RANS neutron spectrum at 3 m away from the target by PHITS simulation.

3. Simulation calculation

3.1 RANS cold moderator design

RANS cold moderator configuration uses a mesitylene cold moderator at 20K with a polyethylene pre-moderator at room temperature in a slab geometry. Mesitylene has good characteristics of non-explosive, easy to handle, good irradiation tolerance, and good neutronic performance [9]. In this study, we considered a realistic configuration of moderator with a 0.3 cm aluminium vessel wall and vacuum layer. The lateral size of the cold moderator is about 10 x 10 cm^2, and the size of the neutron beam extraction hole is 7.5 x 7.5 cm^2. Mesitylene is installed in an aluminium vessel which is surrounded by polyethylene pre-moderator (See Fig. 3(a)). Fig.3 (b) shows the RANS target assembly with the cold moderator using a slab geometry. The cold neutron performance at 2 m in the extraction direction from the target was characterized. So far, we have optimized the thickness of the pre-moderator and the mesitylene to get the highest cold neutron flux.

3.2 Thickness optimization

The PE pre-moderator thickness T_1 and the mesitylene moderator thickness T_2 were free parameters of the optimization. In the calculation for the cold neutron intensity, neutron scattering kernels of mesitylene (20K), polyethylene (296K), water (296K), and graphite (293.6K) were used. The scattering kernel data of polyethylene, water, and graphite are from JENDL-4.0 [10] and scattering kernel data of mesitylene was provided by F. Cantargi, et al. [11]. In the calculation, the detector was placed at 2 m away from the RANS target. [T-Track] tally was used to calculate the neutron intensity within an energy range of 0.1meV to 10meV. The dimension of the surrounding polyethylene was kept as shown in Fig.3 (a).

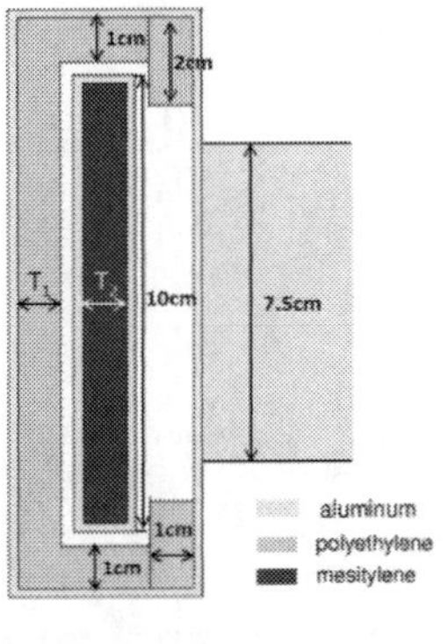

Fig.3 (a). RANS cold moderator with slab geometry.

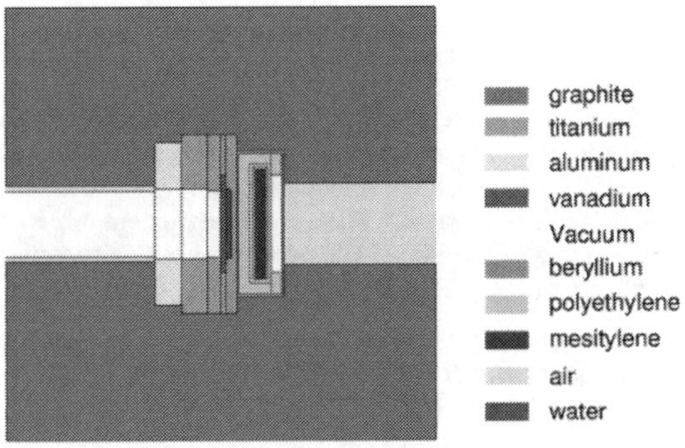

Fig.3 (b). RANS cold neutron moderator geometry.

Cold neutron intensity was calculated by varying pre-moderator thickness from 0 to 5 cm and moderator thickness from 1 cm to 6 cm. Optimization results are shown in Fig. 4.

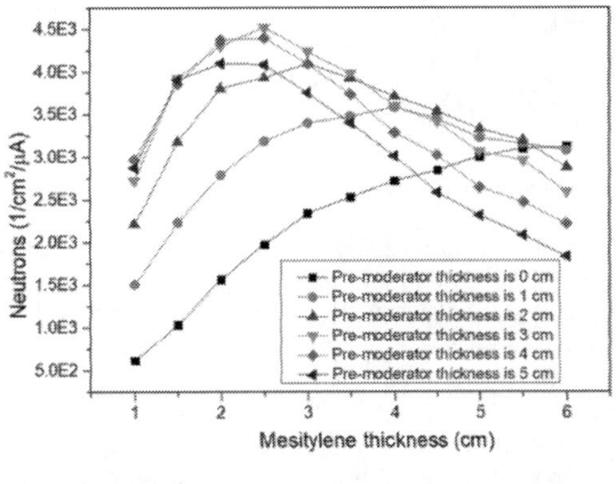

Fig. 4. Cold neutron intensity as a function of pre-moderator and moderator thickness

The pre-moderator is a key to maximize the cold neutron flux. When the pre-moderator thickness increase beyond 4 cm, the efficiency becomes smaller. For a fixed pre-moderator thickness, the cold neutron intensity increases up to a mesitylene thickness of 2.5 cm. The highest cold neutron intensity was found to be 4.5×10^3 neutron/cm^2/μA with a pre-moderator thickness of 3 cm and a moderator thickness of 2.5 cm.

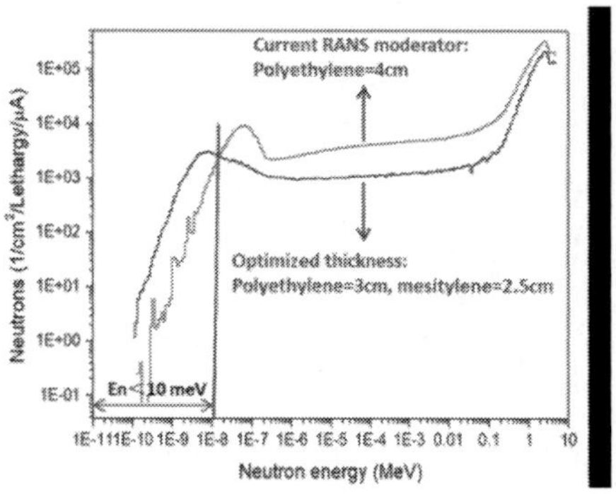

Fig. 5. Neutron spectrum of the RANS current moderator and for an optimized cold moderator

Compared with the RANS current moderator of 4 cm polyethylene and the optimized cold moderator with pre-moderator and mesitylene moderator thicknesses are 3 cm and 2.5 cm, respectively. The optimized moderator gives neutron intensity below 10 meV higher by a 5.6 compared to the current RANS moderator.

3.2 Evaluation of reflector dimension

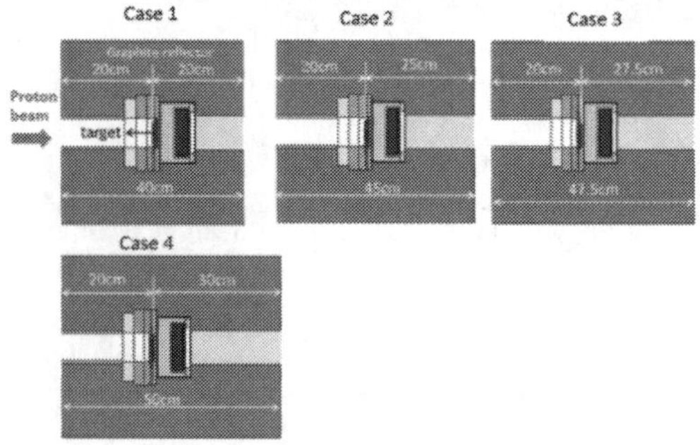

Fig. 6. Four cases of graphite reflector dimension

The cold neutron intensity was evaluated with respect to the reflector dimensions. Case 1 is the current RANS graphite dimension with 20 cm-thick in both forward and backward direction from the neutron production target. For case 2, case 3, and case 4, the graphite thickness in forward direction was increased by 5 cm, 7.5 cm, and 10 cm, respectively while the graphite thickness in backward direction is kept as 20 cm. All the calculation conditions were kept the same as previously in terms of kernel nuclear data, cold neutron energy, and tally. The statistical errors of the calculation are below 2%. The results are shown in Fig. 7.

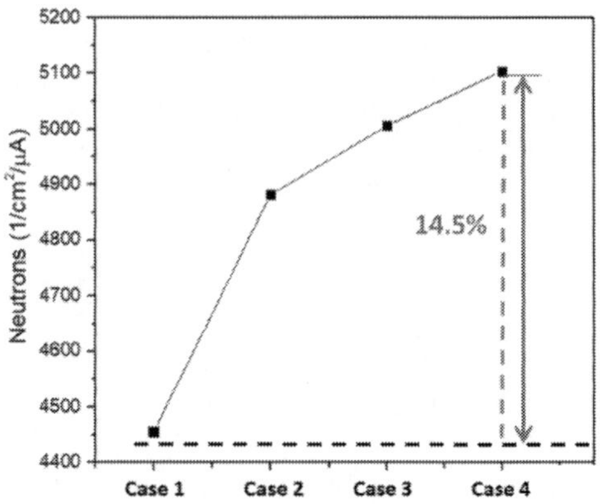

Fig. 7. Cold neutron intensity as a function of the graphite reflector dimension

It shows that by adding thickness in the forward direction, i.e., neutron extraction direction, the cold neutron intensity can be increased accordingly. By adding 10 cm of graphite at neutron emission side, cold neutron intensity could be increased by 14.5% .

4. Conclusions

Currently, we focused on the neutron intensity enhancement. The main conclusions are as follows:

1. By optimizing the thickness of polyethylene and mesitylene, at 2 m away from the target, the cold neutron intensity can be increased by a factor of 5.6 compared with the current moderator model;
2. Increasing the graphite thickness in the neutron emission direction has considerable effects to increase the cold neutron intensity. For RANS case, graphite is going to be remoulded to get higher cold neutron intensity;
3. High energy neutrons whose energy is higher than 1 MeV exist in the spectrum with respect to the slab geometry. So the neutron extraction at other angles will be considered for RANS to decrease the fast neutron background.
4. The estimations of neutron pulse and neutron brightness are under study.

Reference

[1] Takada, Hiroshi, et al. "Materials and life science experimental facility at the Japan proton accelerator research complex I: pulsed spallation neutron source." Quantum Beam Science 1.2 (2017): 8. < https://j-parc.jp/c/en/ >.

[2] Bauer, Günter S. "The European Spallation Neutron Source Project (ESS)." Utilisation and Reliability of High Power Proton Accelerators (1998): 59. <https://europeanspallationsource.se/>.

[3] Fischer, W. E. "SINQ—the spallation neutron source, a new research facility at PSI." Physica B: Condensed Matter 234 (1997): 1202-1208. <https://www.psi.ch/en/media/the-sinq-neutron-source>.

[4] Ageron, Paul. "Cold neutron sources at ILL." Nuclear Instruments and Methods in Physics Research Section A: Accelerators, Spectrometers, Detectors and Associated Equipment 284.1 (1989): 197-199. < https://www.ill.eu/>.
[5] Mason, T. E., et al. "The Spallation Neutron Source in Oak Ridge: A powerful tool for materials research." Physica B: Condensed Matter 385 (2006): 955-960. < https://neutrons.ornl.gov/hfir>.
[6] Kawabata, Yuji, et al. "Characteristics of neutron guide tubes in upgraded JRR-3." Journal of Nuclear Science and Technology 30.10 (1993): 1050-1058. < https://jrr3.jaea.go.jp/>.
[7] Yamagata, Yutaka, et al. "Development of a neutron generating target for compact neutron sources using low energy proton beams." Journal of Radioanalytical and Nuclear Chemistry 305.3 (2015): 787-794.
[8] Niita, Koji, et al. "PHITS—a particle and heavy ion transport code system." Radiation measurements 41.9-10 (2006): 1080-1090.
[9] Natkaniec, Ireneusz, Krystyna Holderna-Natkaniec, and Jurgen Kalus. "Neutron scattering studies of methyl derivatives of benzene selected as potential materials for cold neutron moderators." Physica B: Condensed Matter 350.1-3 (2004): E651-E653.
[10] Shibata, Keiichi, et al. "JENDL-4.0: a new library for nuclear science and engineering." Journal of Nuclear Science and Technology 48.1 (2011): 1-30.
[11] Cantargi, F., et al. "New scattering kernels for some materials of interest as advanced cold neutron moderators." Physica B: Condensed Matter 385 (2006): 1312-1314.

Study on Moderation Properties of Cold Mesitylene using KUANS

Seiji Tasaki[1,*] , *Yuma* Idobata[1], *Yuya* Adachi[1], *Fumiaki* Funama[1], and *Yutaka* Abe[1]

[1]Department of Nuclear Engineering, Kyoto University, Kyotodaigaku-Katsura Nishikyou-ku Kyoto 615-8530, Japan

Abstract. Neutron moderation properties from the cold mesitylene moderator have been studied. Kyoto University Accelerator driven Neutron Source has been used for these experiments. The container of the mesitylene moderator is situated in front of the polyethylene moderator and the change of the time of flight spectrum has been recorded as a function of the temperature of the mesitylene moderator. By fitting the Maxwell distribution to the obtained TOF spectra, the neutron temperature corresponding to the mesitylene temperature is estimated.

1 Introduction

In recent years, compact neuron sources have been developed. In spite of their low neutron flux, they may be useful by specific scientific domains. For study of nano-scale structure in materials using small angle scattering, or reflectometry, cold neutrons are commonly used because the separation between the incident and the diffracted neutron beams are roughly proportional to the wavelength and larger and easier to measure with cold neutrons.

For producing such long-wavelength neutrons, cold neutron sources (CNS) are employed. For most of such CNS, liquid hydrogen or solid CH_4 are used as cold moderator. For compact sources, however, such materials are not suitable because a) they are gas in room temperature and the volume change is quite large when they become liquid or solid, and b) they are explosive in the gas state. Mesitylene having three metyl's around the benzene ring and staying liquid over a large range of temperatures has been suggested as a good moderator for small neutron sources [1].

In the present study, we measured the wavelength distribution from the cooled mesitylene moderator in order to clarify the moderation properties for neutrons of cold mesitylene using Kyoto University Accelerator-driven Neutron Source (KUANS).

In KUANS neutrons are produced by ^{9}Be(p, n)^{9}B reaction using pulsed 3.5MeV-proton beam. The neutrons are moderated by the polyethylene and the moderated neutrons are emitted to the direction perpendicular to the proton beam [1].

2 Measurements

Figure 1 shows schematic view of the experimental set-up for the present measurements.

The mesitylene moderator is situated in front of the polyethylene moderator having the size of 120x120x85 mm^3 in room temperature. The mesitylene container with 25(t)x96(w)x100(h) mm^3 is held in a vacuum container with the size of 43(t)x123(w)x155(h) mm^3. The moderators are surrounded by the graphite reflector.

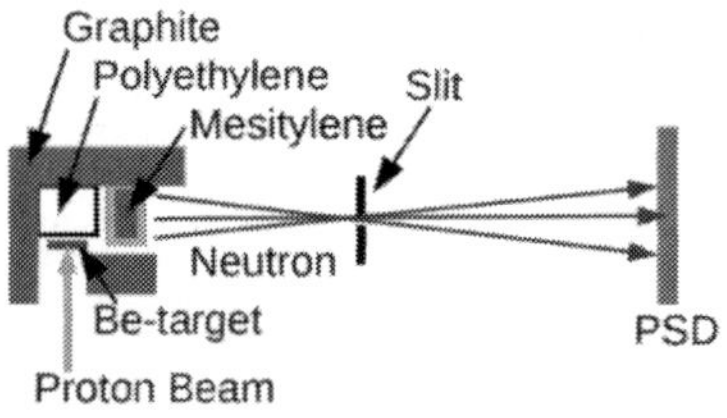

Fig. 1. Experimental set-up of the present study.

The neutrons emitted from the moderator surface go through the 20 (h)x5 (w) mm^2-slit situated at 1710 mm from the moderator surface and are detected with 3He 1-dimensional position sensitive detector (PSD) at 1546 mm from the slit. With the PSD, it is possible to measure 1-dimensional 'pin hole image' over the moderator as well as the time of flight (TOF) spectrum. The TOF spectra presented in this report are taken from the mesitylene area of the PSD data. Total flight length is 3256mm and T0 (time between the system trigger and emission of thermal neutron from the moderator) is 70 μs for KUANS. The repetition rate and pulse width of the proton pulse are 100Hz and 60μs, respectively. Average proton current was about 60 μA. Since the time bin for TOF measurements is 10 μs, 6 successive channels of TOF data are arithmetically averaged.

Mesitylene moderator was cooled with cryostat from room temperature to 28K. During cooling process (about 10hr), the accelerator was operated, and the neutron spectra were measured. The TOF data are summed up for

* Corresponding author: tasaki@nucleng.kyoto-u.ac.jp

https://doi.org/10.1051/epjconf/202023104005

every 30min during when the temperature changes about 10 degree in kelvin. We adopt mean temperature for each TOF data.
The TOF spectra of neutron from the mesitylene moderator for the mesitylene temperature T_M =251, 204, 141, 84, 28K are shown in Fig. 2. Change of the spectrum is small for over 200K which is close to the melting point of mesitylene. As the T_M decreased, the peak intensity is lowered and shifted to the longer wavelength side.

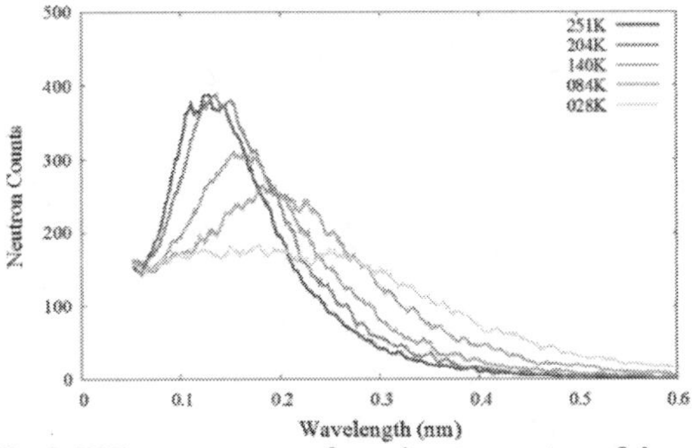

Fig. 2. TOF neutron spectra for various temperature of the mesitylene moderator.

The relative neutron intensity versus wavelength for T_M= 204, 141, 84, 28K comparing to that for 251K is shown in Fig. 3.

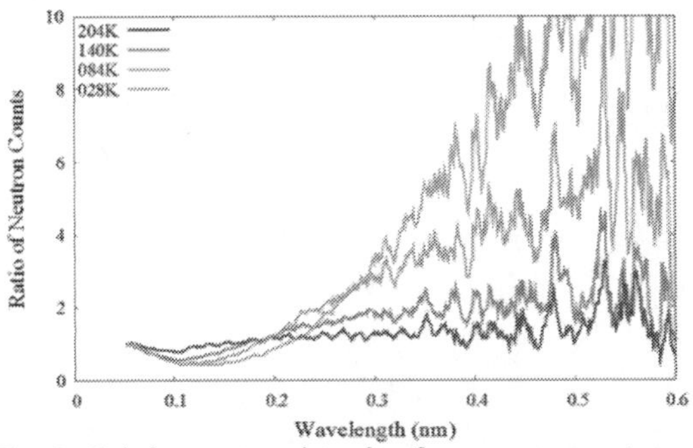

Fig. 3. Relative neutron intensity for some temperature comparing to that for 251K.

For wavelength longer than 0.25nm, the ratio is larger than unity and increases as the wavelength becomes longer and as T_M lowers. At the lowest temperature of 28K and at wavelength of 0.4nm, neutron intensity is six times higher compard to that at 251K.

3 Discussion

In order to obtain the neutron temperature, we fit a Maxwell distribution to the neutron wavelength distribution shown in Fig.2. The aim of the fit is to obtain the temperature of the neutrons T_n in the cold mesitylene moderator. Hence the fitting is carried out for neutrons with wavelength longer than 0.2nm. Examples of the fit results are shown in Fig.4. The fit function is the neutron flux and has the following form for neutron wavelength of λ:

$$\phi(\lambda) = \frac{A}{\lambda^5} \exp\left[-\frac{T_0}{T_n}\left(\frac{\lambda_0^2}{\lambda^2}\right)\right] \quad (1)$$

where λ_0=0.178nm, and T_0=300K are the wavelength and corresponding temperature for 25meV-neutron. The fit parameters are A and T_n, the latter of which represents the moderated neutron temperature.

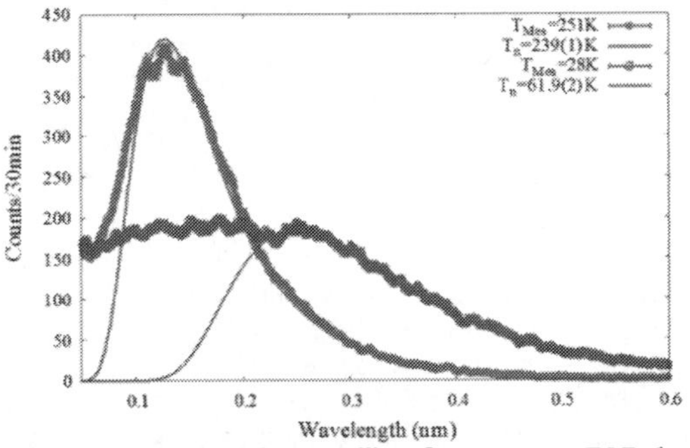

Fig. 4. Examples of Maxwellian fit to neutron TOF data shown in Fig.2 in the 251K (red) and 28K (blue) case.

The red and blue line in Fig.4 represent the spectrum for T_M=251K and 28K, respectively. For the former case T_n=239K, and for the latter T_n=61.9K, respectively. There may be two reasons why T_n is higher than T_M : (1) Because the number of excitation levels around several meV range in mesitylene is not enough, some neutrons are emitted before being fully moderated. (2) Thickness 25mm of the mesitylene moderator container is not enough for full moderation of neutron. Hence the discrepancy between the fit function and TOF spectrum in the short wavelength range is due to the under-moderation of the neutrons.

Repeating the same procedure for other temperature of mesitylene, the relation between T_M and T_n has been obtained. The result is shown in Fig.6.

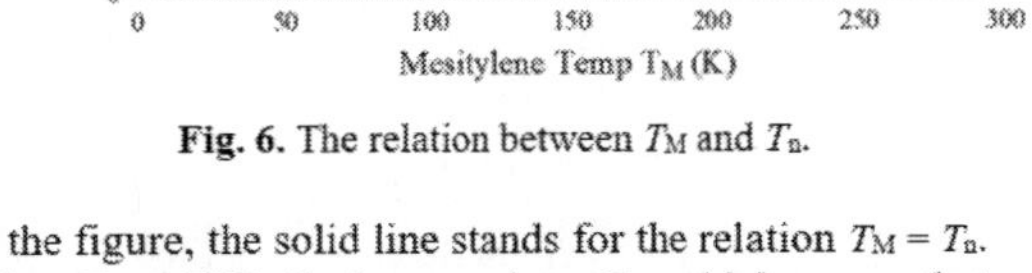

Fig. 6. The relation between T_M and T_n.

In the figure, the solid line stands for the relation $T_M = T_n$. When T_M>100K, T_n changes along T_M, which means that the neutron temperature follows the mesitylene temperature.

On the other hand, when T_M<100K there appears discrepancy between T_M and T_n, which means that at low mesitylene temperature neutrons are under-moderated; their temperature stays above the mesitylene temperature. In spite that, cooled mesitylene enables us to obtain an increase of flux of long wavelength neutrons with the factor 10 at 0.5nm compared to ambient temperature.

The slowing down distance for mesitylene is estimated to be about 6cm, since that for H_2O is about 5cm and is reverse proportional to the hydrogen density ($6.7x10^{22}$, $5.19x10^{22}$ for H_2O and mesitylene, respectively) [3]. The size of the mesitylene container is 25x96x100 mm^3 in the present experiments, and the 'thickness' 25mm is too short for fully moderation. The neutron flux may be improved if the neutron is emitted the 'side' of the container.

4 Conclusions

Neutron TOF spectra from the mesitylene moderator cooled from the room temperature to 28K were measured. Measured spectra have changed corresponding to the mesitylene temperature. As the mesitylene temperature decreases, the peak of the spectrum shifts to longer wavelength side, and the neutron counts in the wavelength range longer than 0.2nm increases. At 0.4nm, the ratio of the neutron counts from the cooled mesitylene against uncooled mesitylene becomes about 6. The neutron counts with the wavelength over 0.2nm, is well reproduced by the Maxwellian distribution. From the fit, neutron temperature is obtained and is almost the same in the temperature range above 100K, and the difference appears under 100K. The neutron flux may be improved by the optimization of the arrangement of the moderator system.

Acknowledgement

We would like to thank Mr. M. Hirose and Prof. T. Nagae for the operation of KUANS and the mesitylene moderator.

References

1. Lourdes Torres, J.R. Granada, *Physica B: Condensed Matter*, **385-386**, Part 2, 2006, pp 1306-1308.
2. S. Tasaki et al., proceedings of Ucans III&IV, *Physics Procedia* ***60***, 2014, pp 181-185.
3. J. R. Lamarsh, *Introduction to Nuclear Reactor Theory*, Addison-Wesley Publishing Company Inc., Massachusetts, USA 1965.

Low-dimensional moderators at ESS and compact neutron sources

Luca Zanini[1,*], *Esben* Klinkby[1,2], *Ferenc* Mezei[1], and *Alan* Takibayev[1]

[1]European Spallation Source ESS ERIC, P.O Box 176, Lund, Sweden

[2]DTU Nutech, Technical University of Denmark, Roskilde, Denmark

Abstract. Low-dimensional moderators were designed for the European Spallation Source (ESS), and the same concepts can be applied to compact sources. At ESS, quasi two-dimensional (2D) high-brightness moderators will serve all the instruments of the initial suite. The design of the moderators is influenced by several factors, such as the layout of the facility, the requirements for beam extraction, and the number of instruments; these constraints led to the choice of the 2D "butterfly" moderator for ESS. In an accelerator-based compact source, such moderators can be designed in an even more efficient way than for high-power sources, taking advantage of the lower heat loads and of the more compact arrangement of the target-moderator system, which gives more freedom in the optimization of the geometrical setup. Some promising design options have been explored.

1 Introduction

The neutron moderators designed for ESS are based on the low-dimensional moderator principles, which are described in Ref. [1] and several references therein. The design of the ESS moderators has been finalized and details are given in Ref. [1]. Since the early development of these moderators, we explored the possibility to use them in research reactors [2], and more recently in compact sources [3]. In [3] it was suggested that the use of low-dimensional moderators is more efficient (in terms of brightness per fast neutron produced in the target) in compact sources than for high-power facilities such as ESS. In this work we further explore the applicability of such moderators in compact neutron sources.

2 Main features of the ESS moderators

ESS will be the most powerful spallation neutron source in the world, with an average beam power of 5 MW and peak power of 125 MW. The facility is designed with 42 beamports surrounding the moderators, covering an angle of 240 degrees, with an average angular spacing between beamports of 6 degrees. Such a grid of beamports is intended to give flexibility in positioning instruments over the lifetime of the facility. All the instruments in the initial scope of ESS will point to the moderators placed at the top of the tungsten spallation target. The requirement that the moderators should provide a bispectral (thermal and cold) neutron spectrum for all the beamports arranged in two sectors of 120 degrees each had a strong influence on the design of the moderator shape: the so-called "butterfly" design was the best identified configuration to fulfil this requirement. The height of the moderator was set to 3 cm, after an optimization process that involved all the initial instruments in the ESS suite, and considered the moderator brightness as well as the brightness transfer from the guides to the sample. Details on the optimization process are given in Refs. [1,4]. Some upgrade possibilities are discussed in Ref. [1].

3 Low-dimensional moderators for compact sources

The energy of the neutrons emitted from the target depends on the primary beam energy; for compact sources, see Ref. [5]: for lower proton energies, and consequently, lower neutron energies, a more compact system can be designed, i.e., with a smaller reflector and a thinner premoderator. Compared to a spallation source, in a compact neutron source the neutron production takes place in a smaller region and this can result in a gain in performance, as the moderator can be placed closer to the neutron production region. This advantage can be exploited very effectively with low-dimensional moderators, especially quasi one-dimensional (1D) moderators. In the case of ESS, the neutron production region occupies a surface close to the one of the (optimized) butterfly moderator [1], which makes it particularly suitable for the use of 2D moderators, giving the possibility to serve a large number of instruments with a high-brightness moderator.

* Corresponding author: luca.zanini@esss.se

An additional advantage for compact sources is due to the lower power, which allows moving the optics closer to the moderator. In the case of ESS the moderator thickness (3 cm) was chosen as the best compromise between source brightness and brightness transfer by the neutron optics [4]. Based on brightness considerations only, a thinner moderator would have been better performing [1]. For a compact source, it should be possible to move the optics closer to the moderator, and therefore to choose a thinner moderator than used at ESS.

For application to compact sources, in order to maximize the brightness, we considered in this work the quasi 1-dimensional tube moderators, i.e., moderators with a cross section of a few cm^2, and with a length of about 20 cm, consisting of aluminium vessels filled with pure parahydrogen at 20 K. Such moderators have a strong directional emission of cold neutrons [1], which will limit the number of beamlines that can be installed.

We did not consider beam transport aspects in this work, but focused only on the brightness of the source.

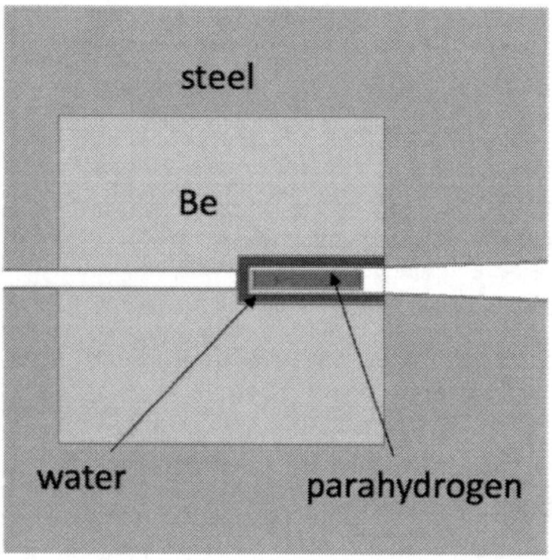

Fig. 1. 1-target geometry configuration.

3.1. Source configurations

We considered a neutron source from a 13 MeV proton beam impinging on a 2-mm thick Be target. The choice of the energy was taken considering the example of the LENS compact source [6]: the 13 MeV energy is at the threshold of production of ^7Be and ^3H in the target, thus limiting the activation of the target and simplifying the maintenance procedures.

We considered two possible geometry setups. In the first configuration (Fig. 1), the beryllium target is placed at the center of a beryllium reflector. A cylindrical aluminium tube with 3 cm diameter, filled with parahydrogen is surrounded by a water premoderator and is placed right after the target. We note that in this basic setup, fast neutrons would be directly seen in the beam extraction channels, a drawback that could be easily avoided by placing the moderator at an angle with respect to the proton beam line. The considered geometry is however suitable for the study of the performance of a tube moderator in such a compact source. In the second configuration (Fig. 2), the proton beam is split and sent to two beryllium targets. The moderator is placed between the two targets. In our example, the moderator is parallel to the proton beam line. A different configuration would be with the moderator perpendicular to the proton beam; this would have the advantage of doubling the neutron beam lines served by the moderator.

The neutron beam extraction channel is in both cases of conical shape, which could in principle accommodate the extraction of neutrons for more than one instrument. However, for simplicity we studied only the brightness at the center of the channel.

The dimensions of the target surface, irradiated by protons, are small (1.5 x 3 cm^2). To run at power levels of the order of tens of kW a larger target would be needed. A larger target surface will result in a decrease in performance, as shown in Section 3.6. Additionally, no engineering details (such as the cooling of target and reflector, and the presence of water and hydrogen pipes) are considered in this model. The reflector is a cube of 60 cm side. Simulations were performed with MCNPX 2.7.0 [7,8]. Neutron production cross sections in Be from ENDF/B-VII were used. The figure of merit for the design optimization was the integrated cold brightness below 5 meV.

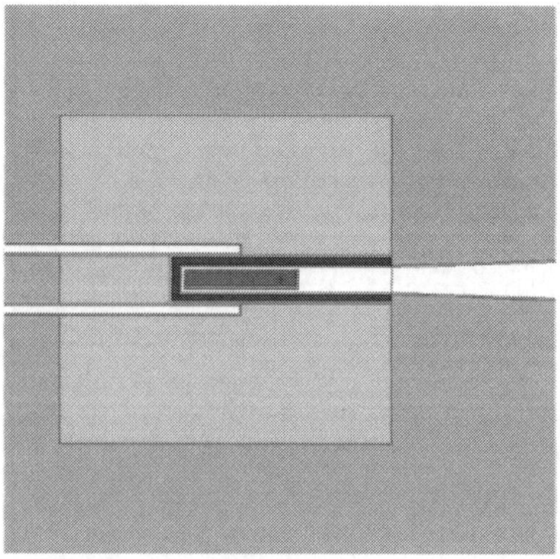

Fig. 2. 2-target geometry configuration.

3.2 Neutron production

Fig. 3 and Fig. 4 show maps of neutron fluences with $E <$ 0.4 eV for the 1- and 2-target configurations, respectively. In both configurations, the density of neutrons is more concentrated at the center of the system, as expected. This has some implications on the design choice of the target-moderator system, because the cold moderator should be placed where the thermal flux is maximum. This obviously favours a 2-target geometry since in this case the moderator is placed at the center of the system, and is more uniformly illuminated by thermal neutrons; a higher brightness is therefore expected from this geometry.

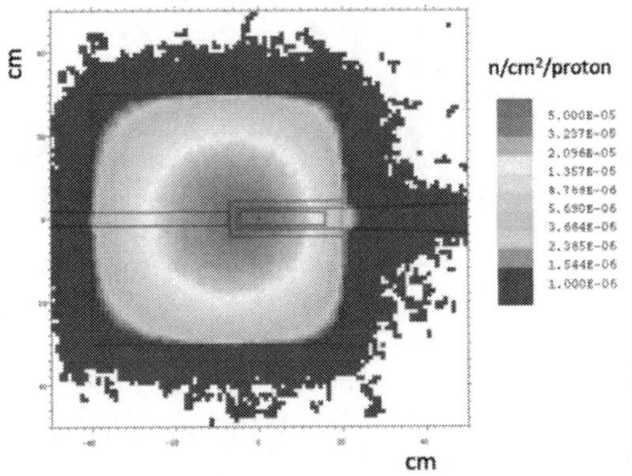

Fig. 3. Low energy ($E < 0.4$ eV) neutron fluence map for the 1-target configuration.

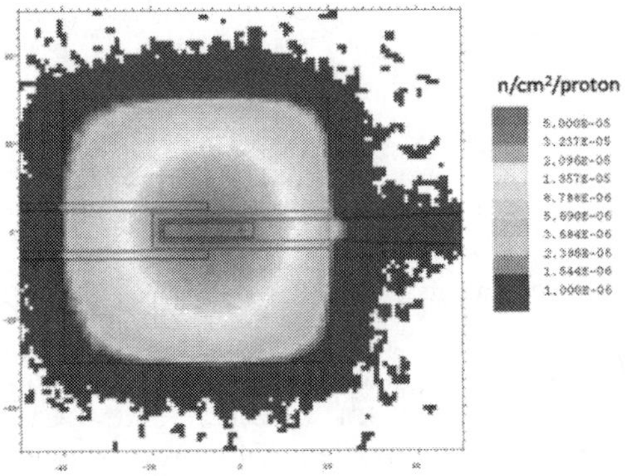

Fig. 4. Low energy ($E < 0.4$ eV) neutron fluence map for the 2-target configuration.

3.3 Moderator length

The source performance for different moderator lengths is shown in Fig. 5 for the 1-target and 2-target configurations. For these calculations, a 1.5 cm thick water premoderator was assumed. In the case of the 1-target geometry, the fact that moderator is not placed in the center limits the optimal moderator length to about 12 cm. For longer moderator lengths, neutron absorption by hydrogen dominates and this reduces the performance of the moderator. On the contrary, for the 2-target configuration the optimal moderator length is of about 20 cm. This confirms that a 2-target configuration is more suitable to be used with a tube moderator. In fact a moderator length of 12 cm is about equal to the mean free path of cold neutrons in pure parahydrogen (11 cm [2]). The optimal moderator length of a parahydrogen tube moderator should be considerably longer than that, if most of the moderator is uniformly illuminated by thermal neutrons; a 20-25 cm length is expected to be optimal, as confirmed by the ESS design [1]. Therefore, the 2-target geometry was adopted for the remaining neutronic studies presented in this paper.

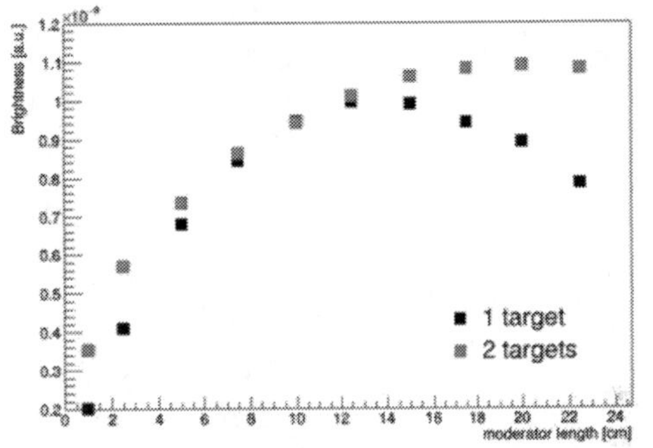

Fig. 5. Effect of moderator length on the source cold brightness (integral brightness below 5 meV) for the 1-target and 2-target geometries.

3.4 Effect of premoderator

We studied the effect of the presence of a water premoderator surrounding the cold moderator for the 2-target geometry. Fig. 6 shows the thermal neutron maps without and with a premoderator (1.5-cm thick). It is clear that having a premoderator increases the concentration of thermal neutrons reaching the cold moderator. Fig. 7 shows the effect on the brightness: the optimal thickness of the premoderator is of 1.5 cm.

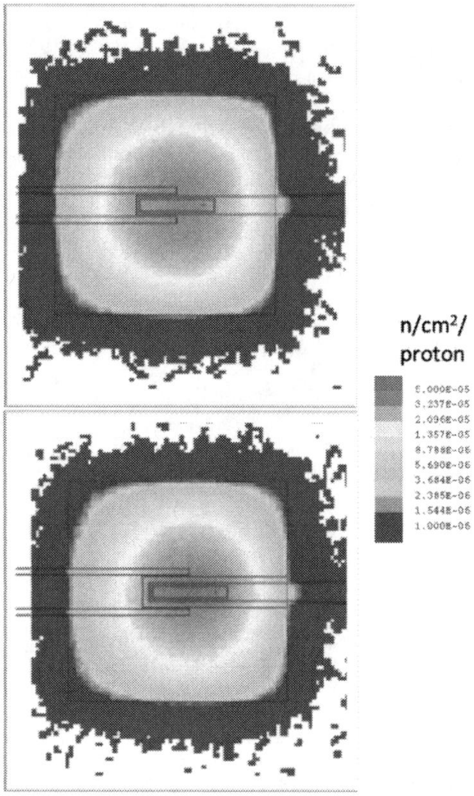

Fig. 6. Effect of the presence of premoderator on the thermal neutron fluence map ($E < 0.4$ eV) in the compact source. Top figure: no premoderator. Bottom figure: 1.5-cm thick premoderator.

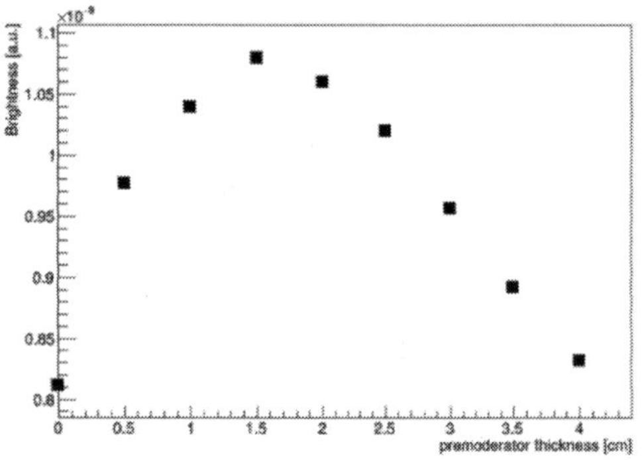

Fig. 7. Effect of premoderator thickness on the source cold brightness.

3.5 Size of Be reflector

Ideally the beryllium reflector surrounding the moderators should be as large as possible. For a compact source however we are interested in reaching an optimal compromise between size (cost) and performance. We considered the variation of the moderator brightness with the size (lateral dimension of a cube of beryllium), in the case with and without premoderator (Fig. 8). The brightness, relative to the maximum achievable with a 1-m size reflector, changes more steeply with the size in the case without premoderator; this is due to the optimal shaping of the thermal neutron map that is achieved by using a premoderator. Fig. 8 indicates that a compact Be reflector with a size between 40 and 50 cm would deliver about 90% of the maximum performance obtained with a large reflector. We therefore see that the advantage of a water premoderator is two-fold: increase of brightness by optimal shaping of the neutron spectrum feeding the moderator, and decrease of the size of the Be reflector. To further reduce costs, it could be worthwhile considering a graphite or lead reflector as a cheaper alternative to beryllium.

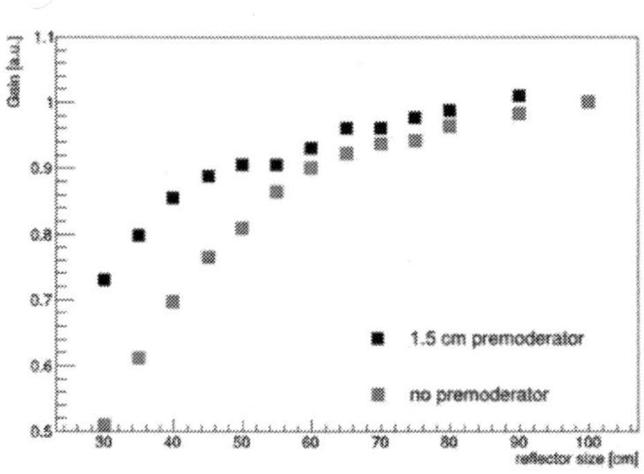

Fig. 8. Dependence of the source cold brightness on the size of Be reflector, with and without water premoderator.

3.6 Size of targets

The results shown so far are for (ideal) very small beryllium targets. However, for operation at power levels of the order of 50-100 kW, the surface of the neutron-producing target must be increased to allow for cooling of the target. Indications are that a surface of 100 cm^2 would be required to allow operation at 100 kW (typically 1 kW/cm^2 is considered the upper limit for water cooling of a conventional flat target [9]). The increase in target size will result in a loss in performance, which is of about 30% at 100 cm^2, compared to an ideal point-like target, see Fig. 9. The decrease is for both 1 and 2 target configurations, but the 2-target configuration remains better performing for all values of target surface.

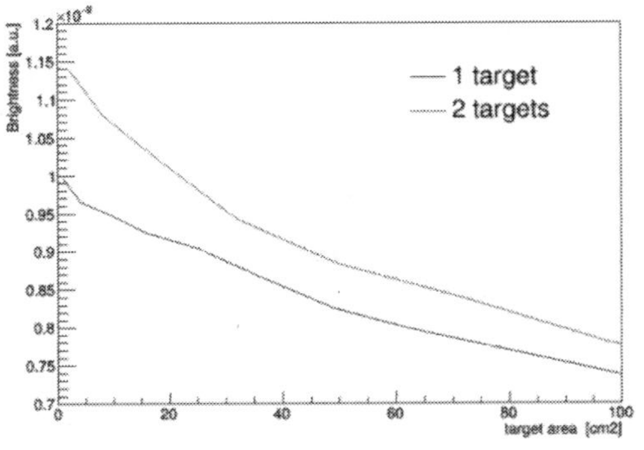

Fig. 9. Target size effect. See explanation in the text.

4 Performance compared with high-power sources

In order to compare the efficiency of two different types of neutron sources, i.e., high-power spallation and compact Be-target sources, we use as a figure of merit the brightness per fast neutron produced in the target. More specifically, as done in Ref. [3], we consider the yield *Y* (number of neutrons per proton on target), which is of about 38 n/p for 2 GeV protons on tungsten, and about 6.2×10^{-3} n/p for the compact source at 13 MeV [3]. The brightness B (E < 5 meV) for ESS is of about 1×10^{-3} n/cm²/s/sr/proton, while for the compact sources is of about 1.1×10^{-6} n/cm²/s/sr/proton. We have therefore a ratio B/Y of about 2.6×10^{-5} for ESS, and 1.8×10^{-4} for the compact source. This is for configuration of the compact source calculated for the 2-target design, with a target surface area of 100 cm² (physics model with no engineering design). The results indicate that a small source with a tube moderator can be one order of magnitude more efficient in slowing down fast neutrons, than a high-power source like ESS. Part of the gain can be attributed to the choice of using a 1D tube moderator, which will allow to serve a smaller number of beamlines (up to 4-6 for each side of the moderator). Part of the gain is also due to the more compact system and the possibility to arrange in a more flexible way the moderator and reflector around the neutron producing target.

If we compare the source performances at operating power of 5 MW for ESS, and (expected) 100 kW for the compact source, the brightness values will be of about 1.6×10^{13} n/cm²/s/sr for ESS, and about 5×10^{10} n/cm²/s/sr for the compact source, i.e., a ratio of about 300. Even if this ratio is expected to increase when calculated for a fully engineered compact source, this level of performance is very promising for a variety of neutron scattering experiments that will not require extremely high fluxes like what is available at ESS.

5 Conclusions

The use of low-dimensional moderators is particularly suitable for compact neutron sources. Such compact systems allow the effective coupling of an optimized moderator and reflector system. The use of a premoderator allows the shaping of the thermal flux feeding the cold moderator, with a consequent increase in performance, as well as the reduction of the size of the reflector. The 2-target geometry that we investigated is particularly promising, and would be convenient for the cooling of the target, as the beam power is split into two targets. The efficiency of these sources, measured by the ratio of the moderator brightness to the number of fast neutrons produced by the target, can be one order of magnitude higher than high-power spallation sources. Thus, use of low-dimensional moderators in compact sources will reduce the gap with high-power sources, giving the possibility to perform high-quality neutron scattering measurements.

References

1. L. Zanini, K. H. Andersen, K. Batkov, E. B. Klinkby, F. Mezei, T. Schoenfeldt, A. Takibayev, Nucl. Instrum. Methods A 925, 33 (2019)
2. F. Mezei, L. Zanini, A. Takibayev, K. Batkov, E. Klinkby, E. Pitcher, T. Schoenfeldt, Journal of Neutron Research 17, 101 (2014).
3. L. Zanini, F. Mezei, K. Batkov. E. Klinkby, A, Takibayev, IOP Conf. Series: Journal of Physics: Conf. Series 1021, 012009 (2019).
4. K. H. Andersen, M. Bertelsen, L. Zanini, E.B. Klinkby, T. Schoenfeldt, P. M. Bentley, J. Saroun, JAC 51, 264 (2018).
5. C.M. Lavelle et al., Nucl. Instrum. Methods A 587 (2008) 324.
6. D. V. Baxter et al, Nucl. Instrum. Methods B 241, 209 (2005).
7. D. Pelowitz, editor, *MCNPX user Manual, Version 2.7.0*, LA-CP-11-0438 (2011).
8. L. S. Waters, et al., AIP Conf. proc. 896, 81 (2007).
9. D. V. Baxter, *CANS TMR design: a personal reflection*, presentation at UCANS-8, Paris, 2019.

Development of new small-angle neutron scattering geometry with ring-shaped collimated beam for compact neutron source

Fumiaki Funama[1,*], *Yuya* Adachi[1], *Seiji* Tasaki[1], and *Yutaka* Abe[1]

[1]Department of Nuclear Engineering, Kyoto University, Nishikyo-ku, Katsura, Kyoto 615-8540, Japan

Abstract. Small-angle neutron scattering (SANS) is an important tool to investigate material properties in nanometer to micrometer scale. The opportunity to conduct SANS experiments is, however, limited because of the low number of available SANS beam lines. Compact neutron sources are expected to play a significant role to increase neutron scattering facilities including SANS beam lines. The problem is that the flux of compact neutron sources can be very low, which makes it difficult to measure scattered neutrons from a sample. A SANS geometry with ring-shaped collimated beam (r-SANS) is developed to conduct SANS experiments at very low flux neutron sources. By using ring-shaped collimated neutrons to hit a large sample, the scattered neutron flux becomes high on the ring center line because the scattered neutrons with each scattered angle overlaps on each point of the ring center line. By setting a ^{3}He point detector on the center line and shielding the surrounding of the small detection area well, high signal to noise ratio experiments are possible. In this paper, we show the concept of this new geometry and a preliminary experimental result of a glassy carbon sample taken with the r-SANS geometry constructed at Kyoto University proton Accelerator Neutron Source (KUANS).

1 Introduction

Small-angle neutron scattering (SANS) aims to study the structure of matter ranging from nanometer to micrometer scales [1]. The mesoscale structure is important to understand the macroscopic behaviour of matter. This technique offers several advantages since neutrons can penetrate into the samples deeply, are sensitive to isotopes and possess magnetic moments. Thus, using SANS and small angle x-ray scattering (SAXS) techniques complementarily, better understanding of the structure of matter in mesoscale is possible.

SANS experiments are normally done in large facilities with neutron sources. The number of SANS beamlines is very limited since nuclear spallation, reaction or fission phenomena are necessary for neutron production, which makes the facilities expensive. In the hope of cutting the cost, shrinking the size of the facilities and making SANS experiments available in smaller scale laboratories, compact neutron sources are being developed. There are several advantages in constructing SANS instruments at compact neutron sources. SANS experiments do not require high spatial and energy resolutions relatively. Because of that, utilizing more scattered neutrons is possible by summing up more neutrons in the same momentum transfer bin. The neutron emission area is much larger than its SAXS counterpart, which means that bigger samples can be measured. The target-station is also relatively easier to access and any shape of slit can be put just in front of the moderator. The residual radioactivity does not remain so much after the experiment especially when a few MeV proton beam is used for the neutron production. By placing the slit near the moderator, environmental noise can be reduced since unnecessary neutrons can be stopped by the slit at the most up-stream part of the instrument. Moreover, the total length of the instrument can be greatly reduced, which enables us to utilize more neutrons during the experiments.

Although SANS with compact neutron sources offer a great potential, several challenges are needed to be addressed. The low neutron intensity of compact neutron sources is a problem for conducting SANS experiments. Several methods and devices were developed to increase the neutron intensity at the sample position. Multi-pinhole SANS geometries use a slit with several pinholes to collimate large area of neutrons and increase the incident neutron beams [2]. Focusing SANS geometries enable us to utilize more neutrons from a pinhole slit using ellipsoidal focusing mirrors [3]. These methods have its own advantages and disadvantages, which depend on different conditions such as the momentum transfer resolution needed in the experiment, the available neutron energy, and the scattering intensity of a sample. However, unlike x-ray production, in order to obtain thermal or cold neutrons, the moderation process is necessary which makes the neutron emission area much larger than x-ray in nature. Moreover, neutrons emitted from the moderator surface have brightness distribution but has no directional dependencies in effect. These intrinsic characteristics can be utilized more to compensate for the lack of neutron flux of compact neutron sources.

In this paper, a new geometry for small-angle neutron

* Corresponding author: funama.fumiaki.26e@kyoto-u.jp

scattering with ring-shaped collimated beams is proposed. The geometry and preliminary experimental results will be discussed.

2 SANS geometry with ring-shaped collimated beam (r-SANS geometry)

A SANS geometry with ring-shaped collimated beams (r-SANS) is developed in order to utilize more neutrons at compact neutron sources whose flux is very low. The schematic diagram of the geometry is shown in Fig. 1. The geometry is composed of a moderator, three ring-shaped slits, a sample and a ^{3}He point detector placed next to each other. The distances between the moderator and the first slit, the first slit and the second slit, the second slit and the third slit, the third slit and the sample, and the sample and the detector are defined by *LM1, L12, L23, L3s* and *Lsd*, respectively. The total instrument length, between the moderator surface and the detector, is defined by *Lmd*. The ^{3}He point detector is covered with a detector shielding material, which is made of B_4C. The shielding window is a hole through which scattered neutrons are able to be detected. The window has radius *rd*. The width of the three ring-shaped slits is *Rw*. The center of all the components are aligned on the ring center line. The nominal scattering angle θ is then determined using *L3s*, *Lsd* and the nominal incident angle, which is determined by the first and third slit ring radii (*R1* and *R2*), *L12* and *L23*. The wavelength, λ, of the scattered neutron is calculated using time of flight method (TOF). These values are used to calculate the momentum transfer of scattered neutrons, Q, using equation (1).

$$Q = \frac{4\pi \sin\theta}{\lambda} \tag{1}$$

By changing the detector position, *Lsd*, we can change the measurable Q range of the scattered neutrons in the measurement. The slits are made of neutron absorber material so that the neutron beam can only pass through the slits. The second slit is included to block neutrons that passes diagonally from the first to the third slit to reduce the smearing effect.

The mechanism for achieving high signal-to-noise (S/N) ratio experiments is explained as follows. The neutron beam only passes through the three rings in the tangent direction which provides a noiseless space on the ring center line. We put a zero-dimensional ^{3}He point detector, which has a higher counting ratio for neutrons and is less sensitive to gamma rays compared to a typical two-dimensional scintillation detector, on the line. Since the detection area of the detector, which is within the shielding window, is small, it is possible to minimize the noise from the environment by providing a good shielding around the detector. By measuring superimposed scattered beam on the beam center line, high S/N ratio experiments are possible. Since the distances of the components from each other are flexible, the total length of the instrument can also be shortened in order to obtain more scattered neutron signals. This geometry can easily be adopted in other compact neutron sources with minimal cost.

Despite the great possibilities that this geometry offers, there are also several limitations. Since the incident neutron beam hits the sample with the designed ring-shape area, a large amount of sample is needed to cover the whole ring area. Considering the usual size of a moderator for a compact neutron source, a square sample size in the range of 5 - 15 cm is necessary. This may be difficult to achieve for biological samples. However, for other fields such as steel industries, which are interested on the average structure of their products, the sample size can be acceptable. Another limitation of the geometry is that anisotropic scattering of the sample is not distinguished because the detected neutrons are coming from different positions of the sample to the ring center line. Moreover, the momentum transfer resolution can be low depending on the size of the rings (*R1*, *R2* and *Rw*), *rd* and *Lsd*. This low resolution may not be significant since achieving more scattered neutron intensity from a sample with a good S/N ratio condition is usually the priority for SANS measurement using very low flux neutron sources. Lastly, the utilized solid angle for this r-SANS geometry is usually small compared with a typical SANS instrument. In principle, however, it is possible to cover a larger range of scattered angle by putting or moving the ^{3}He point detector along the center axis. Since the scatterers are aligned in ring-shape, the observable scattering intensity of unit solid angle is the same as one of the pinhole geometry if the incident beam divergences for both cases are the same. Since r-SANS geometry uses cross-talk neutrons collimated with the three ring-shaped slits, it causes additional scattered neutrons from the sample.

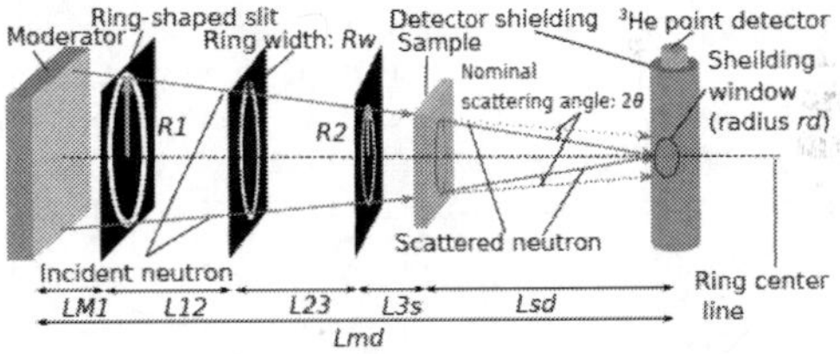

Fig. 1. Schematic diagram of r-SANS geometry

3 Experiment

In order to conduct the test measurements, we built an r-SANS geometry at Kyoto University proton Accelerator Neutron Source (KUANS). KUANS is a compact neutron source with commercially available proton linac (AccSys Technology, Inc.) [4]. The neutron flux of KUANS is one of the lowest among available compact neutron sources for beam experiments [5]. With the same accelerator conditions of r-SANS measurement, the neutron intensity without any collimations was measured with ^{3}He point detector (6 atm, 1 inch) on site. The intensity was just a few hundred n/cm^2/s at 2.5 m away from the polyethylene moderator surface.

The geometry parameters for the experiment are shown in Table 1. The first slit diameter was decided by considering the neutron brightness of the moderator. The third collimator radius is decided by the sample size (5 x

5 cm^2). The distance between the moderator surface and detector, *Lmd*, was 2650 mm, which is a lot shorter than normal SANS set-up, to utilize more neutrons. The distance between the sample and detector, *Lsd*, is decided by the momentum transfer range of interest for our measurement. All ring-shaped slit widths were 4 mm in this experiment. As a sample, a glassy carbon plate (Alpha Aesar, product code: 038021, 1mm (0.04in) thick, type 2) was chosen, which is widely used standard sample for SANS and SAXS measurement [6]. The thickness was 2 mm by piling the two plates up.

Table 1. Parameters in Fig.1 for r-SANS experiment

Parameter	Value
R1	42 mm
R2	21 mm
Rw	4 mm
LM1	45 mm
L12	1010 mm
L23	895 mm
L3s	30 mm
Lsd	640 mm
Lmd	2620 mm
rd	4.5 mm
2θ	0.021 rad

Figure 2 shows the TOF spectra of the direct beam, the scattered neutron and the background measurements. The direct beam of KUANS has a thermal peak at 1.14 Å. The scattered neutron from the glassy carbon sample has a stronger intensity than the background measurement. The background level is 1.58x10^{-3} counts/s (5.68 counts/hour) for neutrons with wavelengths longer than 0.8 Å, which is very low because of the benefits of the r-SANS geometry. With the nominal scattering angle (2θ) and the neutron transmission ratio of the glassy carbon sample, we can obtain the scattering intensity by using the following equation

$$I(Q) \propto \frac{\frac{I_s(\lambda, 2\theta)}{T(\lambda)} - I_b(\lambda, 2\theta)}{I_d(\lambda)} \tag{2}$$

where $I(Q)$ is the scattering intensity of a sample which is proportional to the differential cross section. $I_s(\lambda, 2\theta)$

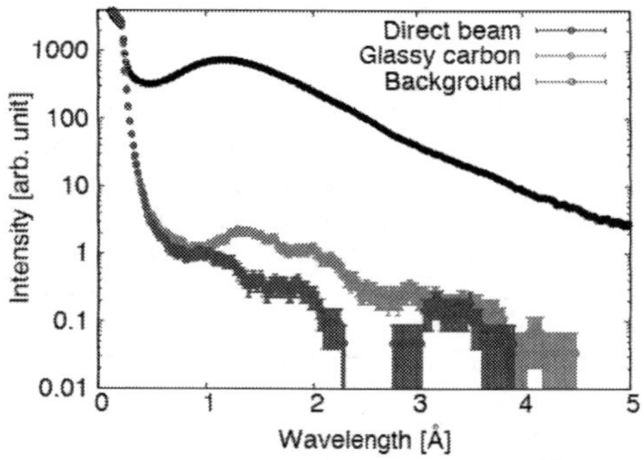

Fig. 2. TOF spectra of the direct beam, the scattered neutron from glassy carbon and the background measurements

and $I_b(\lambda, 2\theta)$ are the scattering intensity of the sample and the background measurements, respectively. $T(\lambda)$ and $I_d(\lambda)$ are the transmission ratio of the sample and the direct beam spectrum.

The measured scattering intensity of the glassy carbon sample is shown in Fig. 3 as well as the reference result of a SANS measurement for the same product sample [6]. The r-SANS experimental result is consistent with the published result although a discrepancy occurs when Q becomes larger than 0.1 [1/Å]. This can be due to the effect of the oblique incident neutrons which pass through the three ring slits but do not have the same scattering angle values with that of the nominal scattering angle. These cross-talk neutron beams can enhance, however at the same time, smear the observed scattering intensity. The development of the analysis method is a subject for future research.

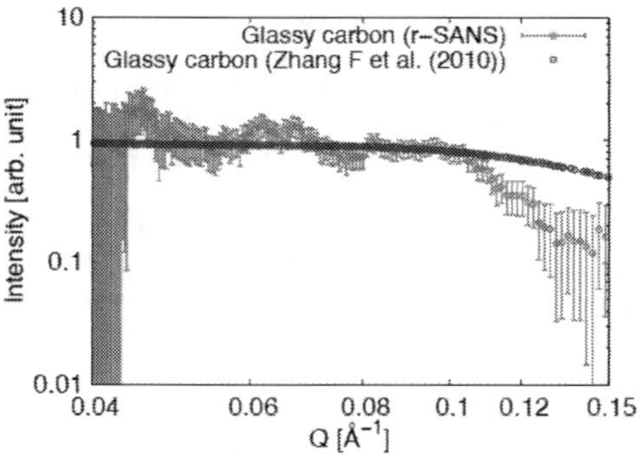

Fig. 3. Scattering intensity of a glassy carbon sample

4 Conclusion

In order to conduct SANS experiments at very low flux neutron sources, we propose a new geometry with ring-shaped collimated beam (r-SANS). This geometry enables us to observe converging scattered neutrons from the sample at the noiseless ring center line with a zero-dimensional ^{3}He point detector. As a test measurement, we constructed an r-SANS geometry at Kyoto University proton Accelerator Neutron Source (KUANS) which has one of the lowest neutron flux among available compact neutron sources for beam experiments. A glassy carbon sample was used for the measurement. Scattered neutrons from the sample was successfully observed and the result was consistent with the published result. However, the result is not statistically enough and does not have a good agreement with available experimental data in the large Q region, $Q > 0.1$ [1/Å]. This disagreement can be due to the effect of the oblique incident neutrons which do not have the same scattering angle with the nominal scattering angle. The estimation of the smearing effect is a subject for the future research in order to improve this new geometry.

The authors are grateful to Mr. Hirose for the operation of KUANS. This work was partially supported by the Seamless Technology Transfer Program through Target driven R & D, Japan Science and Technology Agency (JST).

https://doi.org/10.1051/epjconf/202023105001

References

1. R. J. Roe. *Methods of X-ray and neutron scattering in polymer science*. **739**. New York: Oxford University Press, (2000)
2. C. J. Glinka, J. M. Rowe, J. G. LaRock, J. Appl. Cryst. **19**, 427 (1986)
3. B. Khaykovich, M. V. Gubarev, Y. Bagdasarova, B. D. Ramsey, D. E. Moncton, Nucl. Instrum. Methods Phys. Res., Sect. A, **631**, 98 (2011)
4. S. Tasaki, T. Nagae, M. Hirose, Y. Yamashita, K. Hironaka, Y. Abe, Y. yamagata, Y. Otake, K. Hirota, *Physics Procedia* **60,** 181 (2014)
5. I. S. Anderson, C. Andreani, J. M. Carpenter, G. Festa, G. Gorini, C. -K. Loong, R. Senesi, Physics Reports **654,** 1 (2016)
6. F. Zhang, J. Ilavsky, G. G. Long, J. P. G. Quintana, A. J. Allen, P. R. Jemian, Metall. Mater. Trans. A **41**, 1151 (2010)

Design and Construction of an Imaging beamline at the Nagoya University Neutron Source

Katsuya Hirota[1,*], *Shogo* Awano[1], *Takuhiro* Fujiie[1], *Seiso* Fukumura[1], *Mayu* Hishida[1], *Go* Ichikawa[1,2], *Sohei* Imajo[1,3], *Ikuya* Itoh[1], *Yoshihisa* Iwashita[4], *Masaaki* Kitaguchi[1], *Yoshiaki* Kiyanagi[1], *Yasutoshi* Kuriyama[5], *Koki* Morikawa[1], *Yudai* Niinomi[1], *Hirohiko M.* Shimizu[1], *Kazuki* Tsuchida[1], *Yusuke* Tsuchikwa[1,6], *Yukio* Tsurita[1], *Akira* Uritani[1], *Kenichi* Watanabe[1], *Yutaka* Yamagata[7], *Nana* Yamamoto[1], *Atsushi* Yamazaki[1], *Sachiko* Yoshihashi[1], and *Tamaki* Yoshioka[8]

[1]Nagoya University, Furo-cho, Chikusa, Nagoya, 464-8602, Japan
[2]High Energy Accelerator Research Organization, Tokai, Ibaraki, 319-1106, Japan
[3]RCNP, Osaka University, Mihogaoka, Ibaraki, Osaka, 567-0047, Japan
[4]KyotoUniversity, Gokasho, Uji, Kyoto, 611-0011, Japan
[5]KyotoUniversity, Kumatori, Osaka, 590-0494, Japan
[6]Japan Atomic Energy Agency, Tokai, Ibaraki, 319-1106, Japan
[7]RIKEN, Hirosawa, Wako, Saitama, 391-0198, Japan
[8]Kyushu University, Motooka, Nishi, Fukuoka, 819-0395, Japan

Abstract. The Nagoya University Accelerator driven Neutron Source (NUANS) is constructed at the main campus of the Nagoya University. The electrostatic accelerator is used with the maximum proton energy and intensity of 2.8MeV, 15mA(42kW) respectively. Two neutron beamlines are designed at NUANS. The BL1 is dedicated to BNCT development. The BL2 is designed for research and development for neutron devices and neutron imaging. The neutrons used for the BL2 are generated by using the (p, n) reaction from a thin beryllium target. We constructed a compact target station for the BL2 and measured the neutron transmission image.

1 Introduction

Neutron is a very good probe to investigate the inner structure of materials. Large neutron facilities like J-PARC MLF and SNS have been constructed in this decade, and ESS facility is under construction. These large facilities are very good tools to study in the academic field. However, the opportunity to use them is very low, many neutron scientists use these facilities only once a year. The allocated beam time is usually shorter than 10 days on these powerful machines. The scientists require more neutron beam to study physics, chemistry, and engineering.

One of the solutions to solve this situation is to construct a Compact Accelerator-driven Neutron Source (CANS) [1]. The CANS has some merits compare to large neutron facilities. The constructing cost is lower than larger facilities. The opportunity to use the neutron beam is so higher. All devices from the accelerator to detectors is easy to handle, which is very good for student education. The amount of radiation is smaller than that of the large-scale facilities. Unfortunately, the emitted total neutron is much lower than large facilities. It is very important to consider what you can do with CANS and what you should do with large neutron facilities. We can use these CANS for BNCT(Boron Neutron Capture Therapy), neutron imaging, device development and so on.

*e-mail: hirota@phi.phys.nagoya-u.ac.jp

Nagoya University has constructed CANS using an electrostatic accelerator, Nagoya University Accelerator-driven Neutron Source (NUANS)[2–4]. Two neutron beamlines are designed at NUANS. The first beamline (BL1) is dedicated to BNCT development. The second beamline (BL2) is designed for research and development for neutron devices and neutron imaging. In this paper, we will present the current status of the NUANS BL2.

2 NUANS

Two beamlines are constructed at NUANS. In both beamlines, neutrons are provided by bombarding a high current proton beam to a lithium and beryllium target, respectively. The initial proton beam is provided by a Dynamitron electrostatic accelerator. The 2.8 MeV proton beam used in NUANS has a maximum current of 15 mA. The beam transport for each beamline is selected by using a dipole magnet located at 2.5 m downstream from the accelerator exit. The deflection angles from the initial beam axis to the beamlines are 20° and 70°, respectively. After dividing the beamline, the proton beam hits the neutron generation target.

3 Design and Construction of the BL2

The BL2 is designed as a multipurpose beamline for radiography and device development. The neutron generation target is installed 5m downstream of the proton bending dipole magnet. The layout is shown in a previous paper [2, 3]. Since this beamline uses a high-current proton beam, even if the center of the beam reaches the neutron generation target, in the case of the periphery hits the beam duct, the duct generates heat and eventually the duct breaks down. Therefore, it is necessary to monitor the beam loss so that the periphery of the beam does not hit the duct very much. A circular collimator with a diameter of 50 mm is installed just downstream of the bending magnet to prevent the tail of the proton beam that has passed through the bending magnet from hitting the beam duct. Also, a 4-quadrant slit is installed at a position of 3.8m downstream of the bending magnet to limit the size of the proton beam irradiated to the neutron generation target, and by monitoring the proton current hitting each slit, to confirm the balance of the proton beam in vertical and horizontal directions. Both collimators are water-cooled. A total of 16 thermocouple temperature gauges are attached to the proton beam duct, and the duct temperature is checked. If an abnormal temperature rise is detected, the accelerator beam is automatically stopped. In this way, the proton beam is controlled by inexpensive devices using the circular collimator, the 4-quadrant slit, and thermocouple gauges.

In this BL2, neutrons are generated using the Be(p,n) reaction. Two types of target materials, beryllium and lithium, are candidates for neutron generation targets in this energy region [5]. Because beryllium target is chemically stable and easier to use than lithium, BL2 uses beryllium. The proton beam hit a 40μm thick beryllium foil. This foil is silver brazed to a 5mm thick Vanadium plate. The thickness of the beryllium foil is designed so that the proton beam can penetrate the beryllium foil[6]. Cooling water flows on the back of this neutron generation target so that the heat generated by beam irradiation can be removed.

The radiation shield around the neutron generation target is using polyethylene blocks containing 10 % of B_2O_3, and Pb blocks. Since the estimated maximum neutron energy is less than 1MeV, it is not necessary to shield fast neutrons using iron plates. A multi-layer structure is adopted to shield radiation efficiently. A large amount of neutron flows from the neutron generation target to the upstream side of the proton beam duct, but the neutron shield cannot be inserted into the duct. So the proton beam duct is surrounded by polyethylene blocks containing B_2O_3.

A polyethylene block with a size of 100 × 100 × 100 mm^3 is used as a thermal neutron moderator. In general, water (H_2O) or polyethylene is often used as the thermal neutron moderator in CANS [1, 7, 8]. The reason why polyethylene is used here is that the neutron intensity is not so high, the frequency of carbonization of polyethylene is not high, and it is easier not to use a container to improve the shape of the moderator in the future. Graphite blocks of the 50 mm thickness are used as a reflector around the moderator.

Neutrons emitted from the neutron moderator pass through the hole in the shield and are measured at the experimental port. There are two experimental ports at BL2, BL2-short and BL2-long. The BL2-short and -long are on the same straight line. The length of flight path of the BL2-short is 50 cm and the BL2-long is 500 cm. The corresponding L/D values are 5 and 50, respectively. In L/D = 5, it is generally considered that measurement with sufficient quality is impossible when performing imaging measurement. However, if a measurement of sufficiently thin sample material is possible, measurement with high-intensity neutrons is possible even in such a small neutron source. Currently it measured only at BL2-short, because of the low neutron beam intensity. The effective area of neutron beam at the BL2-short is 10×10 cm^2.

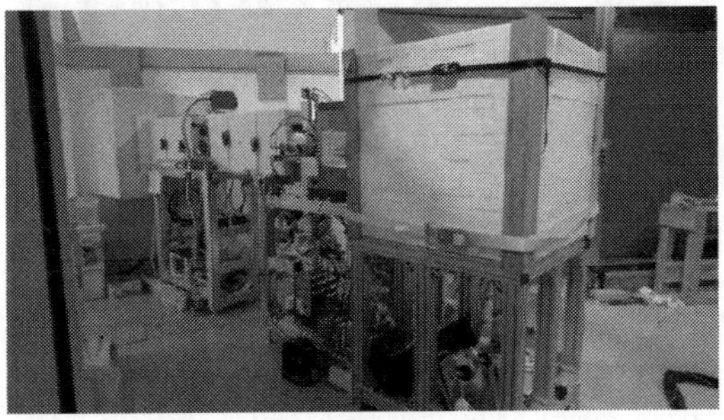

Figure 1. Photograph of NUANS BL2. Left hand side is the proton beam line. The center of the photo is the target shield.

Figure 1 shows the photograph of the NUANS BL2. The left hand side of the photograph is the proton beamline and the center is the target shield. Currently, the accelerator is being adjusted, and the proton beam intensity of 1.5 mA and 4.2 kW, which are the design values, has not been achieved and a thermal neutron beam measurement is used using a proton beam of approximately 0.1 mA. The neutron intensity used in the measurement is 2 $\times 10^4$ n/s/cm^2 at the BL2-short. The assumed thermal neutron energy spectrum is as shown by reference[4]. This accelerator is an electrostatic accelerator and is a DC beam, so neutron energy measurement using TOF measurement is not possible, and the energy spectrum is obtained by calculation. Neutron transmission images are measured with a combination of ^{6}LiF/ZnS(Ag) Scintillator and cooled CCD. The thickness of the Scintillator is 200 μm, which is determined by the balance between detection efficiency and spatial resolution. There is one mirror between the CCD and the scintillator to shield the γ rays, and the light emission image of the scintillator is measured by the CCD image reflected by the mirror. The cooled CCD is BU53LN from BITRAN Corporation [9]. This CCD has 4008x2672 pixel elements of 9 × 9μm, and can acquire images in 16 bits. Among CCDs in the same price range, this CCD camera is characterized by low noise due to the cooling function by the Peltier element, and the element area is large. This CCD camera is coupled to an 85mm F1.4 lens (Nikon Corp.). The lens aperture is usually set to the maximum, and the distance between the camera and the scintillator is changed to adjust the focus.

Figure 2 is a neutron transmission images of an ASTM neutron radiographic sensitivity indicator measured on the

BL2-short, which L/D value is about 5. The exposure time is twenty minutes. In the image on the left, the gaps created by spacers of about 100μm can be measured as lines. In the right figure, the defocus image due to beam divergence and sample thickness can be seen. In the right figure, two black lines are seen, the right line is clear but the left line is blurred. This is due to the beam divergence and sample thickness, because the Cd bars are placed on the front and back of the indicator

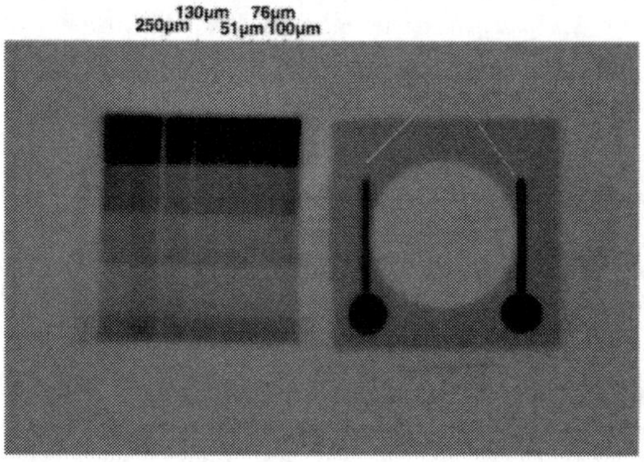

Figure 2. Neutron transmission Image of the ASTM neutron indicator

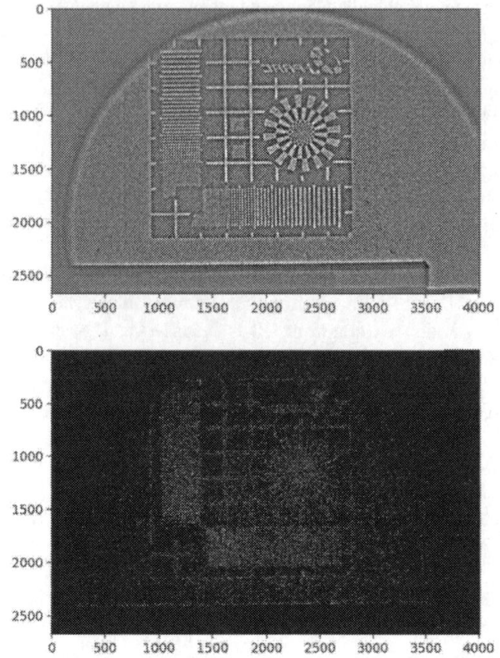

Figure 3. Images before (upper) and after (lower) the Laplacian filter is applied.

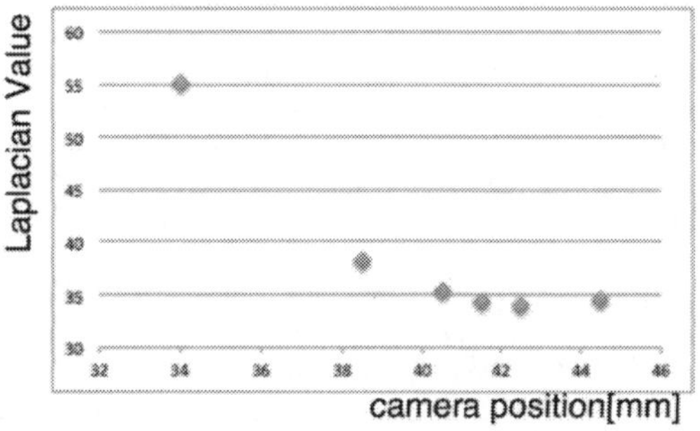

Figure 4. Position dependence of the Laplacian value

4 Focus Adjustment

In parallel with the construction of the neutron source, we are proceeding with a quantitative evaluation of acquired images. The focus adjustment of the neutron camera used for measurement is usually done with the naked eye. The focus position is depending on the person who works and the measurement conditions. To obtain the same result irrespective on the person who performed it, it was decided to evaluate the focus adjustment quantitatively. For focus adjustment, it is necessary to perform edge extraction on the acquired image. Therefore, the image obtained by the CCD camera is subjected to a noise filter for removing the CCD element error and an FFT (Fast Fourier transform) filter for removing the neutron intensity distribution dependency, and then the second derivative of the space (Laplace calculation). Since the Laplacian filter is easily affected by noise, it is necessary to eliminate noise such as dot filtering using a noise filter.

Figures 3 shows the images before (upper) and after (lower) the Laplacian filter is applied. The upper figure shows the result of applying the noise filter and bandpass filter to CCD output. The lower figure shows the image obtained by extracting an edge by the Laplacian filter. This sample neutron transmission image was taken at the E2 port of the KUR research reactor.

Figure 4 plots the calculation of the variance of the Laplacian with changing the camera position. This variance value is predicted to take the minimum value when focused. In this figure these values have a minimum at the position is 42.5mm. This camera position is the same as the focus point checked by our naked eyes. It may be a method that can be used to adjust the CCD camera's focus in CANS, although this method needs a little more detailed confirmation.

5 Summary

The spallation large accelerator facility is versatile and is widely used for precision measurement for academic purposes. On the other hand, a small neutron source using a small accelerator is expected to be used as a familiar neutron source in the future as a device specialized for a specific purpose. Under such circumstances, an accelerator neutron source using low-energy protons with Ep = 2.8

MeV was constructed. BL2 is designed as a multi-purpose neutron beamline, and the thermal neutrons generated by the Be (p, n) reaction and thermal moderator. Currently being adjusted the accelerator, 2×10^4 n/s/cm^2 at the BL2-short port position neutrons are available. The length from the moderator surface to BL2-short is 50 cm and the corresponding L/D value is 5. The neutron intensity will be increased in the future.

As part of the quantitative evaluation of neutron measurement, the focus adjustment during image acquisition is digitized. By performing a Laplacian calculation on the image, the edge of the acquired image was extracted and used for focus evaluation.

References

[1] I.S. Anderson, C. Andreani, J. M. Carpentaer, G. Festa, G. Gorini, C. K. Loong and R. Senesi. Phys. Rep. **654** (2016) 1-58

[2] Katsuya Hirota, **PoS(KMI2017)**(2017)025

[3] Ikuya Ito, D. Furusawa, K. Hirota, G. Ichikawa, Y. Iwashita, M. Kitaguchi, Y. Kiyanagi, Y. Menjo, Y. Miwata, H. M. Shimizu, K. Tsuchida, Y. Tsuchikawa, Y. Tsurita, A. Uritani, K. Watanabe, Y. Yamagata, A. Yamazaki, S. Yoshihashi,**PoS(KMI2017)**68

[4] Y. Tsuchikawa, K. Abo, D. Furuzawa, K. Hirota, G. Ichikawa, S. Imajo, I. Ito, Y. Iwashita, M. Kitaguchi, Y. Kiyanagi, Y. Miwata, K. Saito, H. M. Shimizu, H. Takehista, K. Tsuchida, Y. Tsurita, A. Uritani, K. Watabnabe, Y. Yamagata, A. yamazaki, S. Yoshihashi, JPS Conf. Proc. **22**, 011024(2018)

[5] M.R.Hawkesworth, Atomic Energy Review **15**, No.2 (1977) 169-220

[6] Yutaka Yamagata, Katsuya Hirota, Jungmyoung Ju, Sheng Wang, Shin-ya Morita, Jun-ichi Kato, Yoshie Otake, Atsushi Taketani, Yoshichika Seki, Masako Yamada, Hideo Ota, Unico Bautista, Oinngan Jia. J Rad. Nucl. Chem. (2015) **305**: 787-794

[7] S. Wang, Y. Otake, Y. Yamagata, T. Nagae, H. Fujioka, M. Hirose, Y. Kiyanagi, M. Furusaka, K. Hirota, Physics Procedia **60** (2014) 319

[8] S. Tasaki, T. Nagae, M. Hirose, Y. Yamashita, K. Hironaka, Y. Abe, Y. Yamagata, Y. Otake, K. Hirota, Physics Procedia **60** (2014) 181

[9] https://www.bitran.co.jp

Neutron spectroscopy from 1 to 15 MeV with Mimac-FastN, a mobile and directional fast neutron spectrometer and an active phantom for BNCT and PFBT

Daniel Santos, *Nadine* Sauzet, *Olivier* Guillaudin, *Jean-François* Muraz

Laboratoire de Physique Subatomique et de Cosmologie, Université Grenoble-Alpes -CNRS/IN2P3, France

Abstract. In the frame of direct dark matter search, the fast neutrons producing elastic collisions on the nuclei of the active volume are the ultimate background. The MIMAC (MIcro-tpc MAtrix Chambers) project has developed a directional detector providing the directional signature to discriminate them from the searched events based on 3D nuclear tracks reconstruction. The MIMAC team of the LPSC has adapted one MIMAC chamber as a mobile fast neutron spectrometer, the Mimac-FastN detector, having a wide neutron energy range (10 keV – 600 MeV) working with different gas mixtures and pressures. This presentation will be focused on the MeV range with ^{4}He + 5% CO_2 gas mixture at 700 mbar. A boron coating inside the active volume used for calibration purpose opens the possibility to use the active volume as an active phantom for Boron Neutron Capture Therapy (BNCT) and Proton Fusion Boron Therapy (PFBT).

1. Introduction

To perform a fast neutron spectroscopy without a time flight measurement or without many ^{3}He counters is a real challenge and required in many different domains, such as neutron dosimetry, identification of fissile nuclear material and nuclear physics in general. Some applications require measurements above 10 MeV and a large energy range. Among these applications, we can mention the secondary neutrons in radiotherapy, and the characterization of cosmic neutrons produced in the atmosphere by cosmic particles, going up to 100 MeV.

Neutron spectroscopy at high energies (above 1 MeV) is challenging for the present available detector technologies. Indeed, iterative moderation using neutron capture on converters leads to poor energy resolution and requires hypothesis on the expected neutron energy, the detection in solids through elastic collisions is limited due to the absorption of recoils in the converter, whereas detection in liquid scintillators results in a limited measuring range.

2. Detection Principle

Mimac-FastN is a micro-TPC (Time Projection Chamber) based on a micro-pattern detector coupled to a fast self-triggered electronics [1]. The chamber is filled with a gas that constitutes the converter of fast neutrons into nuclear recoils able to ionize the gas in the active volume. In the present paper, we describe the operation of Mimac-FastN with 2 litres of a gas mixture of 95 % of ^{4}He and 5 % of CO_2 as a quencher, at 700 mbar, to measure neutron energies between 1 MeV up to 15 MeV. The gas mixture and the pressure can be modified depending on the application energy range [11,12]. Fast neutron detection is performed through the tracking of the nuclear recoils that result from nuclear elastic scattering between incident fast neutrons and the gas nuclei. The nuclear recoils lose part of their kinetic energy by ionization in the detector gas. The primary electrons resulting from this ionization process are collected by an electrical field of 160 V/cm through a 25 cm long drift chamber, up to the micro-pattern detector (a square bulk Micromegas [2] with a 512 μm gap, and sides of 10.8 cm). A high electrical field of 10.5 kV/cm between the grid and the anode of the Micromegas produces avalanches, which result in the signal amplification. Resulting secondary electrons are collected on the pixelated anode and the ions drifting toward the grid.

The Mimac-FastN electronic board [3] manages two synchronized types of data. The first one is the energy released in ionization by the nuclear recoil, read through a charge preamplifier connected to the mesh of the micro-pattern detector. This preamplifier, developed at LPSC, has a gain of about 100 mV/pC (that is adjustable depending on the energy range required), and a time constant of 2 ms so that the rise time of the signal is small compared to the electronic decay time. The second type of data is the fired strips of pixels on the anode of the micro-pattern detector (512 strips, 256 in X and 256 in Y), which gives access to the 2D position of the charges.

The data on the grid and on the pixelated anode are read out at a sampling frequency of 40 or 50 MHz, depending on the length of tracks to be produced, and managed by the electronic board. In this way, each nuclear recoil track is sliced in samples. In the gas mixture $^4He/CO_2$ (5%) at 700 mbar and at 40 MHz sampling, each sample has a perpendicular component to the anode of 241 μm (referring to a Magboltz simulation that gives a drift velocity of 9.65 μm/ns in this gas mixture, which leads to a length of 9.65 μm/ns x 25 ns). So the 3D nuclear recoil track is reconstructed thanks to the composition of the 2D picture on the pixelated anode, and the perpendicular component inferred from the electronic sampling. The pixelated anode readout is performed by the 8 MIMAC ASICs, specifically developed by the MIMAC team of the LPSC [4].

Analysing the event-by-event sampled data from the grid and the pixelated anode, the kinetic energy of the incident neutron can be measured.

* Corresponding author: daniel.santos@lpsc.in2p3.fr

The neutron kinetic energy is deduced from the kinetic energy of the nuclear recoil by the following equation:

$$E_n = \frac{(1 + m_R)^2}{4m_R} \times \frac{E_R}{cos^2(\theta_{RN})}$$

being E_n the incident neutron energy, E_R the kinetic energy of the nuclear recoil, θ_{RN} the angle between the nuclear recoil track and the incident neutron direction, and m_R the nuclear recoil mass.

The kinetic energy of the nuclear recoil is determined from the measured ionization energy, corrected from the ionization quenching factor ([7] and [8]) in the considered gas mixture.

The angle θ_{RN} is estimated from the 3D recoil track reconstruction, and from the neutron emitter position that gives the incident neutron direction, see figure 1.

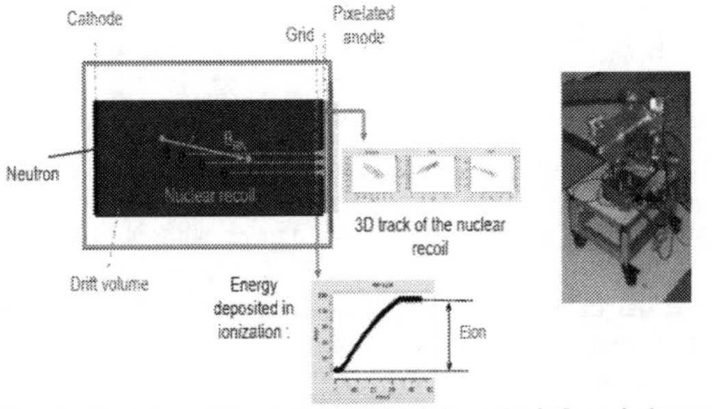

Fig. 1: Drawing of the detector principle at the left, and picture of Mimac-FastN at the right

3. Energy Calibration

The charge profile is measured through a charge preamplifier connected to the mesh and sent to a Flash-ADC (on the electronic board) that digitizes the signal on 4096 channels. The Flash-ADC energy calibration is done through a natural boron coating fixed on the cathode and by means of the following capture reaction of thermal neutrons by the ^{10}B isotope:

$$n + {}^{10}_{5}B \rightarrow {}^{4}_{2}He + {}^{7}_{3}Li + \gamma \quad (94\ \%)$$

$$n + {}^{10}_{5}B \rightarrow {}^{4}_{2}He + {}^{7}_{3}Li \quad (6\ \%)$$

The boron coating consists of an IBS (Ion Beam Sputtering) deposit of 500 nm of $^{nat}B_4C$ on an aluminum sheet, see picture in figure 2 (left). The ^{10}B isotope represents 20% of the natural boron. The coating has a specific shape in order to check the spatial resolution of the boron projection picture on the anode.

The energies deposited in ionization by the ^4He and ^7Li particles are measured on the Flash-ADC, and their tracks are imaged on the pixelated anode.

The figure 2 (right) shows the anode projection of the detector exposed to a 3 MeV neutron field crossing a 5 cm thickness moderator of high density polyethylene (HDPE). The projection on the anode of the first point of all the tracks directed towards the anode, highlights the boron coating, due to neutron captures, that become predominant with respect to the total elastic scatterings on the gas nuclei, at low energies. The presence on this picture of the specific shape of the boron corner proves the uniformity of the electrical field lines in the field cage and gives an estimation of the spatial resolution.

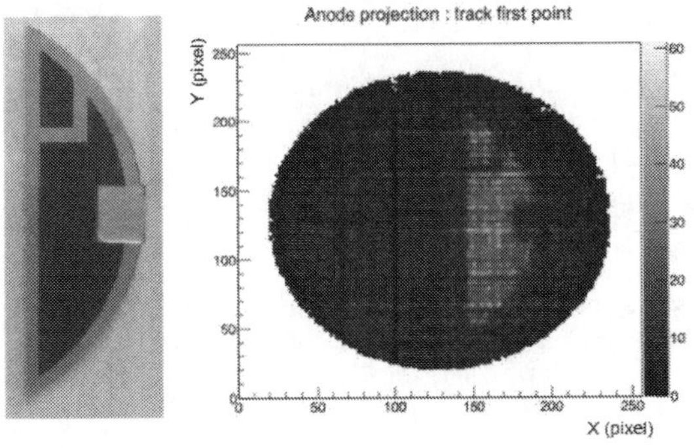

Fig. 2 : On the left, a picture of the B_4C coating used for energy calibration and spatial resolution measurements ; on the right, projection on the pixelated anode of the first point of all the tracks, that highlights the boron shaped coating, in a neutron beam of 3 MeV moderated through 5 cm of HDPE.

A selection of all the tracks whose interaction points are located on the boron coating projection, leads to the energy spectrum of the particles issued from neutron captures on ^{10}B, as presented in figure 3. This measured spectrum can be compared to the ionization energy spectrum calculated with Geant 4, shown in figure 4.

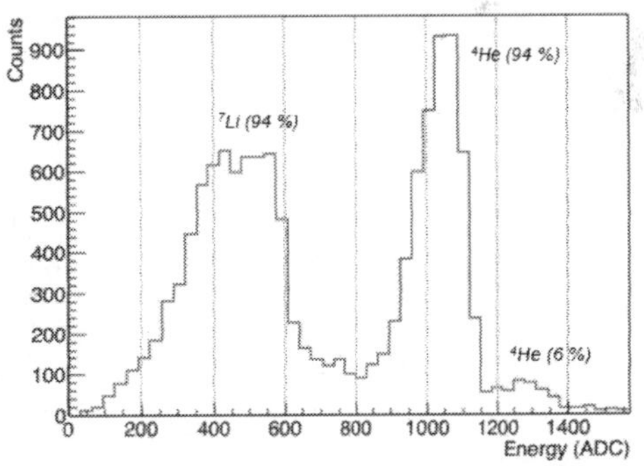

Fig. 3: Measured ionization energy spectrum of the ^4He and ^7Li particles resulting from neutron captures on the boron coating, in a neutron beam of 3 MeV moderated through 5 cm of HDPE.

The Geant4 simulation embeds an ionization-quenching factor, which is defined as the amount of ionization energy deposited by the particle, compared to its kinetic energy ([7] and [8]). In the energy ranges considered above, the mean ionization quenching factor considered by the GEANT4 simulation is 97,5 % for ^7Li, and 98,8 % for ^4He.

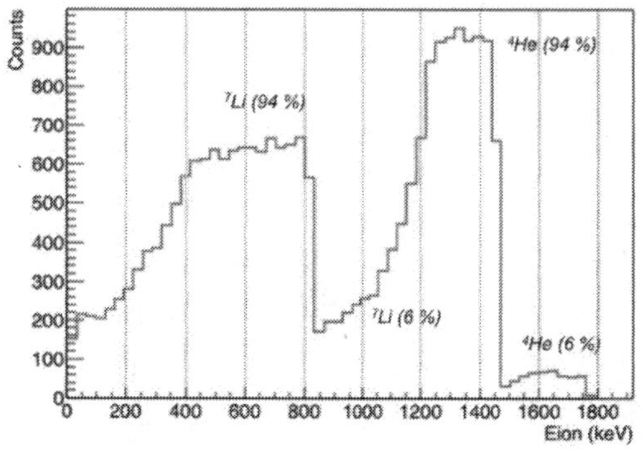

Fig. 4: Simulated ionization energy spectrum of the ^{4}He and ^{7}Li particles resulting from neutron captures on the boron coating.

The energy calibration of the Flash-ADC can be done comparing the simulation with the measurements taking into account the end-points or the mean peak values of the peaks issued from the branching ratio of 94 %. The peak issued from the alpha particle emitted in the case of a branching ratio of 6 % constitutes a checkpoint of the calibration equation.

This measurement opens the possibility to use the detector as an active phantom for BNCT [15] placed in a high flux epithermal neutron field just after the moderator and counting the neutron captures produced at the coating level simulating the tumor and the amount of ^{10}B placed on it by a previous vectorization. In this way a dose calculation can be performed with a controlled amount of boron and with the thermal neutron field available. In addition, the active phantom can be used with a high energy proton beam during a PFBT (Proton Fusion Boron Therapy) counting the alpha particles produced on ^{11}B.

4. Experimental Results

The angular distributions of the ^{4}He recoils with respect to the neutron direction, resulting from elastic scattering with fast neutrons, is a function of the neutron energy.

These angular distributions, see figure 5, show that for neutron energies above 3 MeV, angles above 60° are more likely. This has two implications. The first one is that the most probable kinetic energy of the ^{4}He recoil is 1.1 MeV for a neutron of 15 MeV. In a mixture of ^{4}He/CO_2 (5%) at 700 mbar, a recoil track with this kinetic energy is 3.3 cm long according to SRIM [6], and so remains contained in the drift volume as described previously. From these simulations, we deduce that above 3 MeV, the higher the neutron energy is, the smaller the recoil track will be.

The second consequence is that a 3D geometry is required to detect recoils for neutron energies above 3 MeV. The advantage of a gaseous detector like Mimac-FastN with a cylindric or cubic symmetry geometry, is that recoils can be detected whatever their direction is. The pixelated anode could represent a limitation, if the nuclear recoil tracks are parallel to this plane, since in this case, the 3rd dimension calculated by the electronic sampling will be limited to a few samples. However, knowing the position of the neutron emitter, the chamber can be orientated perpendicularly to the mean emission direction. In such a configuration, if the θ_{RN} angles are above 60°, the tracks' orientations are optimized compared with the anode plane.

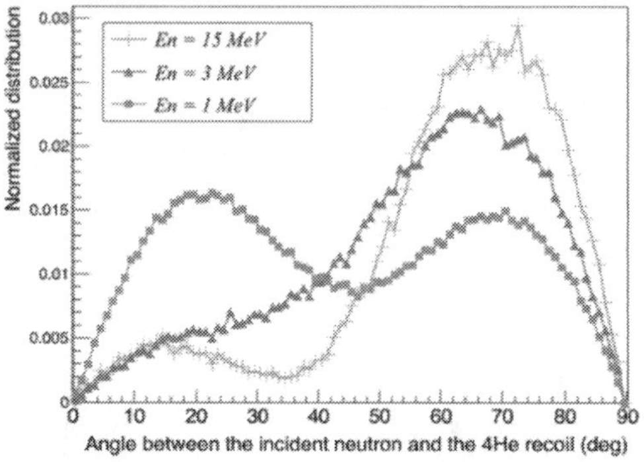

Fig. 5: The simulated angular distributions of ^{4}He recoils with respect to the neutron direction in the laboratory coordinate system, resulting from elastic scattering with neutrons of 1 MeV, 3 MeV and 15 MeV.

Angular distributions in the laboratory frame have been calculated with the Monte Carlo code Geant 4 [5], version 10.1.2, with the physics list QGSP_BERT_HP_LIV, and a chamber filled with a gas mixture of ^{4}He/CO2 (5%) at 700 mbar.

4.1 Experimental results from D(d(220)keV,n)

Spectroscopy of the neutrons produced by the reaction D(d(220 keV,n) has been performed at the GENESIS facility [9], with Mimac-FastN placed at 0° degree with respect to the deuteron beam axis. In this configuration, the D(d(220 keV),n) nuclear reaction produces neutrons of 3.1 MeV.

The target was a solid target, composed of titanium loaded with deuterium, and evaporated on a 3 mm thick copper backing [10].

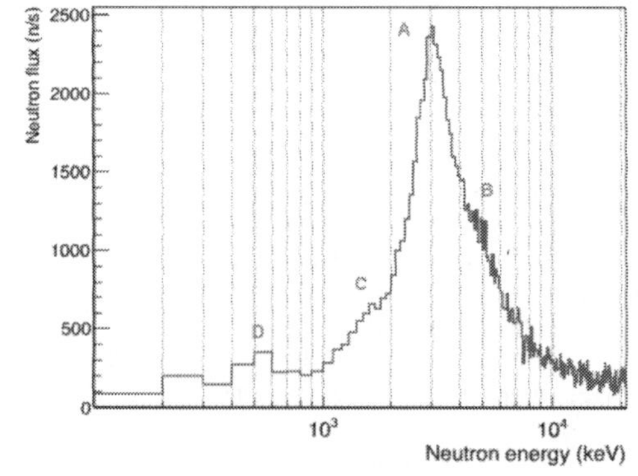

Fig. 6 : Measured neutron spectrum with the reaction D(d(220 keV),n) at GENESIS, with a binning of 120 keV/bin.

The measurement has been done in one hour, with a deuteron current of 100 μA, in the configuration with the longitudinal axis of the detector perpendicular to the beam axis. The detector was positioned at 1 meter from the target. In this directional measurement, we consider the target location as the neutron source position. The spectrometer Mimac-FastN was filled with a gas mixture of 95 % of ^{4}He and 5 % of CO_2 at 700 mbar.

Fast neutrons can be scattered on the concrete walls of the bunker. All the discriminations presented previously have been applied on the data, and the degrees of freedom brought by the 3D geometry have been explored to discriminate part of these scattered neutrons on the walls of the facility, and finally to reconstruct the neutron energy spectrum, plotted in figure 6.

The measured neutron spectrum reveals a polyenergetic spectrum with 4 structures. This is explained by the target composition and the interaction of deuterons with all the components of the target.

The peak **A** is the expected production of neutrons of 3.1 MeV resulting from the reaction D(d(220 keV),n). The shoulder **B** is produced by neutrons resulting from ^{63}Cu(d,n)^{64}Zn and ^{65}Cu(d,n)^{66}Zn with energies between 5.5 MeV and 6.8 MeV, and from ^{48}Ti(d,n)^{49}V with energies around 4.5 MeV. The shoulder **C** results from ^{64}Zn(d,n)^{65}Ga with energies around 1.8 MeV. The structure **D** is a contribution of the residual scattered neutrons on the walls of the accelerator bunker.

These measurements reveal that cross sections of low energy deuterons on copper and zinc are not negligible, despite the lack of data reported in the literature on this subject.

4.2 Experimental results from T(d(220)keV,n)

In the same facility and with the same experimental set-up, spectroscopy of the neutrons produced by the reaction T(d(220 keV),n) has been performed. At 0° degrees with respect to the deuteron beam axis, the T(d(220 keV),n) nuclear reaction produces neutrons of 15.1 MeV.

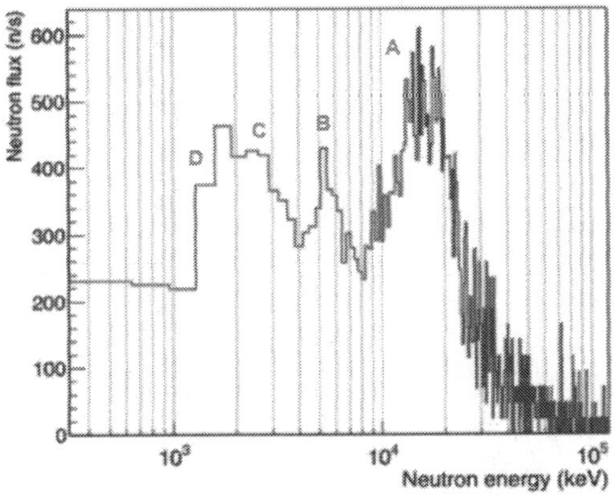

Fig. 7: Measured neutron spectrum with the reaction T(d(220 keV),n) at GENESIS, with a binning of 320 keV/bin.

The target has the same structure as the one described previously with the deuterium loading replaced by a tritium loading.

The measurement has been done in 1 hour, with a deuteron current of 5 μA. The neutron spectrometer longitudinal axis is perpendicular to the beam axis. It is located at 1.7 meters from the target, and the gas mixture is the same as the one used for the reaction D(d(220 keV),n) experiment.

The figure 7 shows the reconstructed neutron spectrum.

5. Conclusions

In this presentation, we have described the ability of Mimac-FastN to measure mono-energetic neutron spectra at 3 MeV and 15 MeV. Using the same gas mixture, with a threshold on the neutron energy as low as 200 keV, this directional fast neutron spectrometer gives a complete polyenergetic neutron spectrum exploring the material and eventual pollutions of the target or neutron sources.

This ability to provide polyenergetic neutron spectrum has already been applied to characterize the angular distribution of fast neutrons produced in a nuclear reaction proposed for a radiotherapy called Accelerator-Based Boron Neutron Capture Therapy (AB-BNCT) [13]. We have recently demonstrated the interesting active phantom application for BNCT (Boron Neutron Capture Therapy) and PFBT (Proton Fusion Boron Theapy) profiting of the calibration device.

With its high spatial resolution 3D track reconstruction associated with its large adjustable measuring range, Mimac-FastN is a versatile instrument that opens new fields for directional neutron spectrometry, for applications such as nuclear matter characterization, detection of target pollution, nuclear cross section measurements or monitoring the neutron production for radioprotection purposes.

Besides this study, preliminary measurements performed at CERF (CERN) [14] with this instrument have recently shown a good potential for spectrometry at neutron energies as high as 200 MeV, a range covering atmospheric neutron production and high energy neutron monitoring.

For more details on this MIMAC-FastN data analysis see our recent publication [16]

6. Acknowledgements

This work has been funded by the "Prematuration" program of CNRS, by the LabEx Enigmass, and by Linksium SATT (Technology Transfer Accelerator Office).

We thank the LPSC accelerator team for the operation of the GENESIS facility labeled CNRS for their support during all the performed experiments.

7. References

[1] D. Santos *et al.*, "MIMAC: A micro-TPC for directional detection of dark matter", EAS Publications Series, vol. 53, pp. 25-31, 2012.

[2] I. Giomataris *et al.*,"Micromegas in a bulk", NIM A, vol. 560, pp. 405-408, 2006.

[3] O. Bourrion *et al.*, "Data acquisition electronics and reconstruction software for real time 3D track reconstruction within the MIMAC project", JINST 6 C11003, 2011.

[4] J. P. Richer *et al.*, "Development of a front end ASIC for Dark Matter directional detection with MIMAC", NIM. A, vol. 620, pp. 470-476, 2006.

[5] S. Agostinelli *et al.*, "GEANT4, a simulation toolkit", NIM A, vol. 506, pp. 250-303, 2003.

[6] J.F. Ziegler and J.P. Biersack, SRIM - The Stopping and Range of Ions in Matter (Pergamon Press New York, www.srim.org, 1985).

[7] D. Santos *et al.*, "Ionization Quenching Factor Measurement of ^{4}He", arXiv:0810.1137v1, Oct. 2008.

[8] O. Guillaudin *et al.*, "Quenching factor measurement in low pressure gas detector for directional dark matter search", EAS Publications Series, vol. 53, pp. 119-127, 2012.

[9] GENESIS facility : http://lpsc.in2p3.fr/index.php/fr/peren-energie-nucleaire

[10] C.Monnin et al., "Characterization of deuteride titanium targets used in neutron generators", NIM A, vol. 453, pp. 493-500, 2000.

[11] D.Maire et al., "Neutron energy reconstruction and fluence determination at 27 keV with the LNE-IRSN-MIMAC microTPC recoil detector", IEEE transactions on Nuclear Science, 63(3) : 1934-1941, June 2016.

[12] Q.Riffard et al., "MIMAC low energy electron-recoil discrimination measured with fast neutrons", JINST, August 2016, Vol.11 Issue 8, p1-1.

[13] M.E Capoulat, N.Sauzet et al., "Neutron spectrometry of the ^{9}Be(d(1.45 MeV),n)^{10}B reaction for accelerator-based BNCT", NIM B, vol.445, pp 57-62, 2019

[14] A.Mitaroff, M.Silari, "The CERN-EU high-energy reference field (CERF) facility for dosimetry at commercial flight altitudes and in space", Radiation Protection Dosimetry, vol.102, n°1, pp 7-22, 2002

[15] Nuclear Physics European CollaborationCommittee (NuPECC) Report, Nuclear Physics for Medicine, 2014: Chap.10, pp46-42

[16] N.Sauzet, D. Santos, O. Guillaudin et al., "Fast Neutron spectroscopy from 1 MeV up to 15 MeV with Mimac-FastN, a mobile and directional fast neutron spectrometer", arXiv:1906.03878, to be published in NIMA.

High-count-rate ^{3}He position-sensitive detector system: NEUNET-HCR

Setsuo Sato[1,*], *Tomohiro* Seya[1], *Hidetoshi* Oshita[1], *Hideto* Kato[2], *Noriyuki* Hikida[2], *Kazuya* Ishizawa[2], *Akinori* Yamaguchi[3], and *Mitsugu* Matsuura[3]

[1]High Energy Accelerator Research Organization, 1-1 Oho Tsukuba Ibaraki, Japan
[2]Canon Electron Tubes & Devices Co., Ltd., 1385 Shimoishigami Ohtawara Tochigi, Japan
[3]CLEAR PULSE Co., Ltd., 6-25-17 Chuo Ota-ku Tokyo, Japan

Abstract. Highly reliable ^{3}He gas detectors are most often used in neutron scattering experiments but have disadvantages such as low count rates and low position resolutions. In this study, a ^{3}He gas detector system called NEUNET-HCR was developed, whose count rate is approximately 10 times that of the current NEUNET system, which was developed for a high position resolution with a low count rate. NEUNET-HCR achieved a maximum count rate of 535 kcps per detector, with a loss of approximately 30%. As the NEUNET-HCR system was developed based on the NEUNET system, various improvements were required to achieve the high count rate. Further, a high count rate was achieved in the NEUNET-HCR system without significantly degrading the position resolution.

1 Introduction

Recently, the industrial use of neutrons has become widespread, and compact accelerator-driven neutron sources have been used to develop new materials. However, the development of neutron detectors that support these experimental facilities has been delayed, and ^{3}He gas detectors are still used in many neutron experiments. ^{3}He gas detectors are the most reliable neutron detectors and are typically utilized as position-sensitive detectors (PSDs) [1, 2]. However, these detectors have low count rates and low position resolutions. Therefore, joint research has begun between Canon Electron Tubes & Devices Co., Ltd. (CETD), which manufactures PSDs, and the High Energy Accelerator Research Organization (KEK), which develops readout circuits for neutron detectors. Consequently, improvements to compensate for these shortcomings have been made. Firstly, we developed a NEUNET system that increased the position resolution, and a position resolution of 2 mm or less in the axial direction was obtained with a PSD 1 m in length and 8 mm in diameter. However, we found that detectors with high count rates are demanded. Thus, we changed the focus of our research to achieve higher count rates. In this study, we succeeded in increasing the count rate to 535 kcps with approximately 30% loss. Even if the maximum count rate for practical usage is 200 kcps, the obtained count rate is approximately 20 times higher than that of the commercial ^{3}He-PSD, which has a maximum count rate of 10 kcps [2].

A high-count-rate readout circuit was developed based on a NEUNET system [1]. The NEUNET system was originally designed based on the assumption that the PSD had a low count rate; therefore, the processing speed needed to be increased. However, when a high-count-rate readout circuit is used for a slow PSD, various characteristics are degraded and there is a risk of double counting. Therefore, the pulse width measurement method designed using another detector system [3] was applied to develop a method for distinguishing between fast and slow PSDs.

2 Readout circuit

The high-count-rate readout circuit was developed based on the NEUNET system. As the NEUNET system was designed for slow gas detectors with high position resolution, numerous improvements were necessary to achieve a count rate 10 times faster than that of the current system.

2.1 NEUNET system

The NEUNET system was developed with the objective of obtaining a position resolution (full width at half maximum, FWHM) of less than 2 mm with a length of 1 m. Figure 1 shows a typical PSD system with an effective length of 1 m, a detector diameter of 8 mm (top), and amplifiers, as well as a NEUNET circuit (bottom) embedded behind a shield. As the data output is obtained through only one network cable, small to large systems can be configured in a scalable manner.

The PSD amplifier circuit should have low noise in order to obtain an FWHM less than 2 mm with a length

* Corresponding author: setsuo.satoh@kek.jp

https://doi.org/10.1051/epjconf/202023105004

of 1 m. Consequently, the amplifier was designed such that it can be installed behind the PSD.

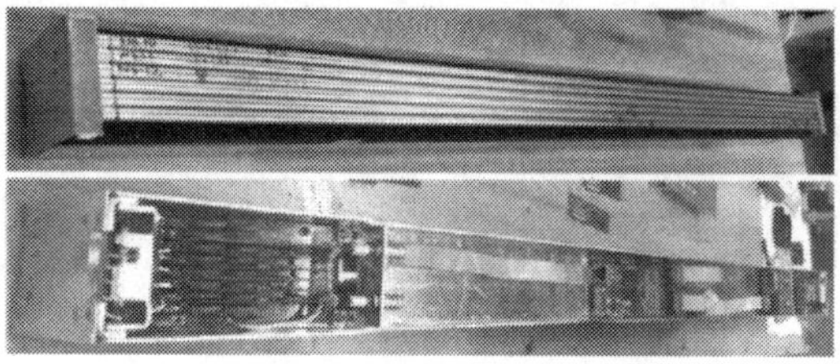

Fig. 1. PSD system of 1 m length and 8 mm diameter.

Figure 2 shows the PSD amplifier board. Charge amplifier (AMP14b) boards are attached to and detached from an amplifier mounting (MAMP14b) board with connectors; these can easily be replaced individually. Nine AMP14b boards are attached to the MAMP14b board: eight boards are used for the actual measurements, and the other board serves as both an electric field shield and backup. The AMP14b is composed of a two-stage amplifier. The first-stage integrated circuit uses a low-noise, high-speed field-effect transistor operational amplifier. The soldered side of the board is almost completely grounded to minimize electric field crosstalk. In addition, input signal guards are installed to minimize charge leakage. The second stage is a pulse-shaping circuit that selects the pulse width that can pass through.

Fig. 2. PSD amplifier board.

Figure 3 shows the NEUNET board, which has 16-channel high-speed analog–digital converters (ADCs) that are processed by a field-programmable gate array (FPGA). The NEUNET board sends the data to a control PC by a network.

Fig. 3. NEUNET board.

Figure 4 shows a block diagram of the NEUNET system. The 16 signals from both sides of the eight PSDs are amplified by AMP14b and converted into digital signals by ADCs on NEUNET. The waveform of the ADC output with the FPGA is analyzed to determine the pulse height as the charge corresponding to the neutron. If the signal is determined to be a neutron signal, the time and detector number are added as event data and sent to the control PC by the network.

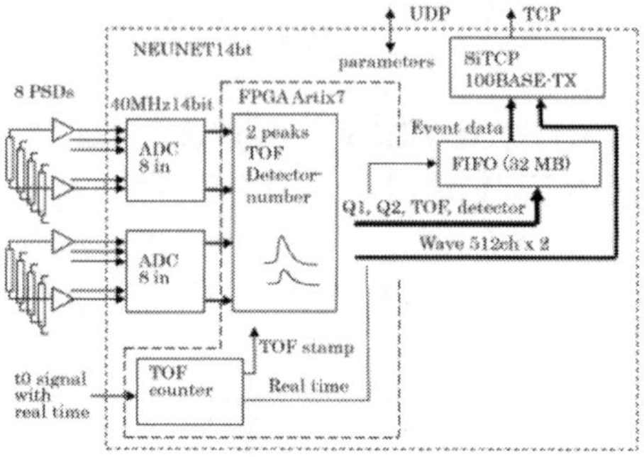

Fig. 4. Block diagram of the NEUNET system.

The ADC resolution was increased from 12 to 14 bits to achieve a high position resolution of 2 mm with a PSD length of 1 m. The position of the PSD was obtained from the ratio between the pulse heights of the charges from both ends, and each pulse height was obtained from the difference between the maximum peak and baseline of the ADC output. As the maximum range of the pulse height distribution was 14 bits, limiting it to 1/4 or more yields 12 bits as the position resolution of 4096 points. When 1 m is adjusted to 3000 points, 2 mm is 1/500 of 1 m; therefore, 2 mm is represented by 6 points and is a usable range.

The ADC used is a pipeline type that tends to diffuse the electrical noise at the input side. Although it has been reduced by ceramic capacitors, etc., it was not completely lost; therefore, a digital filter was programmed. A weighted moving average, which requires fewer circuits, was employed as the digital filter. No filter, 5, 13, and 16 clocks (no weighting factor) were tested. The 13-clock filter yielded the best results and was therefore adopted by default. Figure 5 shows the position distributions with the 13-clock filter and no filter. They were measured under the same conditions; however, the 13-clock filter has a much better position resolution than no filter. Although these techniques often conflict with the high count rate, they are in a trade-off relationship because they are important characteristics that are inherent in the PSD.

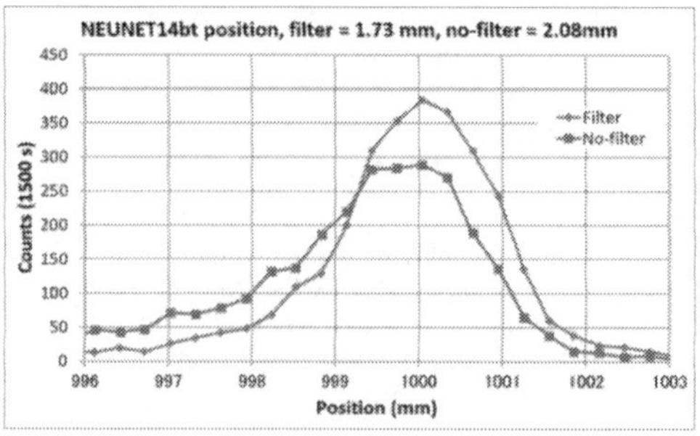

Fig. 5. Position distributions with the 13-clock filter and the no filter.

2.2 NEUNET-HCR system

Several improvements were required in the NEUNET system to ensure that the count rate of the NEUNET-HCR system would be 10 times faster.

• Adoption of a fast pulse-shaping time constant for the AMP14b: The pulse-shaping time constant of the AMP14b is approximately 10 times faster than that of the current board. However, when used with a slow PSD, the risk of double counting is increased and various characteristics are degraded.

• Adoption of a 1000BASE-T network IC: 100BASE-TX was used as the basic network IC; however, by changing to 1000BASE-T, it became possible to exceed the network speed of 12 MB/s.

• Change of the network transfer method: The network transfer method was changed from a handshake method to a one-way method. In the handshake method, the control PC informs NEUNET in advance of the maximum number of transfer bytes, and within that range, NEUNET sends back the data, including the number of transfer bytes, to the control PC. The network receive program specifies the number of received bytes before the transfer; however, the transfer direction is reversed for each handshake and the network flow is degraded. The one-way method simply sends data from NEUNET to the control PC for each neutron. It is suitable for high-speed transfer because it does not stop the network flow. From the measurement of the maximum transfer amount using a test pulse generator, this change improved from 0.5 M cps to 2.6 M cps per NEUNET.

• Deadtime control: When a slow PSD signal is amplified with AMP14b with a fast time constant, it sometimes counts the signal twice (double counting). Therefore, a certain dead time is provided after the pulse detection process in order to prevent double counting. For a fast PSD signal, there is no risk of double counting. Therefore, the dead time has been set to 0 for obtaining the high count rate. When a dead time of 800 ns is set, the count rate does not exceed 450 kcps per detector; however, when the dead time is set to 0 ns, the count rate exceeds 500 kcps. The dead time setting is, therefore, changeable.

• Histogram memory stoppage: The data storage method of a large-capacity memory also hinders speed. The large-capacity memory is indispensable for a smooth network transfer. As NEUNET is not a large-scale circuit, histogram data are stored in the same memory. Although these data are no longer used, they are inherited owing to their requirements in the previous system. Measurement without the histogram access using the test pulse generator improves from 2.1 M cps to 2.6 M cps per NEUNET.

• Pulse width measurement function: A pulse width measurement function is provided to diagnose whether a PSD can increase the count rate. This technology [3] has been developed for other detector systems. Figure 6 shows the pulse width measurement method. The pulse width is obtained by acquiring the pulse height from the pulse waveform. To determine the pulse width, the counting starts from the rising edge of the pulse and stops when it reaches half the height of the pulse.

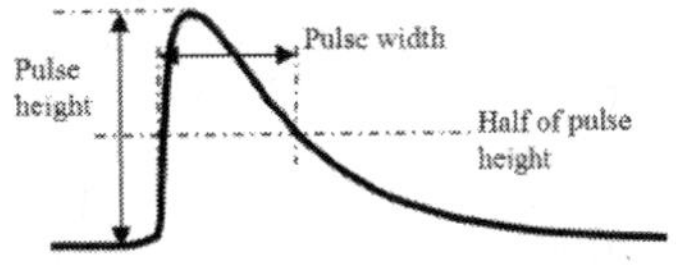

Fig. 6. Pulse-width measurement method.

With these improvements and new functions, the limit of 4 MB/s for transferring 8 PSD data was increased to 21 MB/s. As there are 8 bytes of event data per neutron, an average transfer rate of 325 kcps per PSD is possible.

3 High-count-rate detector

A high-count-rate PSD can be achieved according to measurements of various conditions. Even with nearly the same gas composition and gas pressure, the characteristics often vary greatly with slight differences. If the tube diameter is short and the gas pressure is low, the PSD tends to have a high count rate.

3.1. Pulse width

AMP14b with a short pulse-shaping time constant was selected because the pulse width of the PSD output signal must be narrowed. In general, a signal slower than the pulse-shaping time constant becomes a pulse with a long width and low height, while a signal that is faster becomes a pulse with a short width and high height.

Figure 7 shows the pulse height versus width (PHW) graphs using the pulse width measurement function. The vertical axis is the pulse width (25 ns per channel), and the horizontal axis is the pulse height. Figure 7(a) shows that the entire signal gathers around 280 ns along the vertical axis, and it is a high-count-rate PSD (stopping gas: CF4, ^{3}He gas pressure: 6 atm.). Figure 7(b) shows that the low pulse height signal extends to around 700 ns, and it is a low-count-rate PSD (stopping gas: CF4, ^{3}He gas pressure: 20 atm.). The conditions are the same except that the gas pressure is different. As shown in these examples, the pulse width measurement function can identify a high-count-rate PSD. Because the fast PSD was filled with ^{3}He gas of 8 mm diameter and a pressure of 6 atm, the detection efficiency was estimated to be 19.8 % at the 1 Å neutron at an average of the tube. Similarly, because the slow PSD had a pressure of 20 atm, the efficiency was estimated to be 49.9 %. In this case, because the fast PSD detection efficiency is low, it is suitable for slow neutrons.

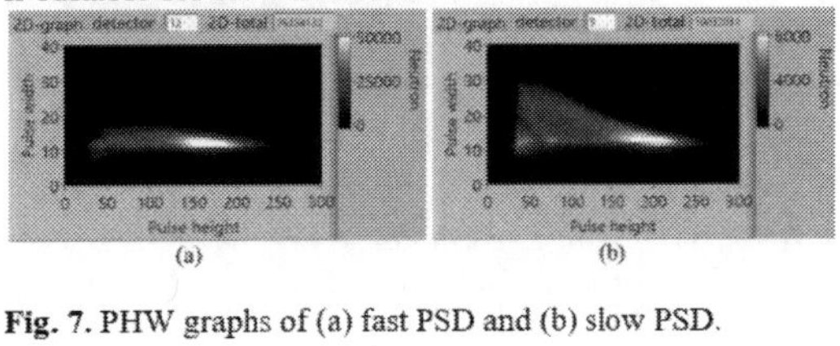

Fig. 7. PHW graphs of (a) fast PSD and (b) slow PSD.

Figure 8 shows the trajectories of a triton and proton after the reaction between a neutron and ^{3}He. The neutron and ^{3}He react, and the proton and triton fly in opposite directions, ionizing the surrounding gas and consuming 746 keV of energy. When flying parallel to a core wire of the PSD, the charge arrives at the core wire almost simultaneously, resulting in a narrower pulse signal with a greater amplitude. When flying perpendicular to the core wire, the charge arrives at the core wire with time dispersion, resulting in wider pulse signal with a lower amplitude. If the generated charge is uneven with a slow PSD, there is a risk of double counting. If the charge transfer speed is faster than the pulse shaping time constant of the AMP14b, these are integrated into one pulse, resulting in a narrow and large-amplitude pulse signal.

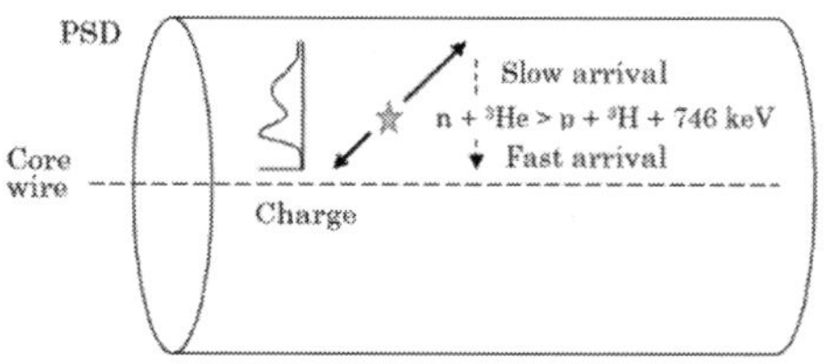

Fig. 8. Trajectories of a triton and proton after the reaction between a neutron and ^{3}He.

3.2 Time constants of AMP14b

The charge amplifier currently of the NEUNET system used at the Materials and Life Science Experimental Facility at the Japan Proton Accelerator Research Complex has an integration time constant of 25 μs and a pulse shaping time constant of 0.5 μs. As the pulse width spreads to approximately 3 μs and neutrons are received at an average of 30 μs with 10 % loss, the maximum average count rate is approximately 30 kcps. Because the setting is slow, good characteristics can be obtained from most PSDs.

For the high-count-rate PSD of NEUNET-HCR, the integration time constant was changed to 1 μs, and the pulse shaping time constant was changed to 0.05 μs, achieving approximately 10 times the original count rate. Figure 9 shows waveforms obtained when the pulse shaping time constant was 0.5 μs and 0.05 μs. In this case, the pulse widths are more than 5 times different.

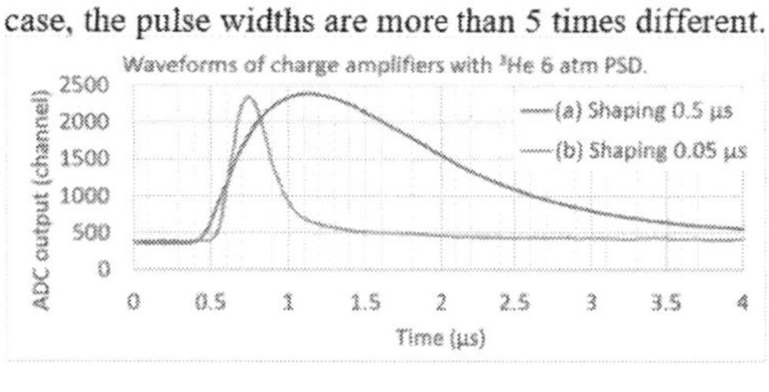

Fig. 9. Pulse shaping time constants of (a) 0.5 μs and (b) 0.05 μs.

4 Experiment with a Neutron Source

The high-count-rate experiment of the NEUNET-HCR system was performed at the B3 beam port at Kyoto University Research Reactor (KUR). The experimented PSD was filled with 14 atm. ^{3}He gas in 8 mm diameter. The detection efficiency was estimated to be 39.2 % at the 1 Å neutron at an average of the tube. As shown in Fig. 10(a), because the number of neutrons was insufficient, the effective thickness of the^{3}He gas was increased by oblique beam injection, and it was measured with varying neutron beam intensity and calibrated with a monitor detector. Figure 10(b) shows the pulse height distribution map. As the beam intensity increases, the beam spreads up to approximately twice the height of the pulse peak because of a pileup. As depicted in the count-rate graph in Fig. 10(c), the maximum count rate of 535 kcps was obtained with a loss of less than 30 %.

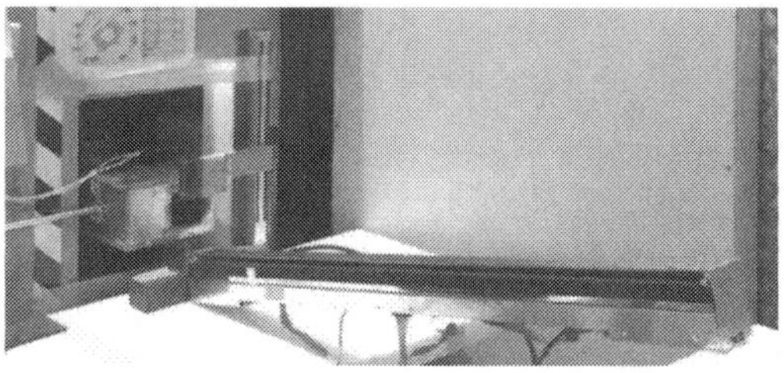

(a)

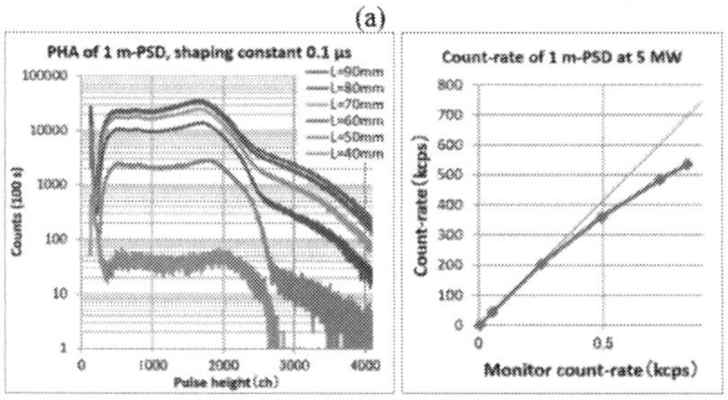

(b) (c)

Fig. 10. (a) Oblique placement of the PSD on the beam. (b) Pulse height distribution. (c) Count rate.

Figure 11 shows the position distribution when neutrons were irradiated at seven points in 15 cm intervals. A pulse-shaping time constant of 0.1 μs yielded a better position resolution than 0.05 μs did. However, the FWHM was less than 6 mm at all points, and the performance as a PSD was maintained.

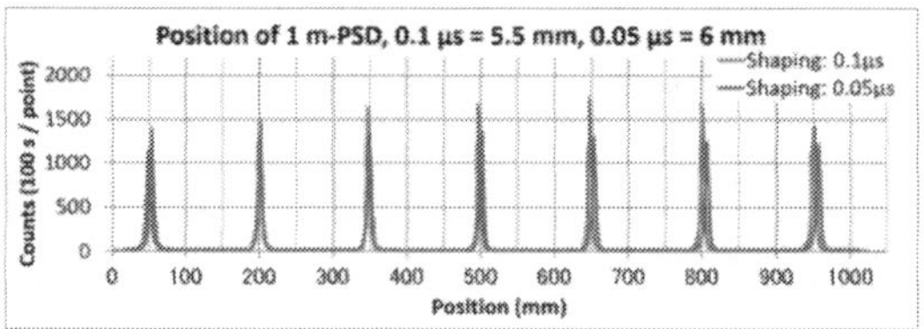

Fig. 11. Position distribution.

5 Conclusion

We developed a novel high-count-rate ^{3}He PSD system. A high count rate of 535 kcps was obtained with a loss of approximately 30%, which is very suitable for use in compact accelerator-driven neutron sources.

The authors wish to thank Dr. Tasaki and Mr. Hirose, who are in charge of the Kyoto University Accelerator-driven Neutron Source (KUANS). As they lent us the KUANS multiple times, we were able to collect a large amount of basic data for the high-count-rate study.

We also thank Dr. Mori, who is in charge of beamline B3 at KUR. We were able to prove that this system can withstand the high neutron count rate of the neutron reactor at KUR.

We also thank Dr. Otake and the staff at RIKEN, who recognized the superiority of this system and introduced it to RANS, a compact accelerator-driven neutron source at RIKEN.

This research was conducted in collaboration with CETD and KEK.

References

1. S. Satoh, S. Muto, N. Kaneko, T. Uchida, M. Tanaka, Y. Yasu, K. Nakayoshi, E. Inoue, H. Sendai, T. Nakatani, T. Otomo, NIMA, **600**, 103–106 (2009), DOI 10.1016/ j.nima.2008.11.054
2. F. Liang, L. Chen, F. Li, G. Jin, Physics Procedia **37**, 1813–1818 (2012), DOI 10.1016/j.phpro.2012.02.508
3. S. Satoh, Physica B, **551**, 401–404 (2018), DOI 10.1016/j.physb.2018.03.011

Characterization of Neutron Beam Applications

Hirohiko M. Shimizu[1,2,*] and *Katsuya* Hirota[1,3]
(on behalf of the Fundamentals Working Group of the Japanese Society for Neutron Science)

[1]Department of Physics, Nagoya University, Furocho, Chikusa, Nagoya 464-8502, Japan
[2]High Energy Accelerator Research Organization (KEK), 1-1 Oho, Tsukuba, Ibaraki 305-0801, Japan
[3]Research Center for Nuclear Physics (RCNP), Osaka University, 10-1 Mihogaoka, Ibaraki, Osaka 567-0047, Japan

Abstract. Analytic framework to quantitatively evaluate the performance of existing neutron scattering instruments would be valuable to find out new possibilities and potentials of slow neutron beams. A conceptual sketch of ongoing effort in Japan is described.

1 Introduction

Quantitative characterization of the analysis capability of the neutron applications is one of the most fundamental information to overlook the potential of expansion of the neutron science. The common measure of the evaluation has not been sufficiently developed for integrating all potential capability and distributing demands to appropriate facilities. We describe the status of the ongoing effort of the quantitative characterization.

2 Sketch of a Characterization of the Analysis Capability of a Neutron Scattering Measurement

We consider the case in which m-kinds of observables are obtained in a neutron scattering measurement and put the set of the observables as $\boldsymbol{y} = (y_1, y_2, \ldots, y_m)$ together with their uncertainties as $\Delta\boldsymbol{y} = (\Delta y_1, \Delta y_2, \ldots, \Delta y_m)$. When each observable is required to be determined with the accuracy of $\boldsymbol{Y} = (Y_1, Y_2, \ldots, Y_m)$ all components of

$$\sigma = \left(\frac{Y_1}{\Delta y_1}, \frac{Y_2}{\Delta y_2}, \ldots, \frac{Y_m}{\Delta y_m}\right) \tag{1}$$

are required to exceed unity. Here we consider the neutron beam use as the combination of "source", "instrument" and "interpretation" as schematically shown in Fig. 1. We

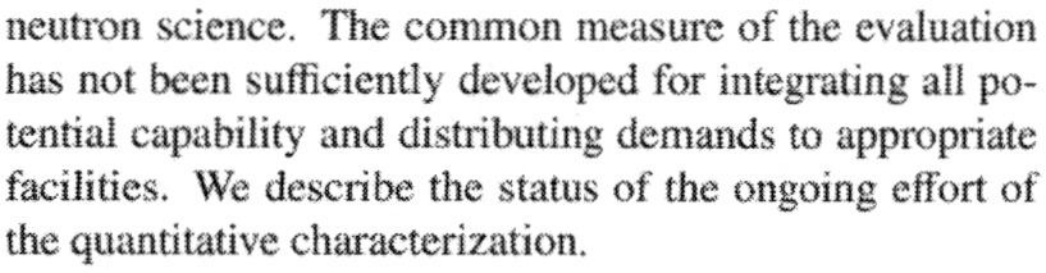

Figure 1. Schematic illustration of the application of the neutron beam driven by accelerators.

use $\Gamma = (\boldsymbol{r}, \boldsymbol{p})$ to represent the phase space of neutrons where $\boldsymbol{r}$ and $\boldsymbol{p}$ are the position and momentum of neutrons. The phase space distribution of neutrons is given as a function of time and can be defined as

$$\frac{\partial N}{\partial \Gamma} = \frac{\partial^6 N}{\partial x \partial y \partial z \partial p_x \partial p_y \partial p_z}, \tag{2}$$

where N is the number of neutrons. The phase space distribution evolves over time to that on the moderator surface ($\partial N/\partial\Gamma_0$), on the entry to the sample ($\partial N/\partial\Gamma_{\rm in}$), on the exit from the sample ($\partial N/\partial\Gamma_{\rm ex}$), which are related as

$$\begin{aligned} M(\Gamma_0, \Gamma_{\rm in}) &: \frac{\partial N}{\partial \Gamma_0} \to \frac{\partial N}{\partial \Gamma_{\rm in}}, \\ M(\Gamma_{\rm in}, \Gamma_{\rm ex}) &: \frac{\partial N}{\partial \Gamma_{\rm in}} \to \frac{\partial N}{\partial \Gamma_{\rm ex}}. \end{aligned} \tag{3}$$

The scattered neutrons are recorded by the detectors as the set of the arrival position and/or timing, which can be described as

$$M(\Gamma_{\rm ex}, \Gamma_{\rm det}) : \frac{\partial N}{\partial \Gamma_{\rm ex}} \to \frac{\partial N}{\partial \Gamma_{\rm det}}. \tag{4}$$

Consequently, the raw data of the measurement can be written as

$$M : \frac{\partial N}{\partial \Gamma_0} \to \frac{\partial N}{\partial \Gamma_{\rm det}}, \tag{5}$$

with $M \equiv M(\Gamma_{\rm ex}, \Gamma_{\rm det}) M(\Gamma_{\rm in}, \Gamma_{\rm ex}) M(\Gamma_0, \Gamma_{\rm in})$. The function of the "instrument" can be considered as the combination of $M(\Gamma_0, \Gamma_{\rm in})$ and $M(\Gamma_{\rm ex}, \Gamma_{\rm det})$. We symbolically use $\mathcal{M} = (M(\Gamma_0, \Gamma_{\rm in}), M(\Gamma_{\rm ex}, \Gamma_{\rm det}))$ to represent the function of the "instrument". The $\mathcal{M}_{\mathfrak{I}}$ is used to represent the $\mathcal{M}$ for a specific "instrument" labeled as $\mathfrak{I}$.

The information of the sample is contained in $M(\Gamma_{\rm in}, \Gamma_{\rm ex})$ and the observables of interest $\boldsymbol{y}$ is to be numerically derived with their uncertainty $\Delta\boldsymbol{y}$ from the $\partial N/\partial\Gamma_{\rm det}$. The achievement of the measurement can be quantified in the $\boldsymbol{\sigma}$, which is the comparison of $\Delta\boldsymbol{y}$ and $\boldsymbol{Y}$. We label the analysis method with $\mathcal{I}$ and the $\boldsymbol{\sigma}$ can be expressed as

$$\boldsymbol{\sigma} = \boldsymbol{\sigma}(\Gamma_0, \mathcal{M}_{\mathfrak{I}}, \mathcal{I}, T), \tag{6}$$

*e-mail: hirohiko.shimizu@nagoya-u.jp

where T is the measurement time. We consider the case in which the systematic uncertainty is smaller than the requirement $\boldsymbol{Y}$ so that all of the component of $\boldsymbol{\sigma}$ can exceed unity if sufficiently long measurement time is spent. For a specific measurement, denoted by ϑ, we put the minimum measurement time to achieve the requirement as

$$\tau_\vartheta = \tau(\Gamma_0, \mathcal{M}_{\Im,\vartheta}, \mathcal{I}_\vartheta), \quad (7)$$

which measures the performance of the combination of the "source" Γ_0, the "instrument" $\mathcal{M}_\Im$, the "interpretation" $\mathcal{I}$ applied for the measurement ϑ.

3 Optimization of Instrument

We consider the case where the "instrument" has n-kinds of adjustable parameters $\boldsymbol{\alpha} = (\alpha_1, \alpha_2, \ldots, \alpha_n)$. The optimized condition can be defined as the $\boldsymbol{\alpha}$ to minimize the τ_ϑ, which can be obtained from the n-ary simultaneous equation

$$\frac{\delta \tau_\vartheta}{\delta \boldsymbol{\alpha}} = \mathbf{0}. \quad (8)$$

This optimization can be generalized to include all auxiliary conditions of the measurement such as the sample choice, sample temperature, sample pressure, applied magnetic field and so on. Here we put the solution of Eq. 8 as $(\boldsymbol{\alpha}_\vartheta)_0$, which is not always achievable. The degree of optimization can be measured in the ratio

$$\mathcal{S}_\vartheta = \frac{(\tau_\vartheta)_0}{\tau_\vartheta}, \quad (9)$$

which indicates the possibility of further improvement if $\mathcal{S}_\vartheta$ is less than unity.

The performance of the "instrument" can be measured by considering the fluence necessary to achieve the requirement of the measurement, which can be written as

$$F_\vartheta = F(\Gamma_0, \mathcal{M}_{\Im,\vartheta}, \mathcal{I}_\vartheta). \quad (10)$$

The optimization condition of the instrument corresponds to minimize F_ϑ.

Each instrument is required to be optimized not only for a single measurement, but also for a set of measurements, denoted by $\mathbb{T}_\Im$. Therefore, the instrument optimization can be formulated by the minimization of

$$F_{\mathbb{T}_\Im} = \sum_{\vartheta \in \mathbb{T}_\Im} w_\vartheta F_\vartheta, \quad (11)$$

where w_ϑ is the weight factor for each measurement ϑ, which will be determined on the basis of the subjectively decision. It would be appropriate to redefine the instrument optimization as

$$\frac{\delta F_{\mathbb{T}_\Im}}{\delta \boldsymbol{\alpha}} = \mathbf{0}, \quad (12)$$

and the degree of optimization as

$$\mathcal{S}_{\mathbb{T}_\Im} = \frac{(F_{\mathbb{T}_\Im})_0}{F_{\mathbb{T}_\Im}}, \quad (13)$$

where $(\boldsymbol{\alpha}_{\mathbb{T}_\Im})_0$ is the solution of Eq. 12.

4 Discussion

The instrument optimization may be generalized to evaluate the degree of optimization of a facility. Most of the neutron facilities have multiple instruments. We put the set of measurements to be carried out in the facility $\Re$ as $\mathbb{T}_\Re$. The weighted sum of fluence required to satisfy the requirement of all measurements can be given as

$$F_{\mathbb{T}_\Re} = \sum_{\vartheta \in \mathbb{T}_\Re} w_\vartheta F_\vartheta. \quad (14)$$

However the optimization of Eq. 14 corresponds to the measurement-by-measurement adjustment of instrument parameters in the facility. It would be more practical to minimize the weighted sum of $F_{\mathbb{T}_\Im}$ given as

$$F_\mathbb{R} = \sum_{\Im \in \mathbb{R}} w_\Im F_{\mathbb{T}_\Im}, \quad (15)$$

where $w_\Im$ is the weight factor for the instrument $\Im$, which will be determined on the basis of the subjectively decision.

The F-value calculated on the basis of common definition will offer a quantitative index to compare instruments or facilities, which will clarify their versatility, uniqueness and significance.

The next step to develop the framework described above is the derivation of the explicit description of the instrument $\mathcal{M} = (M(\Gamma_0, \Gamma_{\text{in}}), M(\Gamma_{\text{ex}}, \Gamma_{\text{det}}))$ for their various usage and also listing of F-values of existing instruments and facilities, which is ongoing in the neutron community in Japan. We expect that this framework will clarify the potential of existing methods, their limitations and possible future extensions to activate advanced application of neutron beam.

Scalable Neutron Imaging Systems at Compact Sources

Knud Thomsen*, *Eberhard* Lehmann, and *Markus* Strobl

Laboratory for Neutron Scattering & Imaging, Paul Scherrer Institut, CH-5232 Villigen PSI, Switzerland

Abstract. Neutron imaging offers unique capabilities not limited to the structural characterisation of materials, components and processes at a large variety of neutron sources and thus is highly qualified to be a prime candidate for instrumentation at compact neutron sources. The peculiar characteristics of neutrons, and in particular the cross sections for their interaction with matter enable imaging results that are directly comparable but, most notably, are effectively complementary to those of wide spread X-ray imaging characterisation. Here, it is highlighted how neutron imaging can significantly add to the value and the versatility of basically all and especially future compact neutron sources. Diverse advantages are identified at all possible sizes and levels of sophistication of compact sources. These areas range from source characterisation and students training to state of the art neutron imaging in 3D to applications and method development, which are possible at compact sources and some of which can take particular advantage of advanced source characteristics such as e.g. several target stations at a single accelerator.

1 Introduction

Neutron imaging (NI) offers a number of unique features qualifying it as a prime candidate for instrumentation at compact neutron sources. The peculiar characteristics of neutrons, in particular their interaction cross sections, allow for image contrast that is in principle comparable but, in particular, complementary to that of easily available X-ray imaging. Neutron imaging, initially on film, was first established as a non-destructive testing tool for industrial use, but through digitalisation and ever increasing resolution capabilities quickly developed into a powerful research tool in many fields of applications.

The most advanced NI facilities, defining the state-of-the-art of the technique, are located at large scale neutron sources where the access routes to beam time are strictly regulated and time consuming.

The establishment of neutron imaging techniques at new compact sources has several key aspects:

- Sustainable in a landscape of a decreasing number of large sources, in particular reactor based
- Potential impact of individual methods to the design and tuning of source properties, which can be optimized according to the different requirements
- Flexibility in the instrument layout (e.g. multiple beams, time structure, variable collimation, sample environment, ...) in order to optimize for method and application
- Easy access to beam time for researchers and industry in particular in the vicinity of the source and in-house, and also for students

1.1 Aspects Relevant to Compact Neutron Sources

Technically, the range for possible installations is vast; this likewise holds for price, sophistication and potential use. Neutron imaging is in principle possible at rather small sources yielding only a relatively low neutron flux. However, e.g. a demanding state-of-the-art detection systems (very high sensitivity, long exposure enabled) is a pre-requisite for impactful application. A very instructive example is the 2W research reactor of the TU Dresden where imaging is pursued (compare image data in Fig. 1).

At the other end of the scale it is possible to take full advantage of a most versatile compact source concept with several independent target stations, like it is intended e.g. in the High Brilliance Source (HBS) project. In such case one can conceive even unparalleled opportunities e.g. for dynamic neutron imaging, like in the idea of crossed neutron beams for stereoscopy with neutrons. Such option would hardly be accessible at any other type of neutron source. However, such options require detailed consideration concerning applications and corresponding instrumentation by the source designers from the very beginning. This should entail a high priority for neutron imaging as a technique with outstanding potential and capabilities and well suited to be established at such neutron sources.

1.2 Current situation for Neutron Imaging at existing Compact Source facilities and Projects

In the overview given in [1] 23 neutron sources are listed and described in some details. From these sources, 11 declare to intend (or perform) some neutron imaging. However, regarding scientific literature or even the

* Corresponding author: knud.thomsen@psi.ch

homepages of the facilities only about 6 display relevant and impactful results.
A very positive example in this regard are the Japanese sources which are additionally networked in the JCANS collaboration [2] and provide a valuable platform to even progress imaging methods relevant for the large scale source of JPARC during its construction and commissioning. And such valuable work continues due to beam availability limitations during large scale user operation.

Some of the new projects like HBS Jülich (D), SONATE Saclay (F), projects in Hungary, and LENS (USA) have intentions to establish neutron imaging capabilities. It is of outmost importance, however, that these projects early on gain access to expertise and practical knowledge in the respective technology of neutron imaging and its applications in order to enable optimum design decisions.

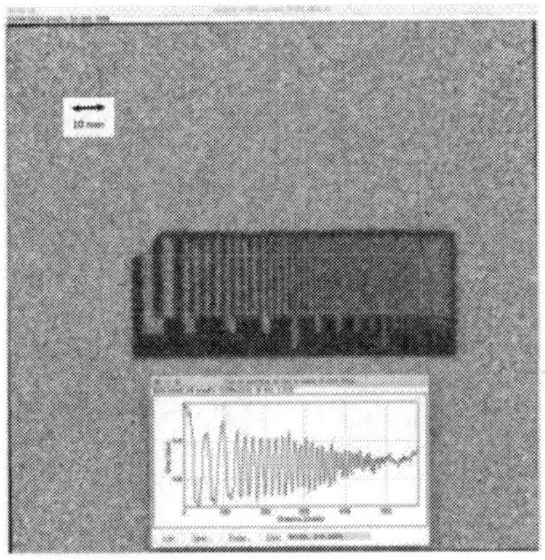

Fig. 1. Example image obtained at 8x10^3 n/cm^{2}s in 30 min (AKR-2, TU Dresden); image courtesy C. Lange, TU Dresden, Germany

2 Imaging as a Key Application at Sources of Different Power and Various Levels of Sophistication

Neutron imaging holds the potential to provide several valuable functions for a small scale source project which include self-supporting applications for the project, industrial use cases and science cases of various nature as well as method development. In the following, some key fields of neutron imaging are briefly introduced.

2.1 Self-supporting Project Applications

2.1.1. Source/Moderator Characterisation

Sources and moderators can be optimized for specific instrumentation and in particular for neutron imaging and even different imaging modalities. Specifically, wavelengths/spectra, sizes, flux, time structure and homogeneity have to be considered and carefully designed. Taking the perspective the other way around, neutron imaging set-ups allow for a detailed characterization, in particular concerning flux, spectrum, divergence and respective homogeneity of emanating neutron beams. Thus pinhole imaging can support efficient facility development and validation of numerical simulation tools. In particular small sources can provide significant flexibility for ongoing optimization of the source layout, and will hence especially provide from such assessment capability.

2.1.2. Training and Preparation Capabilities

Advantageous educational aspects are a direct consequence of these manifold possibilities: students can easily become acquainted with neutron-based instrumentation, make their first steps and can almost immediately obtain useful and illustrative results.

Access at a small source is inherently easier than at big facilities, and hands-on experiments are possible, which cannot be done at more powerful installations. This, in turn allows for a "pathfinder" function where experiments are tried and optimized before expensive beam time at a big source is applied for. As a proof of principle, simple neutron radiography can yield valuable images at fluxes as low as10^4 n/cm^{2}s (see Fig. 1). In parallel to increasing possibilities at larger and more complex, still compact, neutron sources, preparatory experiments can be performed before going to flagship neutron facilities with optimized set-ups.

2.1.3. Public Relations Capabilities

The results of neutron imaging are particularly well suited to provide insights to and generate understanding in the general public for neutrons and neutron science, thus supporting the project and facility. Likewise results and valuable science or industrial applications directly accessible through the observation of high quality images are also exceptionally well suited to generate impact at funding bodies and governance and thus ensure sustainability.

2.2 Industrial Uses

Neutron imaging can provide information of interest to diverse industry and cover a vast range of questions. Quality Assurance topics, in particular improving processing, quality and reliability of products can be claimed to lie at a centre of gravity of relevant investigations in industry. Neutrons deliver especially valuable results e.g. for pyrotechnic components, joining parts, comprising welding, soldering/brazing and, in particular, adhesive bonding (see Fig. 2) but also e.g. processing such as sintering.

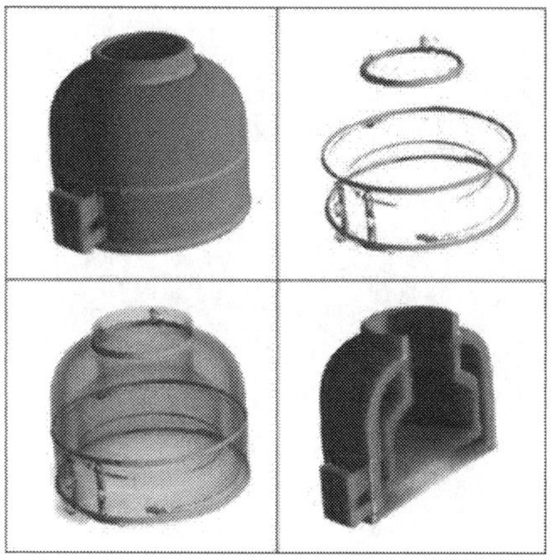

Fig. 2. Example image of assessment of soldering quality, semi-transparent 3D representation highlighting the Boron containing brazing solder distribution.

Further examples include humidity distributions in building materials, but also investigations on the effect of storage on syringes containing medicine, rocket [3] and automotive components, to name but a few.

A specific example of an application at a small scale source is provided in Fig. 3. It concerns a relatively new field, which is to assess relevant characteristics of specimen produced by metal additive manufacturing.

Novel manufacturing technologies like additive ones are often used for sophisticated and expensive parts; typical areas include aerospace and medical industries. For these, special requirements apply also with respect to quality control. Neutron imaging is a tool of choice for such applications.

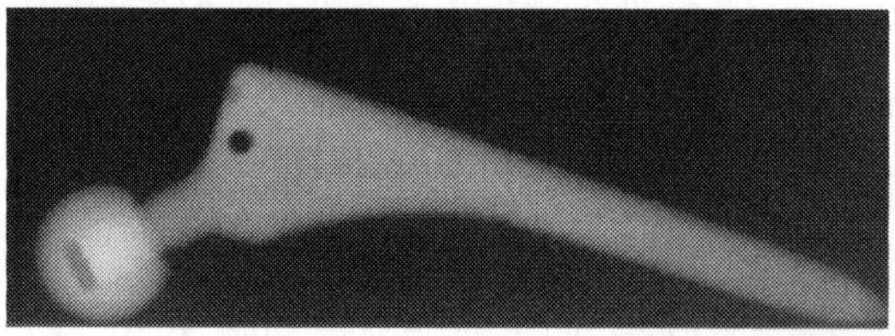

Fig. 3. Example of image of hip replacement, obtained with a moderated 14 MeV D-T source, image courtesy M. Taylor, Phoenix Lab, Madison, USA.

Figure 3 shows the result obtained using a small commercial neutron source, which employs the D-T reaction. Neutrons are moderated and 3 hours of exposure at a thermal flux of approximately 1×10^4 n/cm^{2}s were required to obtain the presented result.

2.3 Science Cases

2.3.1. Engineering materials research

Additive manufacturing is also a dominating topic in engineering materials research and neutron imaging is utilized as post processing characterisation tool supporting process optimisation and simulation of metallic materials and components. The main focus is on bulk metallic materials, which cannot be penetrated by other radiation for imaging purposes, and the achieved porosity and porosity distribution, dependent on process parameters. Novel imaging methods are also used to map strains and phases in the built material, which bear key aspects of mechanical behaviour of the material, material strength and durability under service conditions.

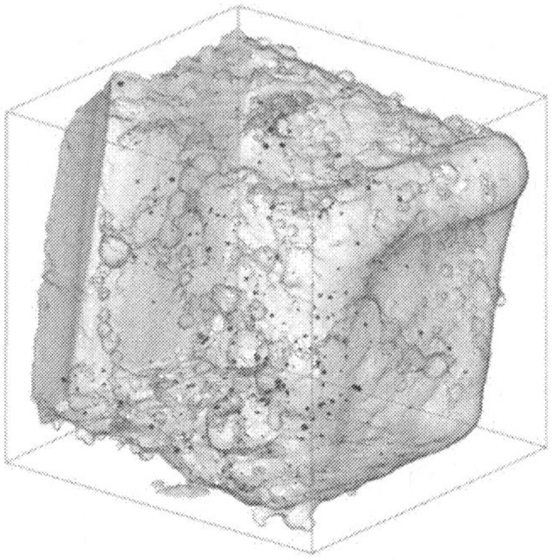

Fig. 4. Example image of porosity in additively manufactured gold (courtesy R. Loge, EPFL) at the sub-10μm scale obtained with the PSI neutron microscope detector (illustrating a far aim for compact sources) [4]

Figure 4 displays a specific example targeting the porosity in additive manufacturing of precious metals, which is of interest for jewellery and watch industries.

Other key topics concern welding, phase transformations and strain in steels or magnetic domain structures in transformer steels as well as hydrogen and hydrogen embrittlement in metals, corrosion and many others more.

2.3.2. Energy Research

The specific high sensitivity of neutrons for some light elements, namely hydrogen and lithium, make neutrons and especially neutron imaging an outstanding research and development tool in the key field energy conversion techniques especially with respect to fuel cells, hydrogen storage and batteries, but also nuclear safety. These also spur industrial interest and applications and examples are numerous in recent literature [5-13].

Many of these investigations are also suited for small sources, in particular in observations of steady states, which can be done in their own right but also establish ideal preparation work for extensions to in operando characterisations.

2.3.3. Soil, Plants and Environment

Recent literature also underlines the distinguished capabilities of neutron imaging in the field of geology, soil and plants and particular the interplay of the latter in the context of water balance and uptake.

Due to the fact that neutron imaging is an excellent tool for the observation of water and liquid flow in porous media a large variety of geological and agriculturally relevant processes and phenomena can be observed with high spatial and time resolution, and especially outstanding means for quantification of related transport phenomena.

2.3.4. Cultural Heritage

Unexpected images of familiar as well as of unknown objects like deciphering the internal make-up of, e.g., ancient sacred statues are considered interesting by most people and important to those researching ancient culture and technology (Fig 5).

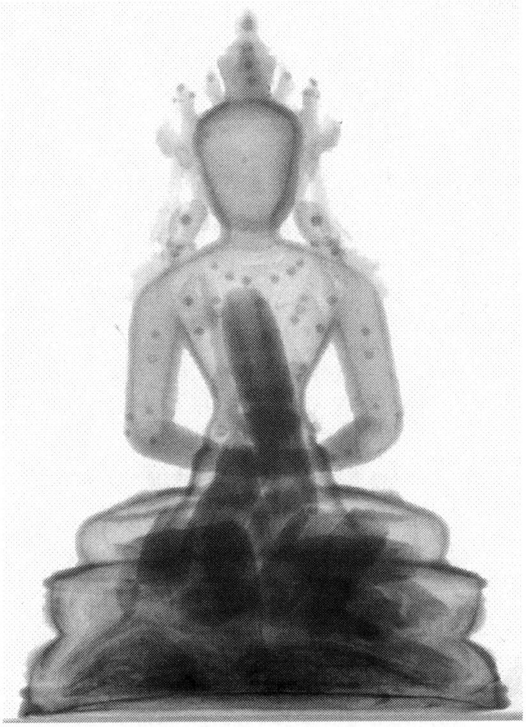

Fig. 5. Example for cultural heritage studies: neutron image of a Tibetan Buddha from the 15th century, showing the filling with scrolls

Neutron imaging is an invaluable tool for the investigation of invaluable cultural heritage as it allows to gain insights in a non-destructive manner, which is often the only possibility to investigate unique and valuable artefacts in details and thus provide unique insights into our past and origin.

It is very often such applications that offer a unique way to demonstrate to the common taxpayer as well as to funding agencies the high potential and manifold usefulness of neutron-based techniques starting at flux levels as low as 1000 n/cm^{2}s.

2.4 Development of new Methods

The development of novel methods at dedicated small sources, potentially even featuring multiple targets and moderators offers an open-ended avenue. Advantages with respect to method development are (i) better availability of beamtime for tests and experiments, (ii) more flexible instrumentation, potentially (iii) down to the neutron source itself.

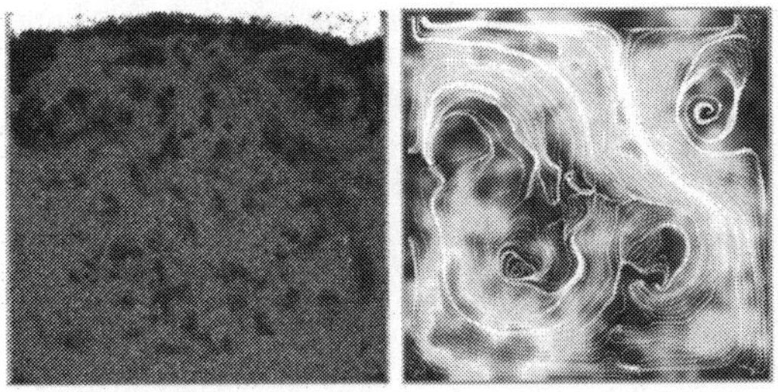

Fig. 6. Example image of recent application aimed at improving imaging- as well as simulation-approaches: moving inclusions in electromagnetically agitated liquid metal, comparison image obtained with neutrons versus fluid dynamic numerical simulation [5]

Outstanding examples are the developments for time-of-flight imaging at JPARC performed at HUNS in Hokkaido, including sophisticated techniques like polarized neutron imaging [15]. Another example might be the test beamline for ESS in Berlin supporting the development of a sophisticated wavelength frame multiplication chopper system not only for neutron imaging [16]. Despite the neutron source BERII being a medium flux reactor source, with regards to time-of-flight through mimicking the ESS time structure, it can be considered a small source in this context.

Some small sources might also provide so far unparalleled opportunities for new concepts and ideas to be tested and realized. A small source with several target stations might for example enable to cross neutron beams from different targets at a sample position. This might provide the combination of different spectra, time structures or techniques. An example in neutron imaging might be found related to recent studies at PSI investigating the movement of inclusions in liquid metal as relevant for industrial manufacturing processes (Fig. 6) and required for the development of reliable simulation tools [5]. The challenge to observe the process with 3D spatial resolution, despite the speed of the process which does not enable tomographic scans, could be solved through neutron stereoscopy utilizing crossed neutron beams.

3 General Considerations & Conclusion

Neutron Imaging is a non-invasive inspection technique with high relevance in several areas of research and industry.

Based on modern digital detection techniques and numerous methodical developments in the past years it has become attractive for research fields like e.g. electro-chemistry, engineering materials, energy applications, construction materials, geology and also archaeometry.

On the other hand, industry can take profit directly in the investigations of advanced manufacturing processes and products, as well as in R&D activities like in the past for Diesel particular filters but also representative inspections like e.g. of adhesive joints, to highlight some prominent examples only.
Nevertheless, state-of-the-art neutron imaging is currently established only at distinct places, where strong neutron sources and dedicated beam lines are available. A list of the neutron imaging installations operated in a user lab are provided in [17], but about 100 facilities used around the globe are attempting to apply neutron imaging [18].

Reasons for the currently low exploitation of this rewarding technique might include:

- Strong neutron sources have been built in the past dedicated to different use cases: nuclear fuel development (irradiation experiments), isotope production, neutron scattering, nuclear engineering education and neutron activation analysis.
- most neutron sources are not mobile or even portable (like X-ray sources) and the inspection of specimen has to be performed "on-site" with often restricted access for important stakeholders of neutron imaging (limiting also awareness)
- neutron scattering communities with coherent applications and purely academic intentions have been dominating the design and applications of research neutron sources for the past decade
- the neutron imaging user community is highly diverse and thus less suited to effectively lobby for their needs
- modern neutron imaging with digital detectors and capable of quantitative also scientific studies emerged delayed to diffraction and spectroscopy techniques

The approach of building compact and tuneable neutron sources offers new challenges in efficient exploitation also for the NI community. It is an outstanding opportunity to build dedicated imaging stations, optimized from the cradle to the grave from the initial source to the detector. At most sources so far, in contrast, a compromise has to be found between the different instruments and stakeholders in the design and particularly the neutronic design of the source.
Today thermal and cold neutron spectra are mainly used for state-of-the-art NI, while fast neutrons are utilized only in few cases and for special purposes. While high-flux pulsed neutron beams are only available at very few powerful spallation sources (JPARC, ISIS, IBR-2, ...) time-of-flight (TOF) methods at continuous sources entail significant flux penalties.
However, very sensitive neutron imaging detectors are meanwhile available and in operation which enable working also with low beam intensities and provide neutron images with reasonable image quality. The combination of tuneable neutron sources with the most suitable and technologically advanced detection system is set to enable valuable and impactful NI also beyond the traditional settings of large-scale facilities.

Aspects that require special (re-)consideration concerning imaging at small sources include:

- all nuclear reactions initially provide high energy neutrons first (on the order of some MeV) and moderation reduces the available flux significantly; can and how can fast neutrons be used efficiently for highly relevant neutron imaging applications?
- what is the optimum source and moderator for specific relevant neutron imaging techniques and can their parameters be tuned even supplying an individual beamline?
- what are the requirements to the time structure and pulsed beam flux to be relevant for advanced ToF imaging?
- are there relevant advanced techniques and in particular imaging methods and applications that can profit from novel concepts such as crossed beams from multiple target stations realized at a single accelerator (compare HBS project [19])?
- while current neutron imaging detector technique serves well the current state of the art in neutron imaging, novel sources and instruments still provide challenges for efficient utilisation. What specific demands are expected in particular from future small neutron sources?

For the moment there are only very few suppliers of compact sources on the market. Some test facilities have been built and upgraded and a few new installations appear to be planned and to advance. None of them is optimized for the exclusive and dominant use for neutron imaging. However, informed design work and applied expertise in neutron imaging are needed to build the best options and to identify challenging applications and use cases matching the particular conditions.

References

1. I. Anderson et al., Research opportunities with compact accelerator driven sources, Physics Report 654, 1-58 (2016)
2. Y. Kiyanagi, Neutron Imaging at Compact Accelerator-Driven Neutron Sources in Japan, J. Imaging, 4, 55; doi:10.3390/jimaging4040055 (2018)
3. G. Bayon, Review on use of neutron radiography at Saclay nuclear research center, Proc.3rd World Conference on Neutron Radiography, pp. 439-446 (1996)
4. V. Manzi-Orezzoli et al., Coating Distribution Analysis on Gas Diffusion Layers for Polymer Electrolyte Fuel Cells by Neutron and X-Ray High

Resolution Tomography, ACS Omega, ID: ao-2019-01763j.R2 (2019)

5. M. Ščepanskis et al., Assessment of Electromagnetic Stirrer Agitated Liquid Metal Flows by Dynamic Neutron Radiography, Metallurgical and Materials Transactions B ISSN 1073-5615 Volume 48 Number 2 Metall and Material Trans B 48:1045-1054 DOI 10.1007/s11663-016-0902-8 (2017)
6. M. Siegwart, et al., Selective Visualization of Water in Fuel Cell Gas Diffusion Layers with Neutron Dark-Field Imaging Journal of The Electrochemical Society 166(2):F149-F157 (2019)
7. M. Makowska, et al., Phase transition mapping by means of neutron imaging in SOFC anode supports during reduction under applied stress, ECS Transactions, 68 (1) 1103-1114 (2015)
8. I. Manke, et al., Characterization of water exchange and two-phase flow in porous gas diffusion materials by H-D contrast neutron radiography, Appl. Phys. Lett. 92, 244101 (2008)
9. M. Strobl, et al., Topical Review: Advances in neutron radiography and tomography, J. Phys. D 42 243001 (2009)
10. Thomas Knoche et al., In situ visualization of the electrolyte solvent filling process by neutron radiography, Journal of Power Sources 331, 267-276 (2016)
11. Y. Wu et al., Effect of serpentine flow-field design on the water management of polymer electrolyte fuel cells: An in-operando neutron radiography study, Journal of Power Sources 399, 254 (2018)
12. M. Cochet et al., Novel Concept for Evaporative Cooling of Fuel Cells: an Experimental Study Based on Neutron Imaging, Fuel Cells 18, 619 (2018)
13. P. Stahl et al., An Investigation of PEFC Sub-Zero Startup: Influence of Initial Conditions and Residual Water, Fuel Cells 17, 778 (2017)
14. P. Boillat, E.H. Lehmann, P. Trtik, M. Cochet Neutron imaging of fuel cells ? Recent trends and future prospects, Current Opinion in Electrochemistry 5, 3 (2017)
15. M. Strobl, H. Heimonen, S. Schmidt, M. Sales, N. Kardjilov, A. Hilger, I. Manke, T. Shinohara, J. Valsecchi Topical review: Polarisation measurements in neutron imaging J. Physics D 52, 12 (2019)
16. M. Strobl, M. Bulat, K. Habicht, The wavelength frame multiplication chopper system for an ESS test-beamline and corresponding implications for ESS instruments, Nucl. Instr. Meth. A 705, 74–84 (2013)
17. E. H. Lehmann, D. Ridikas, Status of Neutron Imaging – Activities in a Worldwide Context, Physics Procedia Volume 69, 2015, 10-17 (2015)
18. http://isnr.de/index.php/facilities/facilities-worlwide
19. HBS Project, Conceptual Design Report, FZ Jülich, ISBN 978-3-95806-280-1 (2017)

Study of a collimation method as a nondestructive diagnostic diagnostic technique by PGNAA for salt distribution in concrete structures at RANS

Yasuo Wakabayashi[1*], *Takao* Hashiguchi[1], *Yuichi* Yoshimura[1, 2], *Maki* Mizuta[1], *Yujiro* Ikeda[1, 3], and *Yoshie* Otake[1]

[1]RIKEN Center for Advanced Photonics, Neutron Beam Technology Team, Wako, Saitama, Japan
[2]Tokyo Institute of Technology, Meguro, Tokyo, Japan
[3]J-PARC Center, Japan Atomic Energy Agency, Tokai, Ibaraki, Japan

Abstract. To meet strong demand for realizing an effective tool to diagnose salt distribution in concrete infrastructures, we have started a development of a new technique using a prompt gamma neutron activation analysis (PGNAA) at the RIKEN accelerator-driven compact neutron source (RANS). So far, by applying PGNAA we have experimentally confirmed that neutrons from RANS can detect small enough amounts of chlorine within the marginal concentration of around 1.2 kg/m ^{3} to involve steel corrosion. In this study, we have proposed two methods to derive the salt depth profile which is critical information of steel corrosion start. The first one utilizes a difference in the intensity ratio of two different γ-ray energies of interest, which is depending on the depth where the neutron capture reaction arises inside the concrete. The second is called the collimator method that measure γ-rays coming through a collimator around detector. Detection of γ-ray associated with ^{35}Cl coming from the assembly of concrete has been also simulated with conditions of neutrons from RANS and a collimator. The feasibility of the method was discussed.

1 Introduction

Concrete infrastructures like a bridge are suffered from chloride attack due to sea wind close to coast and anti-icing agent in mountain area both contain chloride. Chloride ions go into concrete structure and reach to near steel bar. Steel corrosion start from the chloride ion concentration in the range of 1.2 to 2.5 kg/m ^{3} [1], which is called the marginal concentration. As chloride attack progresses, decrease of steel cross section, crack of concrete, rust juice, and so on occur. But when these sign like a rust juice appear, chloride attack is already quit processing. Eventually, serious accident like a collapse of a bridge happen. Therefore, it is important to investigate the chloride ion concentration in concrete structures. Along with chloride ion concentration, to know its distribution between object surface and steel bar in concrete structure is especially important. Hereafter the distribution of chloride ion concentration is called "salt distribution".

Several conventional methods for salt distribution measurement exist [2-4]. They need several steps before measurement. First step is core sampling from structure. Core is moved to experimental room. Then, pre-processing like a cutting, smashing, chemical treatment is applied. After measurement, mortar must be buried into hole after core sampling. In recently, handy device of X-ray fluorescence analysis exist without these steps. But it measures only surface, therefore it finally need core sampling to measure salt distribution. These methods are reliable and accurate, but there are disadvantages such as the limit of sampling location, taking long time, cost, and so on. For these reason, it is difficult to analyze many structures. Non-destructive technique without core sampling is desirable to assay many structures. To meet the demand, we have started the development of on-site non-destructive technique for salt distribution in concrete by prompt gamma neutron activation analysis (PGNAA) at the RIKEN accelerator-driven compact neutron source (RANS) [5]. So far, we have experimentally confirmed that RANS provides suitable neutron fluxes to detect small enough amounts of chlorine within the marginal concentration of around 1.2 kg/m^3 to involve steel corrosion, by applying PGNAA to 4x4x4cm^3 mortar bricks of different chloride ion concentrations from 0.36 to 5.1 kg/m^3 [6, 7].

In this work, we have proposed two methods to measure salt distribution. The first one utilizes a difference in the intensity ratio of two different γ-ray energies of interest, which is depending on the depth where the neutron capture reaction arises inside the concrete. Second one used γ-ray collimator. By collimating γ-ray detection area and neutron beam, γ-ray from target depth are detected. We call it collimator method. Then we carried out trial experiment to show the principle of these methods at RANS [8]. To confirm the feasibility of the concept of the proposed methods, we have performed the simulation with the Monte Carlo simulation code GEANT4 [9] for the production ofγ-rays involved the neutron capture reaction of ^{35}Cl in the

* Corresponding author: y.wakabayashi@riken.jp

https://doi.org/10.1051/epjconf/202023105007

concrete assembly. The γ-rays were produced by injecting a collimated neutron beam into a concrete distributed salt uniformly and detected at concrete surface.

In this paper, we will introduce the feasibility of the proposed methods from the result of trial experiment and simulation.

2 Experiment for proposed methods

Experiment was performed at RANS shown in Figure 1. Neutrons are produced by using the reaction of 7MeV proton whose beam current is 100 μA at maximum with a beryllium target of 300μm thickness. Neutron beam are extracted from a neutron guide, which we call N0, after the target station. N0 have an exit of 160x160 mm^2. By inserting an additional polyethylene collimator from this exit, neutron beam size can be adjusted.

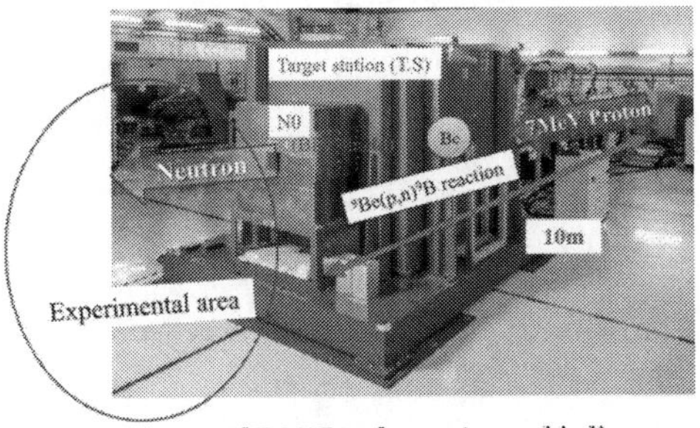

Figure 1 : Picture of RANS and experimental hall

2.1. Experimental Setup

Figure 2 shows the figure of experimental setups. The 30x30 cm^2 concrete bricks with 6-cm or 10-cm thickness were used. A salt sample was placed between concrete bricks and at different depth position from No.1 to No.3. To shield undesirable γ-rays and neutrons derived from RANS, lead bricks and LiF tiles were placed around detectors, concretes, and N0. A neutron collimator made of polyethylene with a hole of φ30mm was inserted into N0. For γ-ray collimators, lead bricks were placed with a gap of 2 cm. Salt of 250 g weight was put in paper box of 8 × 6 × 6 cm^3. Two Ge detectors were placed to view different depth position. In setup (I), they viewed No.1 and No.2 positions. After measurement at setup (I), right side detector was changed to view No.3 position as setup (II). Under these conditions, the γ-ray measurement was done.

2.2 Gamma-ray spectra and γ-ray intensity ratio

Figure 3 shows the salt γ-ray spectra at sample positions of No1, No2, and No3 for the measuring time of 5, 10, and 20 minutes, respectively. The γ-rays derived from the ^{35}Cl(n,γ)^{36}Cl reaction, which are called as ^{35}Cl-γ-rays hereafter, are shown by arrows and appeared in spectra at all positions. This means that the γ-rays can be detected at concrete surface if salt exists within 12 cm depth in principle. Here, to deduce the depth of salt, the γ-ray intensity ratio method is explained. The γ-ray transmission T_γ in concrete is given as,

$$T_\gamma = exp(-d \times f_\gamma), \quad (1)$$

where d is a pass length of γ-ray in concrete, and f_γ is a γ-ray attenuation factor as shown in Table 1. The attenuation factors were obtained by using the database of NIST [10], weight ratios of a standard concrete composition shown in Table 2, and a density of 2.2 g/cm^3. Since transmissions in concrete for different γ-ray energies γ1 and γ2 are different, the ratio of transmissions has a relationship with a γ-ray intensity ratio of $I_{\gamma2}/I_{\gamma1}$ as shown below:

$$T_{\gamma1}/T_{\gamma2} = (A_{\gamma1}/A_{\gamma2}) \times (\varepsilon_{\gamma2}/\varepsilon_{\gamma1}) \times (I_{\gamma2}/I_{\gamma1}), \quad (2)$$

where $T_{\gamma1}$ and $T_{\gamma2}$ are transmission which can be calculated, $A_{\gamma1}$ and $A_{\gamma2}$ are the peak areas in γ-ray spectra, $\varepsilon_{\gamma1}$ and $\varepsilon_{\gamma2}$ are absolute γ-ray detection efficiencies measured with nothing between a standard γ-ray source and a γ-ray detector, $I_{\gamma1}$ and $I_{\gamma2}$ are the known relative intensity ratios [11]. In this work, 517, 788, 1165 keV as γ1 and 1951 keV as γ2 were used. The relative intensity ratios of several major ^{35}Cl-γ-rays are shown in Table 1.

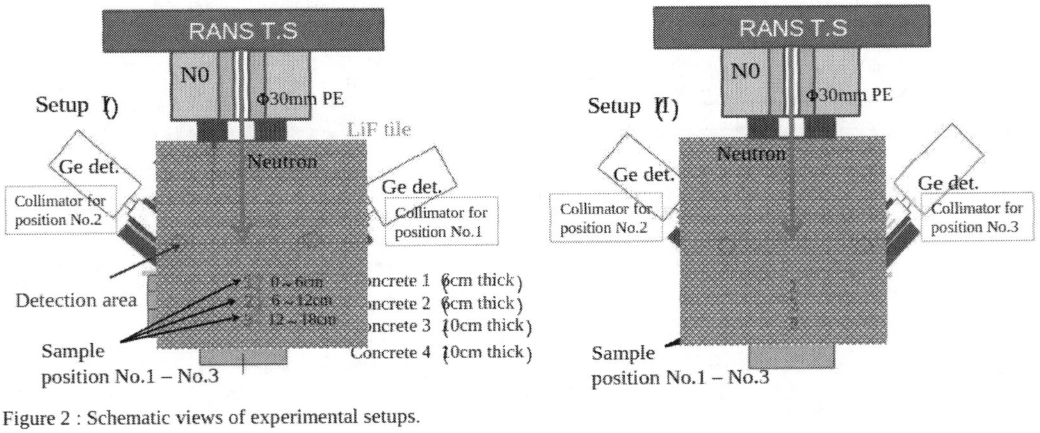

Figure 2 : Schematic views of experimental setups.

 https://doi.org/10.1051/epjconf/202023105007

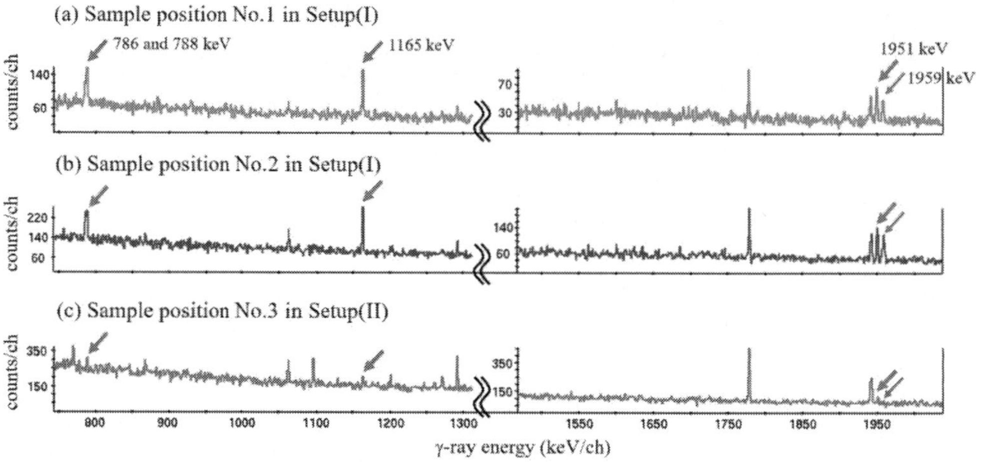

Figure 3 : Measured γ-ray spectra corresponding to sample positions.
(a) sample position No.1 in Setup(I), (b) sample position No.2 in Setup (I), and (c) sample position No.3 in Setup (II).
Arrows are γ-rays from $^{35}Cl(n,\gamma)$ reaction. The 786 and 788 keV γ-rays can't be separated clearly on this spectrum.

2.3 Validation of γ-ray intensity ratio and γ-ray collimator method

We performed the estimation of depth profile by using collimator method and γ-ray intensity ratio as shown in Figure 4 (a). Vertical axis is γ-ray intensity ratio based on the 1951 keV as reference γ-ray, and horizontal axis is γ-ray energy. Solid triangle, circle, and rectangle points in Figure 4 are experimental γ-ray intensity ratios at sample position No.1, No.2, and No.3, respectively, obtained from the right side of equation (2). These points are also plotted with error bars associated with peak fitting which include statistics errors. Solid, dashed, and dotted lines are γ-ray intensity ratios derived from calculations using equation (1) and the left side of equation (2).

In this experiment, salt sample has a thickness of 6 cm and the γ-ray collimators have the gap of 2 cm. Therefore, Ge detectors detect the γ-rays from different pass lengths in concrete like "Distance up" and "Distance down" as shown in Figure 4 (b). Then, we show two calculation lines of "Distance up" and

Table 1 : Major γ-energies, relative intensity ratios, and attenuation factors in capture reaction $^{35}Cl(n,\gamma)^{36}Cl$

Eγ [keV]	Relative intensity ratio [%]	Attenuation factor [cm^{-1}]
517	85.07	0.191
786	38.37	0.158
788	60.83	0.158
1165	100	0.131
1951	71.04	0.100
1959	46.02	0.100
6111	73.96	0.058

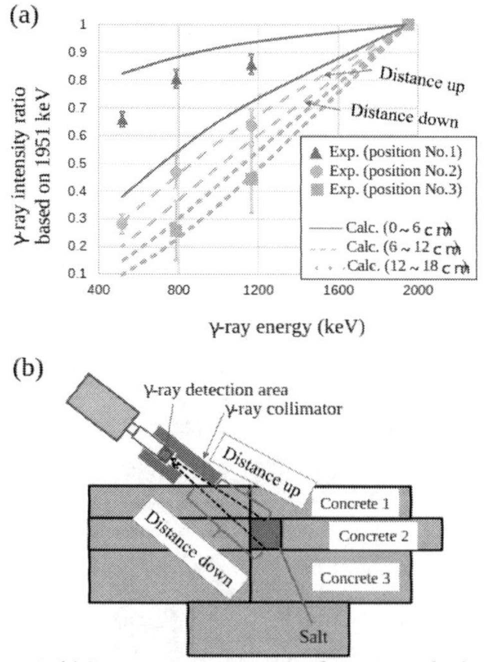

Figure 4 : (a) Gamma-ray intensity ratio of experimental values and calculation based on the 1951 keV γ-ray.
(b) Distance passing through concrete in case of position No.2.

Table 2. Weight ratios of elements contained in a standard concrete and the capture cross sections of thermal neutron of 25 meV.

Element	H	C	O	Mg	Al	Si	K	Ca	Fe	Cl
Cross section for thermal neutron (barn)	0.332	0.00386	0.00019	0.0630	0.230	0.165	2.058	0.429	2.565	33.15
Mix proportion (wt%)	1.03	0.1	54.46	0.22	3.48	34.6	0.22	4.46	1.43	---

"Distance down" for each sample position. From these experimental conditions, experimental values include uncertainties of the sample volumes and the pass lengths, and they are anticipated to be between two lines at each sample position.

Measured γ-ray intensity ratios are in between calculated lines for each sample position. It is indicated that the γ-ray collimator works well to detect γ-rays from each sample position and γ-ray intensity ratios are consistent with pass lengths in concrete for each sample position. Thus, we could conclude that the γ-ray intensity ratio method is consistent with the collimator method, and both are valid for depth profile measurement.

To confirm the experimental feasibility, further experimental with large area concrete of 30x30cm^2 including salt will be performed.

3 Simulation for the proposed methods

To confirm the feasibility of the proposed methods, simulation with GEANT4 [9] for the production of ^{35}Cl-γ-rays in the assembly was performed. In this simulation, from the point of view to detect ^{35}Cl-γ-rays effectively in a certain region at a depth of interest, we identified the optimum angle of the γ-ray collimator and then tried to find a way to estimate chloride ion concentration.

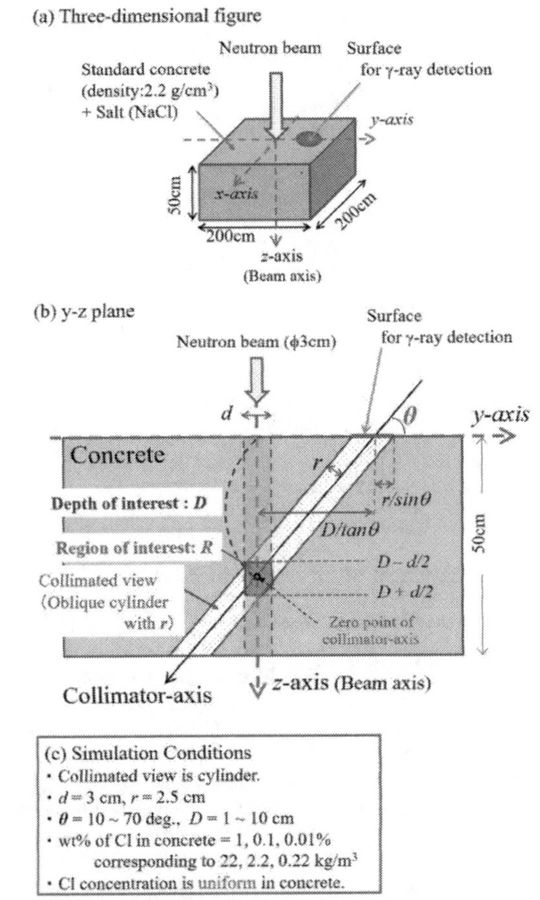

Figure 5 : Simulation geometries and conditions.
(a) Three-dimentional figure of geometries and coordinate.
(b) yz plane. (c) Summary of simulation conditions.

3.1. Simulation conditions

The simulation geometries and conditions of salt containing concrete are shown in Figure 5 (a)-(c), and are explained below. A 200 × 200 × 50 cm^3 standard concrete with salt was used. The compositions in concrete were same as ones shown in Table 2. Salt was homogeneously distributed in the concrete.

We defined that z-axis was the direction of neutron beam axis and x- and y-axes were directions along the concrete surface as shown in Figure 5 (a). In addition, we defined "collimator-axis" which had an angle θ with y-axis, and intersected z-axis as shown in Figure 5 (b). The zero point of this axis was a point across the z-axis. Neutron beam with a diameter of 3 cm and the energy spectrum of RANS [12] was injected to a concrete assembly.

The angle θ was defined as a collimator angle. As a view of a collimator looking into concrete, an oblique cylinder with radius *r* was defined, and called "collimated view" as shown in Figure 5 (b).

A depth of interest, where we aimed to measure the chloride ion concentration, was defined as *D*. A width to delimit a region in collimated view was defined as *d*, and was set to be 3 cm same as the diameter of neutron beam. Then, a region of interest "*R*" was defined as the region where z-axis and collimator-axis intersected and was delimited by *d* and $D \pm d/2$ as shown in Figure 5 (b).

For the detection of ^{35}Cl-γ-rays produced in collimated view, we counted γ-rays passing through the surface with the area of $\pi(r/\sin\theta)^2$ as shown in Figure 5 (a) and (b). Here, it is noted that if we want to view different *D* at the same θ a distance between a center of surface for γ-ray detection and z-axis varies according to $D/\tan\theta$

In this simulation, the amount in wt% of chlorine in concrete was changed from 0.01 to 0.1, and 1% corresponding to the chloride ion concentration of 0.22, 2.2, and 22 kg/m ^{3}, respectively. The collimator angle θ was changed from 10 to 70 degree. The depth of interest *D* was changed from 1 to 10 cm.

3.2 Production and detection rates of γ-rays

Figure 6 shows a contour plot of 35Cl counts in one of simulations as an example. This result was obtained by injecting neutrons of 1×10[illegible] under the conditions of Cl concentration of 2.2 kg/m^3, *D* = 10 cm, and θ= 30 degree. The unit of each point is counts. The region surrounded by dot ellipse is the production of ^{35}Cl-γ-rays in collimated view. The region of interest *R* is indicated by arrow and surrounded by rectangle.

 https://doi.org/10.1051/epjconf/202023105007

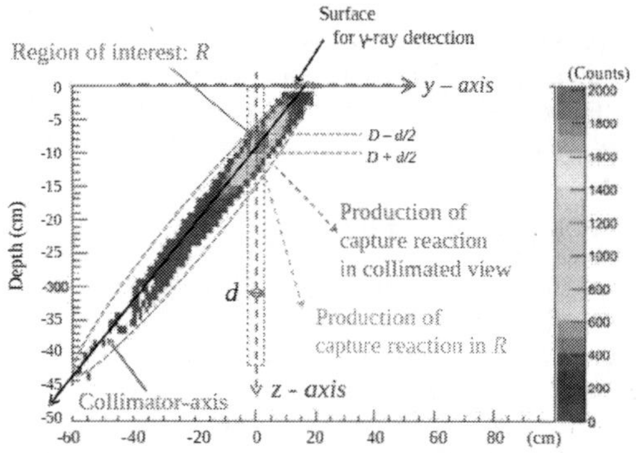

Figure 6 : Contour polt of ^{35}Cl production under simulation condisionts.

Here, the count of production is normalized to the production rate in unit of counts/s/100∝A, where 100 ∝A means the maximum proton beam current. Then, the production rate in collimated view and the production rate in R are plotted in collimator-axis as (a) and (b), respectively, in Figure 7. This production rate is a plot for the 1951 keV γ-ray under the conditions of Cl concentration of 22 kg/m^3, D = 5 cm, and θ= 30 degree.

Next, the production rate is converted to a detection rate that passes through the surface forγ-ray detection. A detection rate at a point in collimated view is calculated by following formula:

Detection rate = (Production rate) × (Ω/4 π)× T_γ, (3)

where Ωis a solid angle and T_γ is a γ-ray transmission in concrete same as in Section 2. Then, the detection rate in collimated view and the detection rate in R become like (c) and (d), respectively, in Figure 7. As obvious, the detection rate along the axis increases as the position becomes close to the concrete surface although the production rate decreases. Positions between R and the surface dominate the detection of the γ-ray. On the other

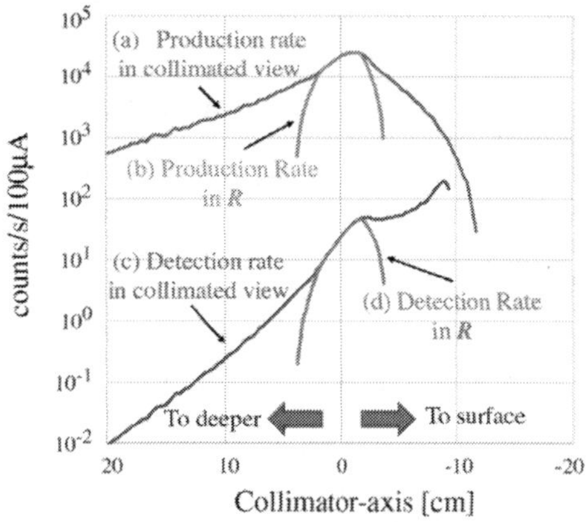

Figure 7 : Comparision with production rates and detection rates under the conditions of Cl concentration of 22 kg/m^3, D = 5 cm, and θ= 30 degree. (a) Production rate in collimated view. (b) Production rate in R. (c) Detection rate in collimated view. (d) Detection rate in R.

hand, the detection rate at a deeper region does not affect so much. Therefore, the key of our study is to identify the contribution of position closer to the surface with respect to that of R.

In following, to determine the optimum collimator angle, the detection rate was used for the estimation of Cl concentration.

3.3 Optimum collimator angle

To estimate optimum collimator angle, the ratio of the detection rates for the 1951 keV γ-ray in R and the total in collimated view were used. For effective detection of the ^{35}Cl-γ-rays in a region of interest, we assumed that an angle at which this ratio become highest should be optimum. Here, the detection rate in R is called RR, the total detection rate is called TR. The ratios of RR divide by TR are shown at the depths of interest D from 1 to 9 cm with θ= 10 to 70 degree in Figure 8. As a result, angles between 40 and 50 degree seem to be optimal for γ-ray detection in R at all depths as shown by the dashed rectangle in Figure 8. Although there is no clear indication, the θ= 50 seems preferable to θ= 40 at a low depth, and the 40 becomes effective at a deep depth.

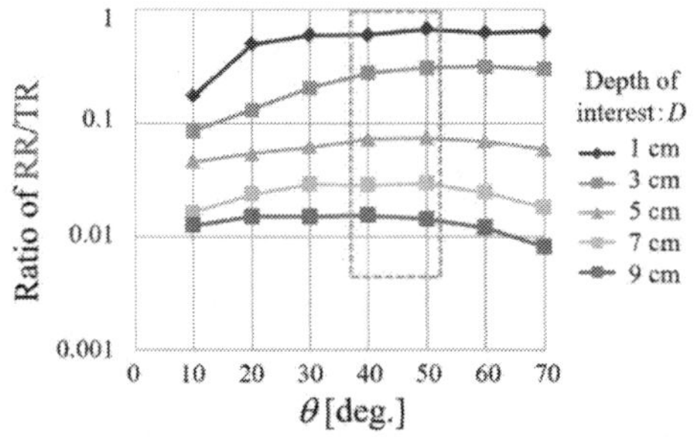

Figure 8 : Ratio of detection rate and total production rate at each depth of interest D for the 1951 keV γ-ray.

3.4 Estimation of chloride ion concentration

Figure 9 shows that RRs and TRs at the depths of interest D from 2.5 cm to 10 cm for the 1951 keV γ-ray for θ= 40 are plotted for chloride ion concentration of 0.22, 2.2, and 22 kg/m^3. It is shown that the ratios (RR/TR) are almost same for the different concentration at the same D. Assuming uniform chloride ion concentration in concrete, this result indicates that chloride ion concentration at D can be deduced from the ratios of RR/TR by obtaing TR experimentally. However, because the ratio is very small at depths larger than 7.5 cm, the estimation should have a large uncertainty. In addition to this uncertainty, we have to take into account uncertainties associated with the collimation. Because a collimated neutron beam brodens after coming out from neutron collimator and a collimated view by using γ-ray collimator also brodens.

From the detection rate in this result, if a chloride ion concentration at D = 5cm is the merginal

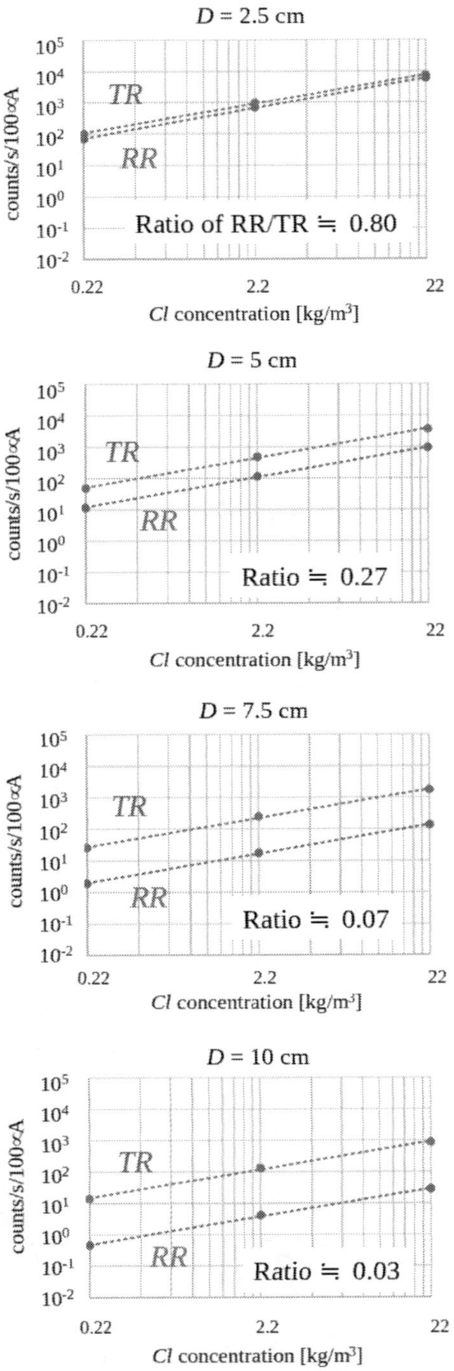

Figure 9 : Detection rate at the collimator angle of 40 degree for the 1951 keV γ-ray at each depth of interest D with chloride ion concentration of 0.22, 2.2, and 22 kg/m^3.

concentration [1] of 1.2 kg/m^3 which corresponds to 0.05 wt%, the *RR* is expected to be around 70 cps. Here, the γ-ray efficiency of the Ge detector with 10 % relative efficiency of 3”x3” NaI is about 0.05 % for the 1951 keV at 5-cm away from a γ-ray source. Then, the count rate of about 0.035 cps is expected at 5-cm away from the concrete surface, and the peak area becomes about 60 counts for about 30 minuetes. This count will be oberved in the γ-ray spectrum because the capture cross section [13] of chlorine is enough higher than other elements in the concrete composition as show in Table 2. In addtion, this measuring time is expected to be shorter than conventional methods. Therefore, it is expected that the proposed methods are feasible to estimate the chloride ion concentration at a depth up to 5 cm. Furthermore, although large unctertainties exist, the density at deeper region could be estimated.

In the future, to evaluate and decrease the uncertainties, we will perfrom the simulation incorporating i) the broadening of a collimated neutron beam, ii) the broadening of a collimated view in concrete for γ-rays, and iii) the chloride ion concentration of non-uniform distribution in concrete. By validating the simulation, we will perform further experiments implementing an actual concrete structures.

4 Summary

Chlorine within 12 cm depth in concrete was detected by NPGA at RANS. The collimator method and the γ-ray intensity ratio seemed to be valid to measure salt distribution in concrete. It was shown that the optimum angle of γ-ray collimator was between 40 to 50 degrees, and the estimation of chloride ion concentration at a depth of interest up to 5 cm seemed to be possible using the ratio of *RR/TR* under assumption of the Cl uniform distribution.

This work was (partially) supported by Photon and Quantum Basic Research Coordinated Development Program from the Ministry of Education, Culture, Sport, Science and Technology, Japan, and Cross-ministerial Strategic Innovation Promotion Program (SIP), “Infrastructure maintenance, renovation and management” (Funding agency: JST).

References

1. JSCE, *Standard Specifications for Concrete Structures-2012 “Design”* (2012) (in Japanese)
2. JIS A 1154. *Methods of test for chloride ion content in hardened concrete* (in Japanese)
3. JSCE-G 574, Concrete Journal, **62**(1), 246 (2006) (in Japanese)
4. H. Kanada, Y. Ishikawa, and T. Uomoto, Concrete Journal, **44**(6), 16-23 (2006). (in Japanese)
5. Y. Otake, et al., J. Dis. Res. **12**(3), 585 (2017)
6. Y. Wakabayashi et al., *Proceedings of the Concrete Structure Scenarios*, JSMS, **17**, 659 (2017) (in Japanese)

https://doi.org/10.1051/epjconf/202023105007

7. Y. Wakabayashi et al., Plasma and Fusion Research, **13**, 2404052 (2018)
8. Y. Wakabayashi et al., *Proceedings of the Concrete Structure Scenarios*, JSMS, **18**, 635 (2018) (in Japanese)
9. J. Allison, K. Amako, J. Apostolakis et al., IEEE Transactions on Nuclear Science. **53**(1), 270 (2006)
10. National Institute of Standards and Technology, XCOM : Photon Cross Sections Database, https://physics.nist.gov/PhysRefData/Xcom/html/xcom1.html
11. National Nuclear Data Center, Thermal Neutron Capture γ's, https://www.nndc.bnl.gov/capgam/
12. Y. Wakabayashi et al., Journal of Nuclear Science Technology, **55**, 859 (2018)
13. K. Shibata, O. Iwamoto, T. Nakagawa, N. Iwamoto, A. Ichihara, S. Kunieda et al., Journal of Nuclear Science Technology, **48**(1), 1 (2011).

Reconstruction on fast neutron CT for concrete structure inspection with a pixel-type detector by applying linear scanning method

Mingfei Yan[1,*], *Yasuo* Wakabayashi[1], *Yoshie* Otake[1,2], *Yujiro* Ikeda[1,3], *Atsushi* Taketani[1], *Takao* Hashiguchi[1], *Sheng* Wang[2,1], *Binbin* Tian[2], *Takaoki* Takanashi[1], *Tomohiro* Kobayashi[1], and *Baolong* Ma[2,1]

[1]RIKEN center for advanced photonics, RIKEN, Wako, Saitama, Japan
[2]School of Energy and Power Engineering, Xi'an Jiaotong University, Xi'an, China
[3]J-PARC Centre, Japan Atomic Energy Agency, Naka-gun, Ibaraki, Japan

Abstract. Concrete structure has been widely used in bridges and highways, however, its performance will be deteriorated after long term serving or suffering disaster. Since fast neutron has strong transmission ability and is sensitive to water content in the concrete structure, it can provide an effective probe to inspect the inner structure of concrete with non-destructive way. Thus, we propose a fast neutron imaging and reconstruction system of 3D CT for concrete structure inspection with Riken accelerator-driven compact neutron source (RANS) using a fast neutron pixel-type detector, which has 8×8 pixels. To have a good space resolution on the reconstructed image, a rotation + linear scanning method is devised and is used to collect the projection data from experiment or calculation. In this paper, reconstruction for a concrete object containing both iron bars and acrylic bars for simulation of water has been conducted. As a result, 3D image of 1cm diameter bar is reconstructed by the sparse reconstruction algorithm.

1 Introduction

Concrete structure has been widely used in the modern buildings (such as bridges, high ways and so on). It is important to inspect the health condition of the concrete structure, especially when the buildings are in service for a long time or have suffered from natural disaster (like earthquake). Usually, the technologies suitable for non-destructive inspection for concrete structure include ground penetrating radar method, ultrasonic wave method, radiation method (X-ray and neutron) and so on[1]. In the radiation method, radiation imaging can directly provide the radiography and tomography images for concrete structure. At present, the X-ray imaging has been widely used in situ non-destructive inspection. Compared with x-ray, neutron has strong transmission ability in medium, and is sensitive to water content, therefore, it can show evident advantages in concrete sutrcture inspection. However, neutron imaging is still under developing [2-3]. In addition, because the crosstalk phenomenon is serious in fast neutron imaging detector, its space resolution is not good as thermal neutron detector, which ranges from ~mm to ~cm[4]. If better space resolution in fast neutron imaging is expected, special method for collecting images should be employed. Thus we have started to develop a fast neutron imaging and reconstruction system of 3D CT for concrete structure inspection in situ with compact neutron source.

Basically,the algorithms involving CT image reconstruction can be divided into analytic and iterative classes. The famous analytic algorithm is filter back projection (FBP) algorithm, which is widely used in medical and industrial CT at present, and its reconstruction results is regarded as the golden standard. However, when the number of projection data is not enough, especially in sparse sampling case, a large number of artifacts will appear on the reconstructed images. However, reconstruction with iterative class algorithms usually can get better results in this case. As a kind of iterative class algorithm, the sparse reconstruction algorithm[5] utilizes the sparse characteristic of the image, which can significantly decrease the experiment time to collect the projection data.

2 Compact neutron source and fast neutron pixel-type detector

RIKEN accelerator-driven compact neutron source (RANS) was developed in 2013, which utilizes 7MeV

* Corresponding author: mingfei.yan@riken.jp

proton to bombard 300 μm thickness beryllium target to produce neutron. As is shown in figure 1, the main components of RANS includes proton accelerator section and target station section. The accelerator section mainly consists of ion source, radio-frequency quadrupole (RFQ) and drift-tube-linacs (DTL). The target station mainly contains beryllium target, cooling water, polyethylene moderator, graphite reflector and layered shieling materials. The moderator can be removed from the side of the target station when implementing fast neutron experiment.

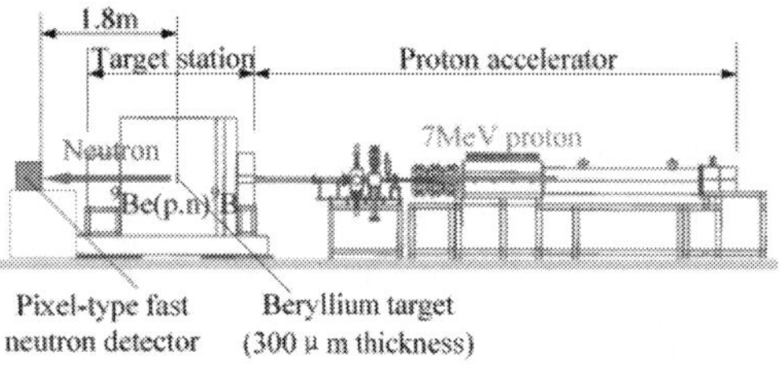

Fig. 1. Structure diagram of RANS.

The maximum current of RANS is 100 μA, and the corresponding energy spectrum at the location of 1.8m far away from the target is shown in figure 2, which is simulated by Geant4 code[6]. In the simulation, a neutron generation function is employed to calculate the neutron spectrum produced from the ^{9}Be+p reaction with 7 MeV protons[7]. The peak intensity of fast neutron is nearby 10^6 $cm^{-2}\cdot s^{-1}$ at the location of 1.8m far away from the target.

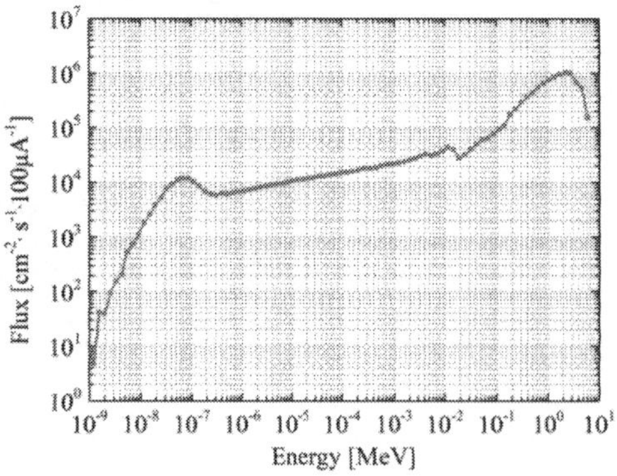

Fig. 2. Energy spectrum at 1.8m far away from the target (without moderator).

To achieve non-destructive inspection for concrete structure by fast neutron imaging method, our team has developed a fast neutron pixel type detector (shown in figure 3), which has 8×8 pixels, and the size of each pixel is 2cm×2cm×5cm (thickness). It utilizes the BC408 plastic scintillator as the neutron to recoiled proton convertor and scintillator, multiple-pixel photon counter (MPPC) as the light to electric signal convertor. As the electric signal produces, the number of counts can be recorded by the counter after passing though the amplifier and comparator. The detector is triggered by the accelerator signal, which is helpful for eliminating the counts produced by lower energy neutron.

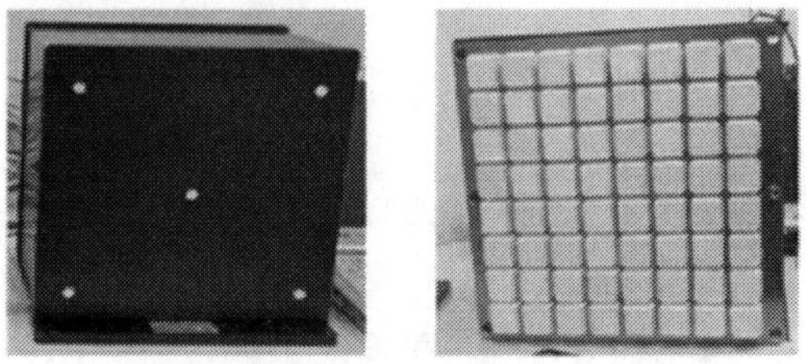

Fig. 3. Fast neutron pixel-type detector

3 Principle of CT and image reconstruction

Neutron intensity attenuates in the matter as Lambert-Beer law:

$$I(x,y)=I_0(x,y)e^{-\int_L \mu(x,y)dl} \tag{1}$$

where $I_0(x,y)$ and $I(x,y)$ refer to the incident and transmitted intensities respectively; $\mu(x,y)$ denotes the macroscopic neutron cross section of the matter. The principle of CT image reconstruction is calculating $\mu(x,y)$ from $I_0(x,y)$, $I(x,y)$ and the geometry configuration of CT system.

CT image reconstruction with iterative class algorithms can be regarded as solving the linear equation set:

$$\mathbf{P}_{M\times1}=\mathbf{W}_{M\times N}\mathbf{U}_{N\times1} \tag{2}$$

$$i.e.\quad \begin{bmatrix} p_1 \\ p_2 \\ \vdots \\ p_M \end{bmatrix} = \begin{bmatrix} w_{11} & w_{12} & \cdots & \\ w_{21} & w_{22} & \cdots & \\ \vdots & \vdots & \ddots & \vdots \\ w_{M1} & w_{M2} & \cdots & \end{bmatrix} \begin{bmatrix} \mu_1 \\ \mu_2 \\ \vdots \\ \mu_N \end{bmatrix} \tag{3}$$

where $\mathbf{P}=[p_1,p_2,\cdots,\quad]$ is termed as projection vector; $\mathbf{W}=(w_{ij})\in\mathbf{W}^M\times\mathbf{W}^N$ is named as system matrix, in which w_{ij} describes the contribution of the *ith* ray to the *jth* pixel, and it can be calculated based on the model shown in figure 4; $\mathbf{U}=[\mu_1,\mu_2,\cdots$ is image vector to be reconstructed.

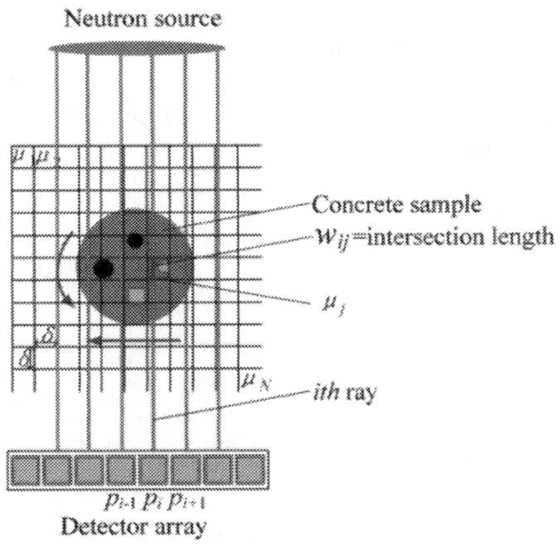

Fig. 4. System matrix calculation model

Usually, rank(**W**) = rank(**W**|**P**)<N (**W** and (**W**|**P**) are termed as the coefficient matrix and augmented matrix for linear equation set respectively), and equations (2) and (3) are undetermined and ill-posed, thus the exact solution can't be obtained. Therefore, the key point of CT image reconstruction with iteration class algorithms lies in constructing the method to solve the undetermined equation.

In this paper, the sparse algorithm based on total variation (TV) minimization—adaptive steepest descent-projection onto convex sets (ASD-POCS) is employed to implement the image reconstruction, which describes the minimization:

$$\|\mathbf{P}-\mathbf{WU}\|_2^2+\alpha\|\mathbf{\Psi}(\mathbf{U})\|_n \tag{4}$$

where $\|\mathbf{P}-\mathbf{WU}\|_2^2$ is fidelity term; $\|\mathbf{\Psi}(\mathbf{U})\|_n$ is sparse constrait term; $\|\mathbf{\Delta}\|_n$ is l_n -norm of vector $\mathbf{\Delta}$ (n=0, 1, 2) ; α is sparse constraint coefficient. If $\mathbf{\Psi}$ is image gradient transformation, thus :

$$\|\mathbf{\Psi}(\mathbf{U})\|_1=\sqrt{\left(\mu_{s,t}-\mu_{s-1,t}\right)^2+\left(\mu_{s,t}-\mu_{s-1,t}\right)^2} \tag{5}$$

where s and t are the pixel indexes on the reconstructed image. Equation (5) can also be termed as total variation of image. ASD-POCS algorithm employs simultaneous algebraic reconstruction technique (SART) to solve the minimization of fidelity term, and adaptive steepest descent method to minimize sparse constraint term.

4 Projection data simulation and reconstruction results

The configuration of fast neutron CT is shown in figure 5. In our research, a cylindrical concrete containing two iron bars and two acrylic bars is as the simulation phantom. Since the pixel size of fast neutron detector is 2cm, to improve the space resolution on the reconstructed image, the linear scanning method is proposed. In scanning porcess, the concrete sample is rotated and translated step by step in each rotated angle to take the projection data for image reconstruction.

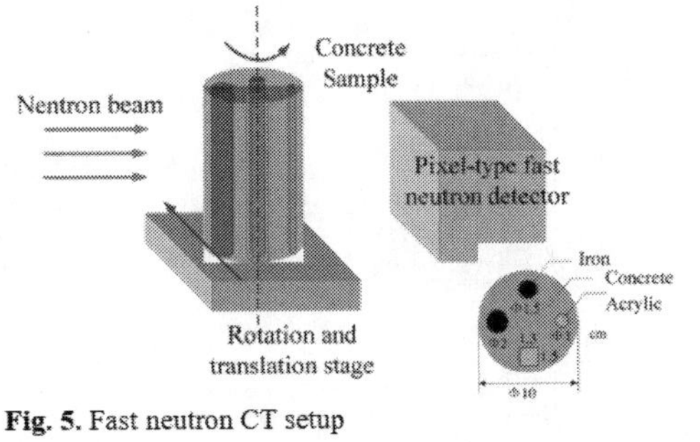

Fig. 5. Fast neutron CT setup

The simulation is implemented by Geant4 code, and the main parameters are shown in table 1.

Table 1. Main simulation parameters of fast neutron CT

Parameter	value
Physics	QBBC
Neutron spectrum	Shown in figure 2
Record of detector	Energy deposition of recoiled proton
Number of rotated angle and step	6/30°(180°)
Translation times and step in each rotated angle	260/1mm

By means of the simulated radiography images of concrete sample with or without iron and acrylic bars, the projection data for image reconstruction can be obtained. Combing with the system matrix and projection data, the image reconstruction is implemented by ASD-POCS algorithm, and the results is shown in table 2. In the reconstruction, we have changed the translation step size and pixel size on the reconstructed image. It can be observed that the accuracy of reconstructed image basically can be improved as the translation step decreasing, especially in the case that the pixel size on the reconstructed image is 1.25 mm. The reason is that the undermined degree of equation (2) can be improved by increasing the number of projection data, especially in the case that the equation is seriously undetermined. It also can be seen that decrease of pixel size (increase of the number of pixels) on the reconstructed image basically will decrease the accuracy of reconstructed image. That is because increasing the number of unknown number in equation (2) may result in it transiting from overdetermined state to undetermined state. In general, the iron bar with 1.5 cm and 2 cm diameters, acrylic bar with 1 cm diameter and 1.5 cm^2 cross section can be identified by the reconstruction after 6 times rotation scanning with 5mm translation step in each rotated angle. It demonstrate that the proposed scanning method can be applied to fast neutron CT to improve the space resolution of imaging.

Table 2. Reconstructed results under different pixel size on reconstructed image and translation step in each rotated angle

		Pixel size/number of pixel on reconstructed image		
		5mm/20×20	2.5mm /40×40	1.25mm /80×80
Translation step	5 mm	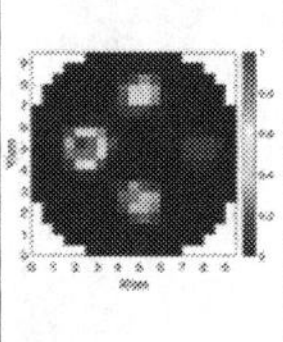	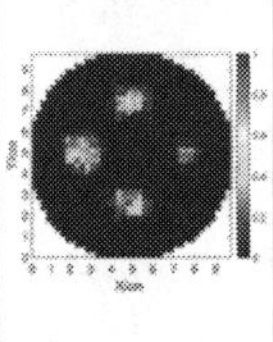	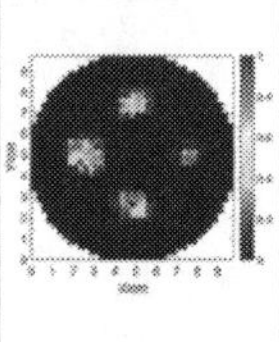
		Pixel size/number of pixel on reconstructed image		
		5mm/20×20	2.5mm /40×40	1.25mm /80×80
Translation step	2 mm	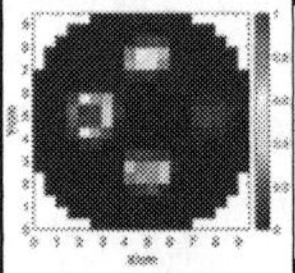	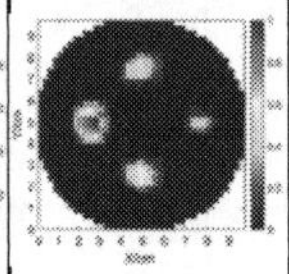	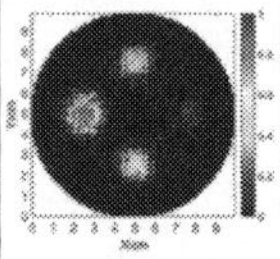
	1 mm	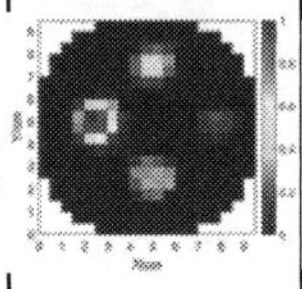	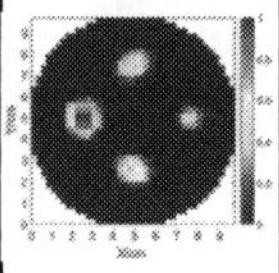	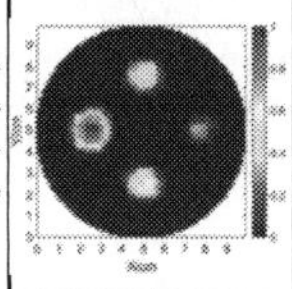

5 Conclusion and outlook

This paper has proposed a rotation+linear scanning method for fast neutron imaging with a pixel-type detector. It is shown that iron bar with 1.5 cm and 2 cm diameters, acrylic bar with 1 cm diameter and 1.5 cm^2 cross section can be identified by the reconstruction after 6 times rotation scanning with 5mm translation step in each rotated angle.

The quantitative evaluation of the accuracy and space resolution of reconstructed image is in progress. And the real experiment will be performed on RANS. We are also planning to employ the area model to calculate the system matrix for the fast neutron CT to get a better space resolution.

This work is supported by a *kakenhi* Grant-in-Aid for JSPS Fellows (18F18763) from Japan Society for the Promotion of Science.

References

1. K. U. R. Sardar, Z. Ibrahim, S. A. Memon, M. Jameel, Constr. Build Mater. **107**, 58-86 (2016)
2. Y. Otake, Y. Seki, Y. Wakabayashi, Y. Ikeda, T. Hashiguchi, Y. Yoshimura, H. Sunaga, A. Taketani, M. Mizuta, Y. Oshima, M. Ishida, J. Disaster Res. **12**, 585-592 (2017)
3. Y. Seki, A. Taketani, T. Hashiguchi, S. Wang, M. Mizuta, Y. Wakabayashi, Y. Otake, Y. Yamagata, H. Baba, K. Kino, K. Hirota, S. Tanaka, Nucl. Instrum. Methods Phys. Res. A **870**, 148-155 (2017)
4. R. Zboraya, R. Adamsb, Z. Kis, Appl. Radiat. Isotopes **119**, 43-50 (2017)
5. M. F. Yan, H. S. Hu, Y. Otake, A. Taketani, Y. Wakabayashi, S. Yanagimachi, S. Wang, Z. H. Pan, G. Hu, Meas. Sci. Technol. **29**, 055404 (2018)
6. S.Agostinelli, J. Allison, K.Amako, J. Apostolakis, H. Araujo, P. Arce, M. Asai, D. Axen, S. Banerjee, et. al, Nucl. Instrum. Methods Phys. Res. A **506**, 250-303 (2003)
7. Y. Wakabayashi, A. Taketani, T. Hashiguchi, Y. Ikeda, T. Kobayashi, S. Wang, M. F. Yan, M. Harada, Y. Ikeda, Y. Otake, J. Nucl. Sci. Technol. **55**, 859-867 (2018)

EPJ Web of Conferences **231**, 05009 (2020)
UCANS-8 https://doi.org/10.1051/epjconf/202023105009

Thermal neutron intensity measurement with fission chamber in current, pulse and Campbell modes

*Shakir Zeinalov**, *Olga Sidorova*[1], *Pavel* Sedyshev[1], *Valery* Shvetsov[1]
Youngseok Lee[2], *Uk-Won* Nam[3]

[1]Joint Institute for Nuclear Research, 141980, Dubna Moscow region, Russia
[2] National Fusion Research Institute, Daejon, Republic of Korea
[3] Korea Astronomy & Space Science Institute, Daejon, Republic of Korea

Abstract. In thermal nuclear reactors, most of the power is generated by thermal neutron induced fission. Therefore, fission chambers with targets that respond directly to slow neutrons are of great interest for thermal neutron flux measurements due to relatively low sensitivity to gamma radiation. However, the extreme conditions associated with experiments at very low cross section demand highly possible thermal neutron flux, leading often to substantial design changes. In this paper we report design of a fission chamber for wide range (from 10 to 10^{12}n/cm^{2}sec) measurement of thermal neutron flux. Test experiments were performed at the first beam of IBR2 pulsed reactor using digital pulse processing (DPP) technique with modern waveform digitizers (WFD). The neutron pulses detected by the fission chamber in each burst (5 Hz repetition rate) of the reactor were digitized and recorded to PC memory for further on-line and off-line analysis. New method is suggested to make link between the pulse counting, the current mode and the Campbell technique.

1 Introduction

In recent decade digital pulse processing (DPP) is considered as very powerful alternative to traditional analogue pulse processing (APP) in experimental nuclear physics research. Mainly it happened due to a diversity options were provided in data analysis and the flexibility of data treatment, which very unlikely can be provided by APP. In this work we demonstrated the power of DPP implemented in thermal neutron counting with fission chamber (FC) in the various modes developed in the reactor instrumentation, namely the counting, the current and the Campbelling techniques. We demonstrated how to implement DPP making direct link between the counting and the current mode measurement with the thermal neutron sensor based on gas filled detectors. It should be noticed that absolute thermal neutron flux measurement in the wide intensity range was very unlikely possible with the same mode. The main problem was in the response function of the sensor, which has limit in the width about few tens of ns. When the distance between pulses became comparable with this limit, then separation of the individual pulses was impossible. In that case measurement usually was possible to perform in the current mode, using the proportionality of the measured current to the neutron intensity. The main problem is how to make link between the counting and the current modes. In this work we propose a solution which we believe able to provide the link.

2 Experimental Setup and Results

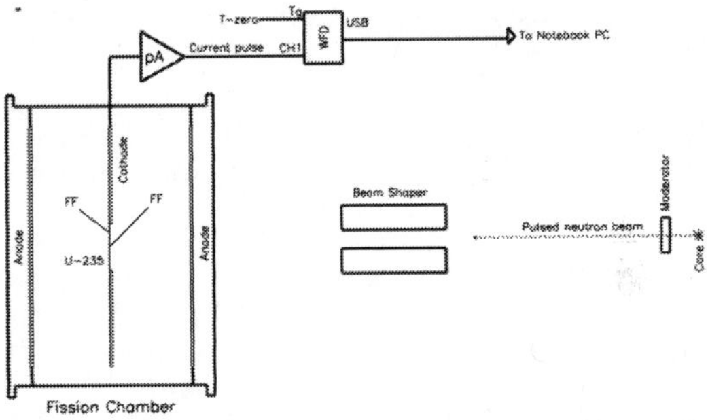

Fig. 1. Fission chamber with associated preamplifier, WFD and beam shaping equipment

In reported experiment we used the fission chamber with ^{235}U target (99,999% enrichment) mounted on the common cathode of a twin parallel plate ionization chamber shown in Fig. 1. The target thickness was 0.5 mg/cm^2 of uranium smeared on the both sides of aluminium backing foil. Ionization chamber was operated at constant flow 50 ml/min of working gas P10. Negative high voltage (500 V) was applied to the common cathode allocated between two anodes at distances of 40 mm from each of them. Due to thickness of the target and backing foil only one of two fission fragments (FF) was detected on thermal neutron induced fission event. Working gas pressure was stabilized at 1.05 bar using a gas pressure controller. Fission chamber was located at distance 27 m from the pulsed nuclear reactor IBR2 with the following main parameters: average power 1.8 MW, fast neutron pulse duration 200 us, the pulse repetition rate 5Hz, the thermal neutron intensity at the target position was about 10^6 n/cm^2 per

* Corresponding author: zeinal@nf.jinr.ru

 https://doi.org/10.1051/epjconf/202023105009

second. Fission fragments (FF) were decelerated in the working gas of FC producing free electrons which were drifted towards the anodes under the applied uniform electric field, inducing electric current pulses on the cathode circuit. The cathode current was amplified and recorded by the WFD, triggered by the IBR2 "T-zero" signal. The WFD sampling frequency (200 kHz) and the number of samples (4096) were selected in order to cover the full thermal neutron burst duration about 20 ms. Sequential reactor bursts were digitized and recorded to the notebook computer's hard disk for further off-line data analysis. One of the recorded reactor bursts presented in Fig.2.

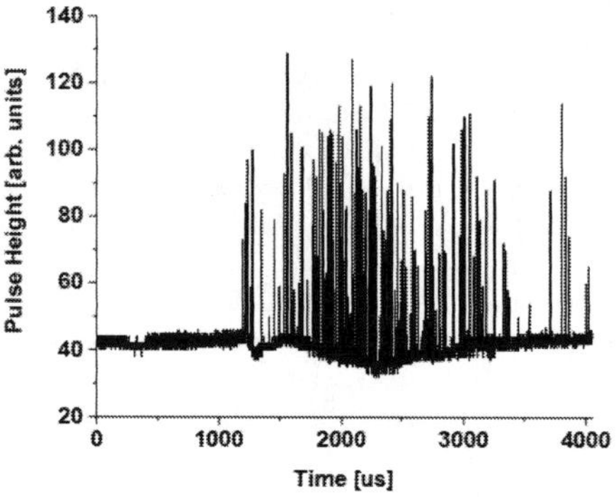

Fig. 2. Recorded waveform of single burst of the IBR-2. Full range is 20500 us

It should be noticed that because of the overlapped pulses baseline was distorted and must be restored using special electronic modules to stabilize the base line for proper counting when the APP was implemented. Use of the DPP avoided the problem in the elegant way, using sequential flow of the reactor bursts. To remove the baseline fluctuation we have used the following method: each recorded waveform of the burst was subtracted from the waveform of previous one, creating waveform shown in Fig. 3. It was easy to see that resulted distribution composed of bipolar waveform with removed baseline and the data obtained was satisfying to the second Campbell theorem [1,2]. According to the theorem the variance of the current's AC component from the randomly distributed in time (Poisson distribution) current pulses is proportional to the average pulse rate and the mean square charge per pulse:

$$\langle I(t)\rangle = \int_0^{\infty} h(t)dt \cdot \sum_{i=0}^{m} \langle n_i\rangle \cdot \langle q_i\rangle \qquad (1)$$

$$\langle I(t)^2\rangle - \langle I(t)\rangle^2 = \int_0^{\infty} h(t)^2 dt \cdot \sum_{i=0}^{m} \langle n_i\rangle \cdot \langle q_i^2\rangle \qquad (2)$$

where, $h(t)$ the circuit's response to a single pulse of a unit charge, $\langle q_i\rangle$ the mean charge per pulse, $\langle q_i^2\rangle$ the mean square charge per pulse, $\langle n_i\rangle$ the average pulse rate in a fixed interval of measurement. In current measurement $\langle I(t)\rangle^2 = 0$, because the current waveform $I(t)$ was residue of two sequential waveforms. So the Eq. 2 reduced to Eq. 3:

$$\langle I(t)^2\rangle = \int_0^{\infty} h(t)^2 dt \cdot \sum_{i=0}^{m} \langle n_i\rangle \cdot \langle q_i^2\rangle \qquad (3)$$

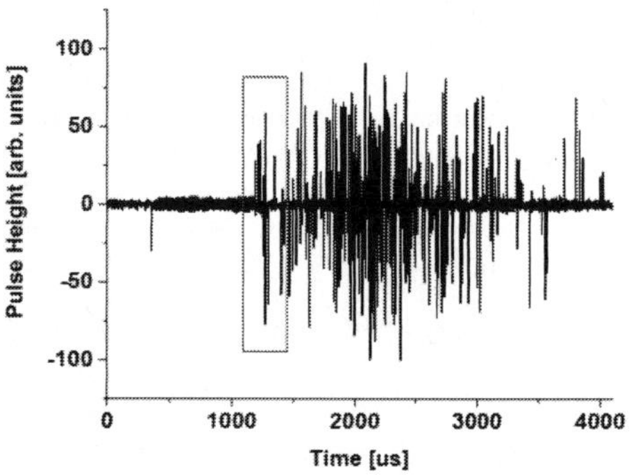

Fig. 3. Resulted waveform after second waveform was subtracted from the first one.

To investigate the Campbell theorem experimentally we divided whole measurement consisting of more than 160000 of reactor bursts into the pieces of 12000 bursts. For each reactor burst we measured the number of detected neutrons, which was fluctuating from burst to burst making distribution presented in Fig.5. There were main group of bursts with the average neutron number n_m=168 per

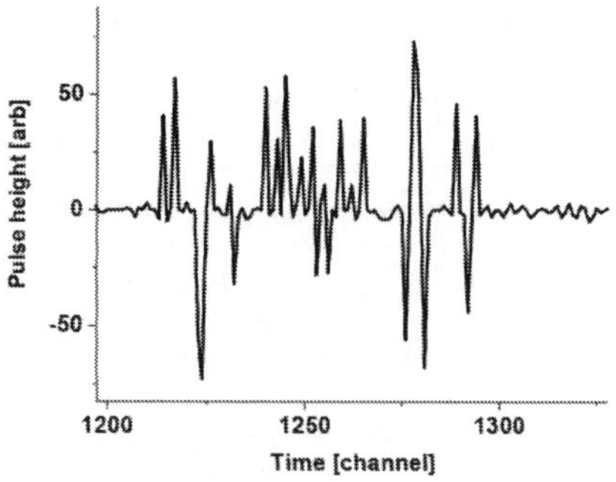

Fig. 4.:The enlarged area from the Fig. 3 to make individual pulses visible

burst and the FWHM=23.9 and a small group with n_l=131.7 and FWHM=21.3. Two group caused by the effect of the threshold in peak search algorithm implemented for FC pulse counting. The neutrons were calculated according to number of peaks above the threshold using constant fraction time marking algorithm, described in our previous publications [3]. From the data presented in Fig. 4 one can estimate the time intervals

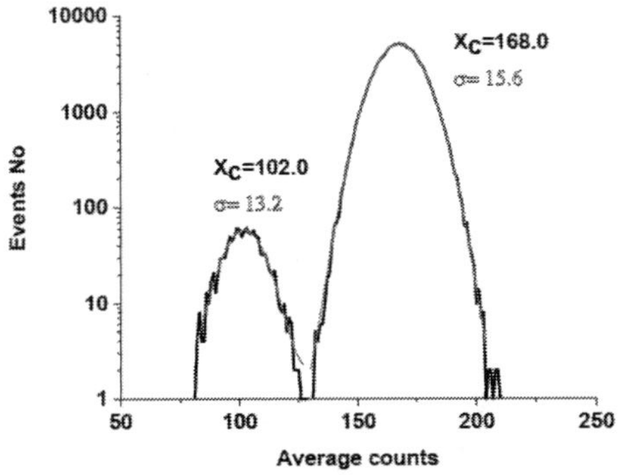

Fig. 5. Distribution of a number of neutrons detected in the set of 200 bursts .

between the sequential pulses presented in Fig. 6 which demonstrates the numbers of events with time difference between successive pulses in the range 15us up to 1250 us. This distribution demonstrates variable counting rate of Poisson distribution inherent to pulsed sources. One of the important features available with measuring technique using DPP is simulating the high counting rates from the set of measurements made at relatively low counting rate.

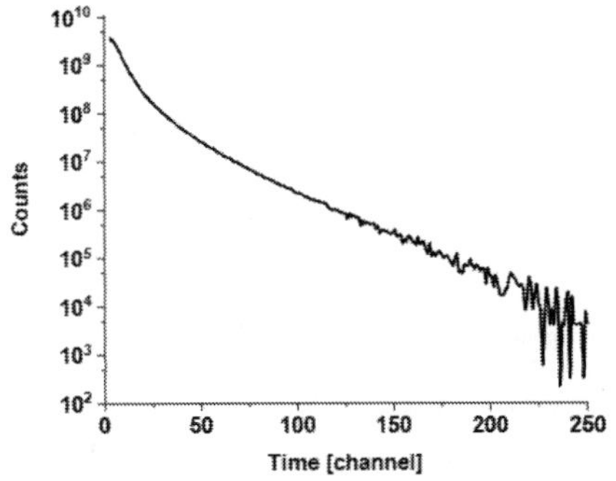

Fig. 6. Distribution of time intervals between successive pulses

In current measurement the full set of data was divided into the smaller subsets, containing 20000 burst each. Taking into account the number of pulses in one burst 168 the number of pulses in $1.6*10^5$ burst would be $\sim 3.0*10^7$. Summing whole pulses into single burst produce single burst, containing $3.0*10^7$ pulses in 20 msec. The most of the pulses overlapped simulating the situation of practically continuous current flow trough detector as demonstrated in Fig. 7. Such situation is exactly the same as it would be with a detector exposed to the neutron flux producing intensity of a detector $1.5*10^9$ pulses/sec. Using this approach we were able to simulate comparison of the pulsed, the current and the Campbelling techniques with the same data set. From the average pulse count in the single burst we managed estimate the total number of pulses in the acquired data set. To simulate the current mode measurement we summed 160000 successive reactor bursts into single distribution as demonstrated on Fig. 8. Fig.9 demonstrates the share between partially moderated neutrons and gamma radiation plus part of the fast neutrons passed the moderator without interaction from plutonium's fission in the reactor core, used to evaluate the FC relative sensitivity to the prompt neutrons and gammas. Fast prompt fission neutrons produced pulses in the chamber mainly by inducing the fission with cross section about 2 b, whereas prompt fission gammas produce pulses in the FC by directly ionizing the FC's working gas atoms. If the target containing about $6*10^{18}$ uranium atoms (double sided 0.5 mg/cm ^{2}) irradiated by the fast neutron flux 10^{12} n/cm^2 would produce

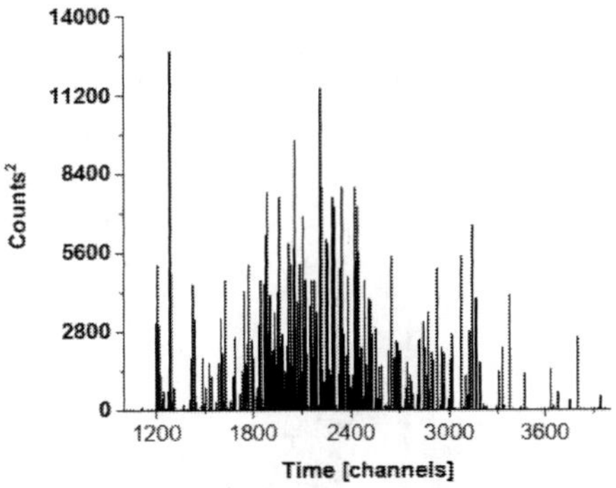

Fig. 7. Autocorrelation function calculated after subtraction of two successive reactor bursts

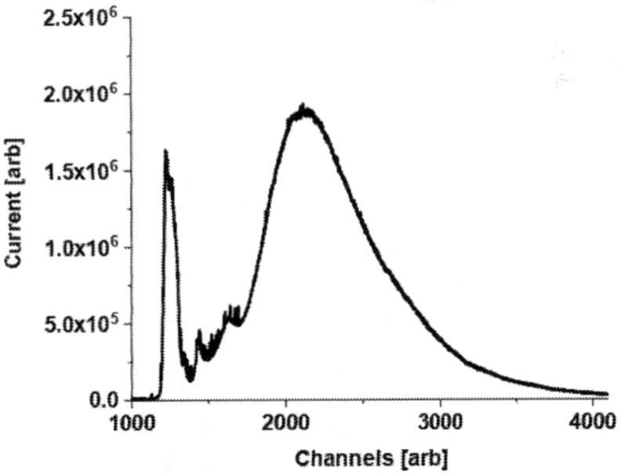

Fig. 8. FC pulses summed for 160000 reactor bursts.

$2*10^6$ pulse/cm^2 in the FC. According to the plot in Fig. 9 prompt fission gammas plus neutrons escaped the moderator produce by factor 31.08 pulses less. For the thermal neutron region the neutron capture cross section about250 times higher, meaning that the neutrons would produce by factor 250 more pulses than gammas and the intensity of pulses from gammas would be at least 31.08 times less. From these rough speculation the sensitivity of FC to gamma radiation would be ~7000 less than sensitivity to thermal neutrons.

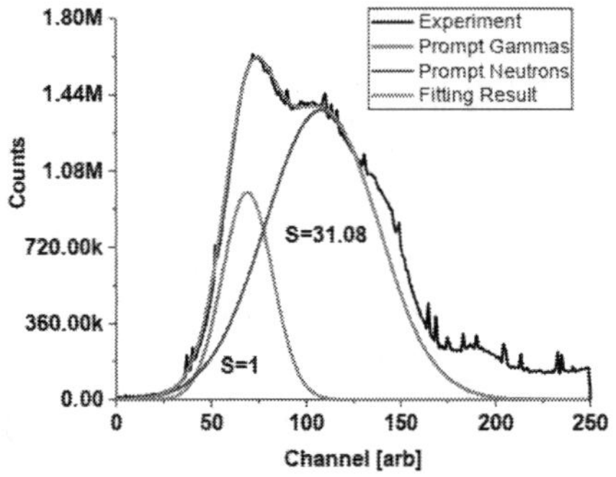

Fig. 9. Radiation from the reactor burst with decomposition into the neutron (partially moderated/resonance) and gamma plus fast neutrons, escaped the moderator, represented by Gaussians. S indicates relative area under the Gaussians.

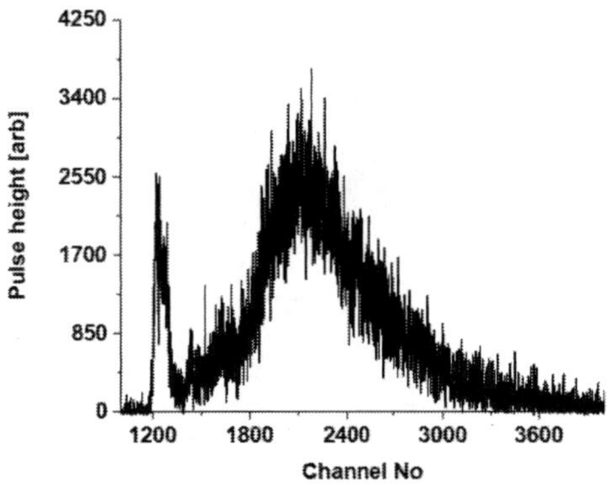

Fig. 10. Group waveform composed of 100 sequential waveforms;

Finally in Fig.11 we have presented results of the current mode and the Campbelling techniques of the pulse rate measurement using FC. Each point was obtained by summing the pulses from the sequential reactor bursts incrementally: the second group burst pulses were added to first one, the third group of burst pulses were added to the second one and so on. Each group consisted of waveforms

acquired in 100 sequential bursts. For the analysis the waveforms were summed creating almost continuous distribution as shown on the graph of the Fig. 10. Then for the current mode measurement the components of the distribution from Fig.10 was integrated producing the number proportional to the pulse intensity in the given group. For the Campbelling technique the components of distribution was squared before integration then the square root of integral was considered as measure of the flux intensity for the given group. Results for the current and Campbelling technique were plotted on the left of Fig. 11 both demonstrating very good linearity.

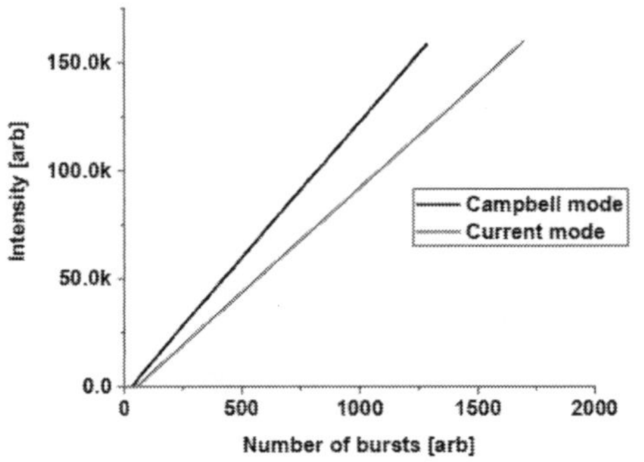

Fig. 11. Comparison of intensity measurement in Campbell and current modes demonstrates very good linearity

3 Conclusions

In this paper we report results of out-of-core thermal neutron flux measurement on pulsed nuclear reactor IBR2 using the counting, the current and Campbelling techniques. Counting mode was used to make direct link between the number of detected neutrons and the current flowing through the FC. Pulsed neutron source provided unique possibility to achieve a variable neutron flux density in the same data set. The neutron induced signals result in pulses with much greater charges than pulses from gammas, the mean square signal will weigh the neutron component by the square of the ratio of neutron signal neutron- to gamma ray induced signal [2]. In our work we find this ratio not less than 7000 from rough estimate of experimental data.

1. N.R. Campbell, V.J. Francis, "Theory of Valve and Circuit Noise, JIEE, Vol. 93. Part III, 1946; DOI:10/1049/ji-3-2.1946.0009.

2. G.F. Knoll, *Radiation detection and measurements*, (Wiley, New York, 2000), p. 522, Chapter 14, IV, Reactor Instrumentation).

3. L. Svetov, Sidorova O., Sh. Zeinalov. *In proc. of XXIV International Seminar on Interaction of Neutrons with Nuclei*, Dubna, May 24-27, 2016, JINR E3-2017-8, p. 430.

https://doi.org/10.1051/epjconf/202023105010

Hybrid boron-10 gaseous detector for slow and fast neutron simultaneous detection

Stanislav Potashev[1,*], *Aleksandr* Drachev[2], *Yury* Burmistrov[1], *Sergey* Karaevsky[1], *Aleksandr* Kasparov[1], *Vasily* Ponomarev[1], and *Gennady* Solodukhov[1]

[1]Institute for Nuclear Research, Russian Academy of Sciences, 117312 Moscow, Russia
[2]State Research Institute for Chemistry and Technology of Organoelement Compounds, 119991 Moscow, Russia

Abstract. A hybrid position-sensitive neutron detector with active ^{10}B-layer and proportional chamber filled by the gas mixture Ar+25%CO_2 is being studied. The detector is used on the neutron source based at the linear accelerator. Recording two pairs of pulse heights which are produced in two gas gaps by ^{7}Li and ^{4}He nuclei between the anode and two cathodes gives a possibility to determine the X and Y neutron coordinates. This detector permits time-of-flight measurements. A significant difference in pulse height spectra for slow and fast (several MeV) neutrons was observed in experiment. In recent experiments with ^{10}B-detector we found that (1) the pulse heights depend on the neutron energy and (2) the position of the maximum in the pulse height spectrum varies monotonically versus the maximum neutron energy.

1 Introduction

There are many applications of neutron position-sensitive detectors in physics [1], technology [2] and biology [3]. The choice of a kind of detector depends on the energy range of neutrons. To investigate the atomic structure in superconductors thermal neutron diffraction can be used [4]. Small angle scattering is used for the investigation of alloys [5], magnetic structures [6], protein molecules in biology [7].

Usually, ^{3}He- and $^{10}BF_3$-tube counters and their assemblies are widely used in a diffraction setup and in small angle scattering. The position sensitive proportional chamber filled by ^{3}He under high pressure [8] or $^{10}BF_3$ is more rare and expensive device. A proportional chamber and a assembly of tube counters filled by the ^{3}He gas have a high efficiency and low sensitivity to γ-rays. But they operate under a high pressure and have a thick entrance window and may experience leakage of expensive gas.

Also, scintillation detectors with the embedded ^{6}Li are developed [9-11]. They are used to detect slow neutrons. Detection of fast neutrons is important in radiation protection and in many branches of physics. As a rule, a scintillation detector of a recoil proton is used for the detection of fast neutrons [12]. However, it has a high sensitivity of the scintillation detector to γ-rays and x-rays which makes it difficult to detect neutrons. Besides that, the substance of scintillation detector distorts the neutron distribution. The pulse height is not proportional to an ionization loss.

Recently, novel hybrid detectors for slow neutrons and for neutron flux measurements have been developed [13, 14]. They have sensitivity only to neutrons. Therefore, they are suited to study rare nuclear reactions. The detector is based on the ^{10}B solid layers combined with a proportional ion chamber [15, 16]. The neutron cross section on the ^{10}B nucleus has a very important feature, namely, that it sharply increases between 1 and 7 MeV up to ~1 b. Therefore, the ^{10}B-detector, unlike the ^{3}He-detector, can also be used as a fast neutron detector [17].

In Section 2 we discuss the concept and design of our detector. In Section 3 we discuss recent experiments with slow and fast neutrons. We conclude with perspectives of hybrid ^{10}B-detectors.

2 Position-sensitive ^{10}B-detector

The detector principle is as follows: ^{10}B nucleus captures a neutron and forms the nucleus $^{11}B^*$ in the excited state, which decays in two ways as shown below

$$n + {}^{10}B \rightarrow {}^{11}B^* \rightarrow {}^4He(1470keV) + {}^7Li(841keV) + \gamma(476keV)$$

branch ratio is 0.936 for 0.025 eV; 0.3 for 1.8 MeV; 0.5 for 2.5 MeV

$$n + {}^{10}B \rightarrow {}^{11}B^* \rightarrow {}^4He(1770keV) + {}^7Li(1013keV)$$

branch ratio is 0.064 for 0.025 eV; 0.7 for 1.8 MeV; 0.5 for 2.5 MeV.

* Corresponding author: potashev@inr.ru

 https://doi.org/10.1051/epjconf/202023105010

The 3-μm ^{10}B layer was deposited on an inner surface of a 1-mm glass plate. The 0.2-μm Al layer was deposited on the boron layer and it forms the first cathode of the detector. A thin layer of a semiconducting polymer was deposited for the cathode protection. The concept and design of detector is demonstrated in Fig.1. Only one nucleus ^{4}He or ^{7}Li produced by the thermal neutron can penetrate to the active gas volume overcoming the remaining layers of boron and aluminum. One nucleus or both nuclei can enter the sensitive gas volume if a neutron has energy of several MeV. The space between the first cathode and the anode of 20-μm tungsten rhenium wires coated by gold serves as the first sensitive gap. The spacing between wires is equal to 2 mm. The second sensitive gap is the space between the anode and second cathode. Both gaps are equal to 2 mm. The neutron event occurs only if at least one nucleus gives ionization signal in the second sensitive gap.

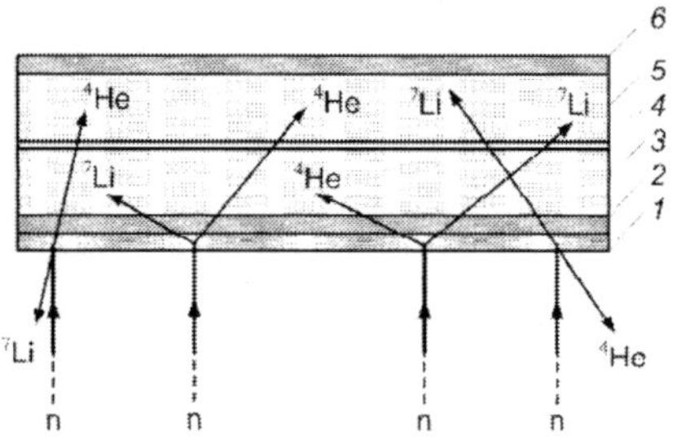

Fig. 1. ^{10}B 2D detector concept. *1* – ^{10}B-layer, 3 μm; *2* – Al-layer, 0.2 μm; *3* – 1-st sensitive gas gap of 2 mm thickness; *4* – WRe-wires, 20 μm; *5* – 2-nd sensitive gap of 2 mm thickness; *6* – 2 mm thickness fiberglass.

Various gas mixtures can be used to fill the detector. The cheap gas mixture Ar+25%CO_2 was used in experiments. It provides a long-time operation of the detector without aging. The charge division method is used to determine the X- and Y-coordinates of the neutron [18]. For this aim, anode wires are connected to a circuit of distributed resistors. Partial pulse heights X_1 and X_2 from two ends of the circuit are used to determine the X-coordinate. Analogously, the pads of the second cathode are connected to a circuit of distributed resistors. Partial pulse heights Y_1 and Y_2 from two ends of another circuit of distributed resistors are used to determine the Y-coordinate. A pulse height resolution of 15% and a spatial resolution has been obtained at 700 V. The detector operates in a proportional mode at low gas gain. This provides small fluctuations in an ionization loss and, consequently, a high resolution in pulse height [19].

To determine a detector spatial resolution a cadmium mask with three 1 mm-slits was used. Slits were in the distances separated by 20 mm from each other. The moderator of the neutron source was configured to the flux enriched by thermal neutrons. The ^{10}B detector was located at a 4 m distance from the source centre. The ^{3}He detector served as a flux monitor. The coordinate distribution on X-axis and Y-axis is shown in Fig.2. The spatial resolution is ~2 mm along X-axis, and ~4 mm along Y-axis.

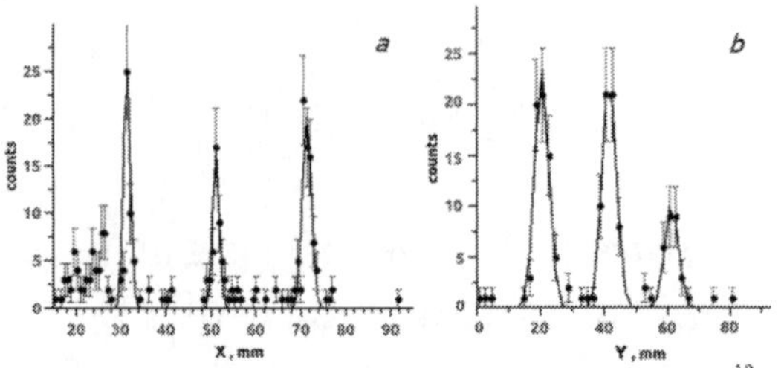

Fig. 2. Neutron coordinate distribution measured in the ^{10}B detector at U = 700 V with Cd-mask along: (a) X-axis; (b) Y-axis.

3 Experiments

3.1 Thermal neutrons

An important task is to measure the thermal neutron flux from the output channel of the moderator at various distances from its center. The ^{10}B- and ^{3}He-detectors was located along an axis of the channel. The ^{3}He-detector was moved along X-axis perpendicularly to the channel axis and it was located behind the ^{10}B-detector. A profile of the flux of the neutron source was measured by using both detectors. They were located at several distances from the moderator center. For example, they were placed at a distance of 204 cm from the center. The measured distribution of the neutron flux by ^{10}B-detector is shown in Fig.3(a). It was compared with the one which was measured with using of the moving ^{3}He-detector (see Fig.3(b)). It can be note in the Fig.3(b) that the neutron flux is accompanied by a large background of scattered neutrons. It can be explained by a fact that the ^{3}He tube detector is sensitive to neutrons from all directions. At the same time, the ^{10}B-detector has a much lower sensitivity to the background of scattered neutrons.

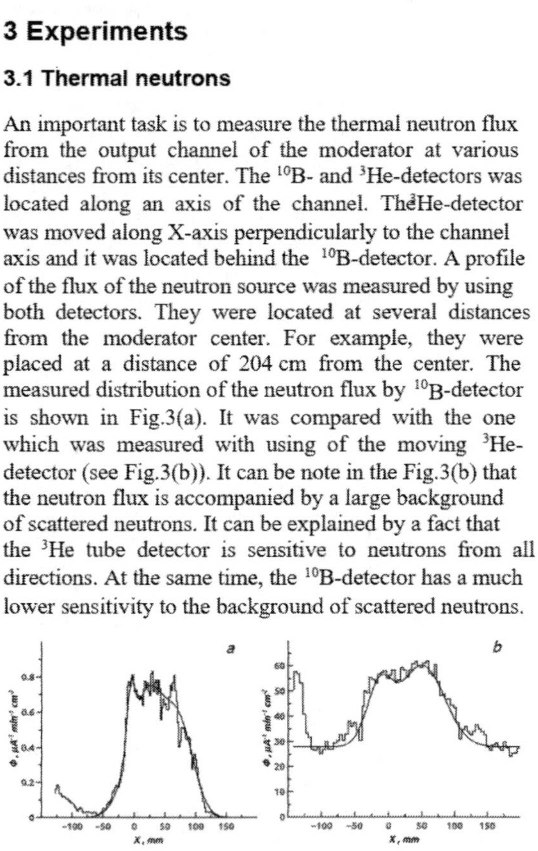

Fig. 3. Spatial neutron distribution from the moderator channel along the X-axis which was measured by: (a) ^{10}B-detector at U = 700 V; (b) the moving ^{3}He tube counter at U = 1400 V.

The ^{10}B-detector gives a possibility to measure a time-of-flight. Time-of-flight measurements were performed with ^{10}B-detector and ^{3}He tube counter. The time-of-flight base for the ^{3}He tube counter was ~7 m, for the ^{10}B-detector – 3 m. Therefore the time-of-flight scale must be substituted by the corresponding energy one. The neutron energy spectra obtained by the time-of-flight technique with the ^{10}B-detector and ^{3}He tube counter are shown in Fig.4. Discrepancy in the shape of spectra can be explained by difference in the material and thickness of entrance windows.

The accuracy and energy range of the time-of-flight measurement depends on the distance. The need for a large distance contradicts the requirement of a large solid angle. Therefore, as a rule, the time-of-flight measurement is precise enough only for thermal or epithermal neutrons.

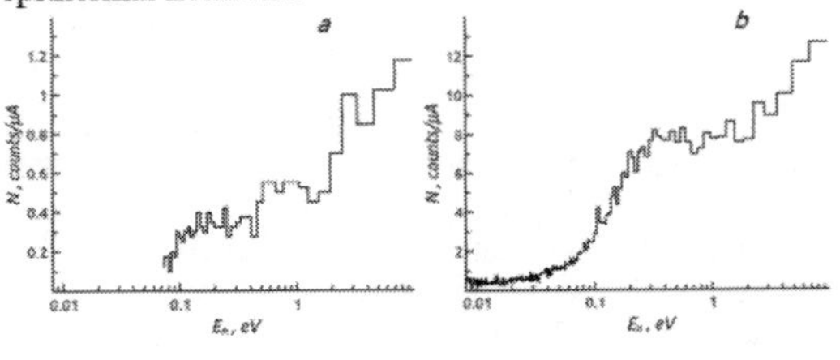

Fig. 4. Neutron energy spectra obtained by the time-of-flight technique. They were measured by: (a) ^{10}B-detector; (b) ^{3}He-tube counter.

3.2 Simulations for thermal and fast neutrons

Simulations gave that the position of maximum in the ionization loss spectrum corresponds to approximately 0.43 MeV for the thermal neutron as it is shown in Fig.5(a). This made it possible to calibrate spectra on an energy scale. Simulations of the ionization loss spectrum for the 3 MeV energy are presented in Fig.5(b).

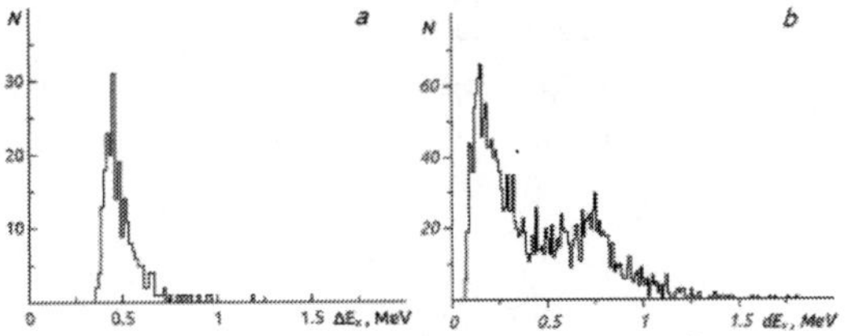

Fig. 5. Simulations of ionisation loss spectra with ^{10}B detector in the first gas gap for: (a) thermal energy; (b) 3 MeV energy.

An average ionization loss of the ^{4}He nucleus decreases with a increasing energy. An average ionization loss of the ^{7}Li nucleus increases with a increasing energy in our case. Note that, for a comparison with experiment the energy threshold must be taken into account.

3.3 Fast neutrons

An ionization loss signal from a thermal neutron is mainly caused by the emitted ^{4}He nucleus because ^{7}Li nuclei cannot penetrate into the gas gap. But if a neutron has an energy of several MeV the shape of pulse height spectra changes crucially. The ^{7}Li nuclei becomes capable to penetrate into the sensitive gas gaps [19].

To perform fast neutron experiments the moderator has been removed. The ^{10}B-detector was located at a distance of 310 cm from the center of the beryllium target at an angle of 43 with respect to the beam. The experimental ionization loss spectra for two gaps of detector are shown in Fig.6.

Neutrons with energies of several MeV transfer an additional energy to the emitted ^{4}He or ^{7}Li nuclei. The ^{4}He nucleus leaves less energy in the gas gap and causes the left peak in the spectrum. The ^{7}Li nucleus thus obtains enough energy to penetrate to the gap and leave a lot of energy in it [17].

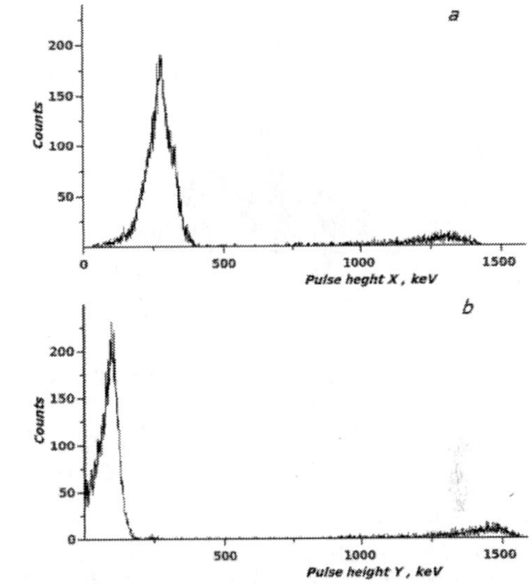

Fig. 6. The experimental ionisation loss spectra for the fast neutrons of several MeV with ^{10}B detector for: (a) the 1-st gap; (b) the 2-nd gap.

Photoneutrons are produced on beryllium target. The yield of neutrons measured in the ^{10}B-detector depends on the electron beam energy as it is shown in Fig.7.

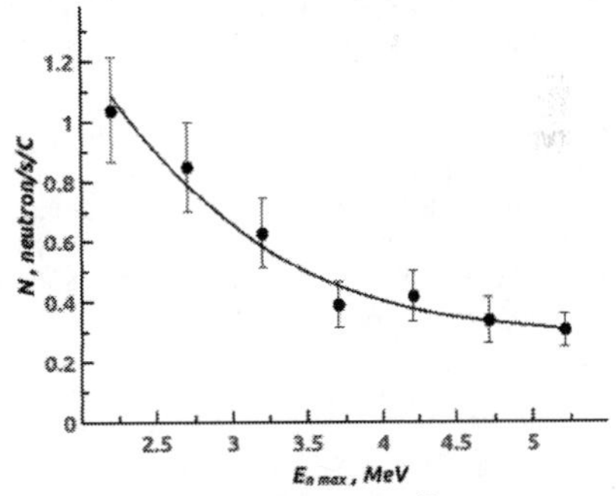

Fig. 7. The neutron counting rate in the ^{10}B-detector per sec per coulomb of the accelerator beam charge versus maximum energy of neutrons.

The number of detected neutrons per second per coulomb of the accelerator beam charge reduces with the maximum neutron energy.

The moderator was installed again for further experiments. But the moderator of the neutron source was configured to the flux which was enriched by fast neutrons of several MeV energy. The beryllium target was placed in the moderator center and channel was opened. The collimator of 3 cm in diameter was installed in the channel. The ^{10}B detector was located at a distance of 119 cm from the target center at the output from the moderator channel. In order to remove slow neutrons from the flux, a 5 mm thick cadmium shield was placed in front of the detector. Two-dimension spatial distribution of neutrons with energies above the cadmium boundary (>0.55 eV) and the neutron maximum energy 7 MeV is shown in the Fig.8.

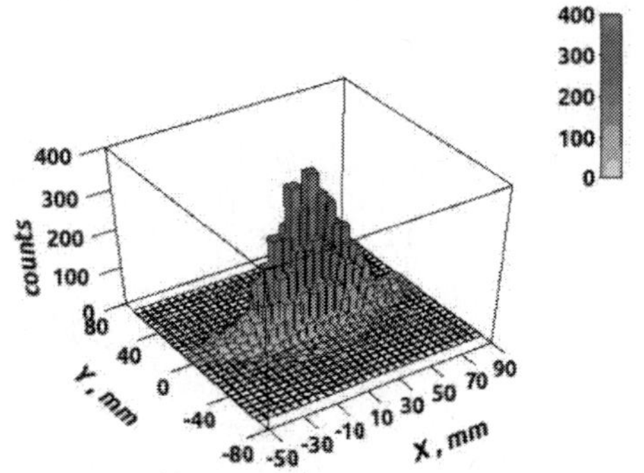

Fig. 8. The neutron two-dimension spatial distribution with the energy above 0.55 eV and the maximum energy 7 MeV.

At the same time, the ^{3}He-detector which was placed behind ^{10}B-detector detected only rare events. The efficiency of the ^{3}He-detector to fast neutrons is extremely low because of the fast neutron cross section on ^{3}He nucleus is very small.

Ionization loss spectra which is produced in the first gap are shown in the Fig.9. The maximum neutron energy $E_{n\ max}$ was from 3 to 7 MeV.

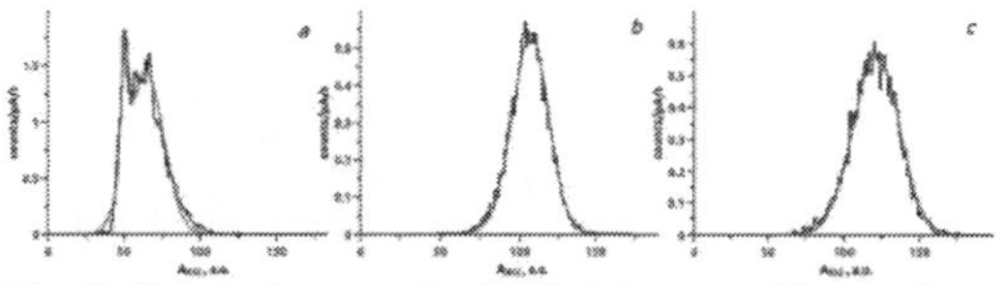

Fig. 9. Energy loss spectra in first gas gap. The maximum neutron energy $E_{n\ max}$ was: (a) 3 MeV; (b) 5 MeV; (c) 7 MeV.

We have found that the ionization loss and, hence, energy loss in the ^{10}B detector gaps depends on the maximum neutron energy $E_{n\ max}$. The position of the maximum P_{max} in the pulse height spectrum varies monotonically versus the maximum neutron energy $E_{n\ max}$. This is demonstrated in Fig.10.

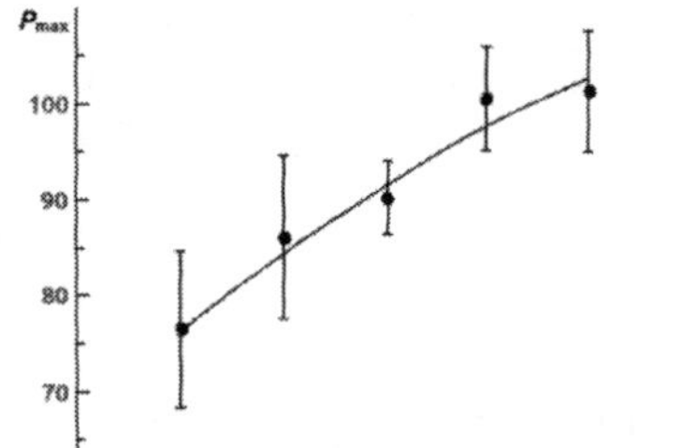

Fig. 10. The position of the maximum P_{max} in the pulse height spectrum versus the maximum neutron energy $E_{n\ max}$.

The one layer detector has a ~$3\cdot10^{-2}$ efficiency for thermal neutrons and ~10^6 efficiency for several MeV neutrons. The efficiency could be enhanced by using many ^{10}B layers. The detector will weakly distort the measured neutron field and it will have a negligible sensitivity to γ-quants.

4 Conclusions

The developed novel hybrid ^{10}B detector demonstrates the possibility of separating thermal and fast neutrons. Experimentally measured pulse heights depend on the maximum neutron energy in the MeV-range. This allows to control the neutron flux energy. The obtained spatial resolutions for thermal neutrons along the X- and Y- axis were ~2 mm and 4 mm respectively.

The detector has a very low sensitivity (< 10^7) to other types of radiation including γ-rays and weakly distorts neutron field due to small scattering and absorption in gas, aluminum and silicium of which such detector is built. The hybrid ^{10}B-detector is suited to study rare nuclear reactions.

We are grateful to L.V. Kravchuk for the interest to this work, and to S.V. Sabinin and V.V. Tishin for the technical support.

References

1. J. M. Tranquada, Guangyong Xu, I.A. Zaliznyak, J. Magn. Magn. Mater. **350**, 148 (2014)

2. E. Kartini, M. Manavan, M.F. Collins, M. Avdeev, Physica B, Condens. Matter. Phys. **551**, 320 (2018)

3. M.M. Castellanos, A. McAuley, J.E. Curtis, Comput. Struct. Biotech. J. **15**, 117 (2017)

4. V.L. Aksenov, Physics-Uspekhi **45**, 645 (2002)

5. S.I. Potashev, V.P. Zavarzina, A.A. Afonin, Bull. Lebedev Phys. Inst. **44**, 271 (2017)

6. M.V. Avdeev, M. Balasoiu, V.L. Aksenov, V.M. Garamus, J. Kohlbrecher, D. Bica, L. Vekas, J. Magn. Magn. Mater. **270**, 371 (2004)

7. D.A. Jacques, J. Trewhella, Protein Science **19**, 642 (2010)

8. J. Fried, J.A. Harder, G.J. Mahler, D. Makowiecki, Nucl. Instrum. Methods in Phys. Res. Sect. A **478,** 415 (2002)

9. T. Nakamura, K. Toh, N. Tsutsui, M. Ebine, A. Birumachi, K. Sakasai, J. Instrum. **12,** C12025 (2017)

10. M. Katagiri, T. Nakamura, M. Ebine, A. Birumachi, Nucl. Instrum. Methods in Phys. Res. Sect. A **573,** 149 (2007)

11. M.G. Strauss, R. Brenner, F.J. Lynch, C.B. Morgan, IEEE Trans. Nucl. Sci. **28,** 800 (1981)

12. R. Beyer, E. Grossea, K. Heidel, J. Hutsch, A.R. Junghans, J. Klug, D. Legrady, R. Nolte, S. Rottger, M. Sobiella, A. Wagner, Nucl. Instrum. Methods in Phys. Res. Sect. A **575**, 449 (2007)

13. M. Kohli, F. Allmendinger, W. Haubler, T. Schroder, M. Klein, M. Meven, U. Shmidt, Nucl. Instrum. Methods in Phys. Res. Sect. A **828**, 242 (2016)

14. S. Uno, T. Uchida, M. Sekimoto, T. Murakami, K. Miyama, M. Shoji, E. Nakano, T. Koike, K. Morita, H. Satoh, T.Kamiyama, Y. Kiyanagi, Physics Procedia **26**, 142 (2012)

15. S. Potashev, Yu. Burmistrov, A. Drachev, S. Karaevsky, E. Konobeevski, S. Zuyev, J. Phys.: Conf. Ser. **798**, 012160 (2017)

16. S.I. Potashev, Yu.M. Burmistrov, A.I. Drachev, S.Kh. Karaevsky, G.V. Solodukhov, J. Surf. Invest.: X-ray, Synchrotron Neutron Tech., **12,** 627 (2018)

17. S. Potashev, Yu. Burmistrov, A. Drachev, S. Karaevsky, E. Konobeevski, A. Kasparov, I. Meshkov, S. Zuyev, A. Afonin, V. Marin, V. Ponomarev, G. Solodukhov, KnE Energy & Physics **12,** 627 (2018)

18. S. Karaevsky, S. Potashev, A. Drachev, Yu. Burmistrov, Bull. Russ. Acad. Sci.: Phys. **82,** 748 (2018)

19. R.A. Boie, J. Fischer, Y. Inagaki, F.C. Merritt, H. Okuno, V. Radeka, Nucl. Instrum. Methods in Phys. Res. Sect. A **200,** 533 (1982)

AUTHOR INDEX

AUTHOR INDEX